Cognitive Technologies

Matthieu Cord · Pádraig Cunningham (Eds.)

Machine Learning Techniques for Multimedia

Case Studies on Organization and Retrieval

With 98 Figures and 20 Tables

Springer

Editors:

Prof. Dr. Matthieu Cord
UPMC University
CNRS (UMR 7606)
Lab. LIP6
104 Avenue du Président, Kennedy
75016 Paris, France
matthieu.cord@lip6.fr

Prof. Dr. Pádraig Cunningham
University College Dublin
Belfield
School of Computer Science &
Informatics
Dublin 2, Ireland
padraig.cunningham@ucd.ie

Managing Editors:

Prof. Dov M. Gabbay
Augustus De Morgan Professor of Logic
Department of Computer Science
King's College London
Strand, London WC2R 2LS, UK

Prof. Dr. Jörg Siekmann
Forschungsbereich Deduktions- und
Multiagentensysteme, DFKI
Stuhlsatzenweg 3, Geb. 43
66123 Saarbrücken, Germany

ISBN: 978-3-642-44362-6 ISBN: 978-3-540-75171-7 (eBook)

Cognitive Technologies ISSN: 1611-2482

ACM Computing Classification: I.2, I.4, I.5, H.3, H.5

Cover design: KünkelLopka, Heidelberg

Printed on acid-free paper

9 8 7 6 5 4 3 2 1

springer.com

Preface

Large collections of digital multimedia data are continuously created in different fields and in many application contexts. Application domains include web searching, cultural heritage, geographic information systems, biomedicine, surveillance systems, etc. The quantity, complexity, diversity and multi-modality of these data are all exponentially growing.

The main challenge of the next decade for researchers involved in these fields is to carry out meaningful interpretations from these raw data. Automatic classification, pattern recognition, information retrieval, data interpretation are all pivotal aspects of the whole problem. Processing this massive multimedia content has emerged as a key area for the application of machine learning techniques. ML techniques and algorithms can 'add value' by analysing these data. This is the situation with the processing of multimedia content. The 'added value' from ML can take a number of forms:

- by providing insight into the domain from which the data are drawn,
- by improving the performance of another process that is manipulating the data or
- by organising the data in some way.

This book brings together some of the experience of the participants of the European Union Network of Excellence on Multimedia Understanding through Semantics Computation and Learning (www.muscle-noe.org). The objective of this network was to promote research collaboration in Europe on the use of machine learning (ML) techniques in processing multimedia data and this book presents some of the fundamental research outputs of the network.

In the MUSCLE network, there are multidisciplinary teams including expertise in machine learning, pattern recognition, artificial intelligence, information retrieval or image and video processing, text and cross-media analysis. Working together, similarities and differences or peculiarities of each data processing context clearly emerged. The possibility to bring together, to factorise many approaches, techniques and algorithms related to the machine learning framework has been very productive. We structured this book in two parts to follow this idea: Part I introduces the machine learning principles and techniques that are used in multimedia data processing and analysis. A comprehensive review of the relevant ML techniques is first presented. With this review we have set out to cover the ML techniques that are in common use in multimedia research, choosing where possible to emphasise techniques that have sound theoretical underpinnings. Part II focuses on multimedia data processing applications, including machine learning issues in domains such as content-based image and video retrieval, biometrics, semantic labelling, human–computer interaction, and data mining in text and music documents. Most of them concern very recent research issues. A very large spectrum of applications is presented in this second part, offering a nice coverage of the most recent developments in each area.

In this spirit, Part I of the book begins in Chap. 1 with a review of Bayesian methods and decision theory as they apply in ML and multimedia data analysis.

Chapter 2 presents a review of relevant supervised ML techniques. This analysis emphasises kernel-based techniques and support vector machines and the justification for these latter techniques is presented in the context of the statistical learning framework.

Unsupervised learning is covered in Chap. 3. This chapter begins with a review of the classic clustering techniques of k-means clustering and hierarchical clustering. Modern advances in clustering are covered with an analysis of kernel-based clustering and spectral clustering, and self-organising maps are covered in detail. The absence of class labels in unsupervised learning makes the question of evaluation and cluster quality assessment more complicated than in supervised learning. So this chapter also includes a comprehensive analysis of cluster validity assessment techniques.

The final chapter in Part I covers dimension reduction. Multimedia data is normally of very high dimension, so dimension reduction is often an important step in the analysis of multimedia data. Dimension reduction can be beneficial not only for reasons of computational efficiency but also because it can improve the accuracy of the analysis. The set of techniques that can be employed for dimension reduction can be partitioned in two important ways; they can be separated into techniques that apply to *supervised* or *unsupervised* learning and into techniques that either entail *feature selection* or *feature extraction*. In this chapter an overview of dimension reduction techniques based on this organisation is presented and the important techniques in each category are described.

Returning to Part II of the book, there are examples of applications of ML techniques on the main modalities in multimedia, i.e. image, text, audio and video. There are also examples of the application of ML in mixed-mode applications, namely text and video in Chap. 9 and text, image and document structure in Chap. 10.

Chapter 5 is concerned with visual information retrieval systems based on supervised classification techniques. Human interactive systems have attracted a lot of research interest in recent years, especially for content-based image retrieval systems. The main scope of this chapter is to present modern online retrieval approaches of semantical concepts within a large image collection. The objective is to use ML techniques to bridge the semantic gap between low-level image features and the query semantics. A set of solutions to deal with the CBIR specificities are proposed and these are demonstrated in a search engine application in this chapter.

An important aspect of this application is the use of an *active* supervised learning methodology to accommodate the user in the learning and retrieval process. This chapter hence provides algorithms in a statistical framework to extend active learning strategies for online content-based image retrieval.

Incremental learning is also the subject of Chap. 6 but in the context of video analysis. The task is to identify objects (e.g. humans) in video and the learning methodology employs an online version of AdaBoost, one of the most powerful of the classification ensemble techniques (described in Part I). The incremental methodology described can achieve improvements in performance without any user interaction since learning events early in the process can update the learning system to improve later performance. The proposed framework is demonstrated on different video surveillance scenarios including pedestrian and car detection, but the approach is quite general and can be used to learn completely different objects.

Face detection and face analysis is the application area in Chap. 7. These are tasks that humans can accomplish with little effort whereas the development of an automated system that accomplishes this task is rather difficult. There are several related problems: detection of an image segment as a face, extraction of the facial expression information and classification of the expression (e.g. in emotion categories). This chapter considers several machine learning algorithms for these problems and recommends Bayesian network classifiers for face detection and facial expression analysis.

In Chap. 8 attention turns to the problem of image retrieval and the problem of query formation. Query-by-example is *convenient* for an application development perspective but is often not practical in practice. This chapter addresses the situation where the user has a mental image of what they require but not a concrete image that can be passed to the retrieval system. Two strategies are presented for addressing this problem, a Bayesian framework that can home in on a useful image through relevance feedback and a process whereby a query can be composed from a visual thesaurus of image segments.

In Chap. 9 we return to the problem of annotating images and video. Two approaches that follow a semi-supervised learning strategy are assessed. The context is news videos and the task is to link names with faces. The first strategy follows an approach analogous to machine translation whereby visual structures are "translated" to semantic descriptors. The second approach performs annotation by finding the densest component of a graph corresponding to the largest group of similar visual structures associated with a semantic description.

Chapter 10 is also concerned with multi-modal analysis in the classification of semi-structured documents containing image and text components. The development of the Web and the growing number of documents available electronically has been paralleled by the emergence of semi-structured data models for representing textual or multimedia documents. The task of supervised classification of semi-structured documents is a generic information retrieval problem which has many different applications: email filtering or classification, thematic classification of Web pages, document ranking, spam detection. Much progress in this area has been obtained through recent machine learning classification techniques. The authors present in this chapter the different classification approaches for structured documents. A generative model is explored in detail and evaluated on the filtering of pornographic Web pages and in the thematic classification of Wikipedia documents.

Music information retrieval is the application area in the final chapter (11). The main focus in this chapter is on the use of self-organising maps to organise a music collection into so-called "Music Maps". The feature extraction performed on the audio files as a pre-processing step for the self-organising map is described. The authors show how this parameterisation of the audio data can also be used to classify the music by genre. The produced 2D maps offer the possibility to visualise and to intuitively navigate into the whole collection. Some nice technological developments are also presented, demonstrating the practical interest of such research approach.

Many original contributions are introduced in this book. Most of them concern the adaptation of recent ML theory and algorithms to the processing of complex, massive, multimedia data. Architectures, demonstrators and even technological products resulting from this analysis are presented. We hope that our abiding preoccupation to make connections between the different application contexts and the general methods of Part I will serve to stimulate the interest of the reader. Despite that, each chapter is self-contained with enough definitions and notations to be read in isolation.

The editors wish particularly to record their gratitude to Kenneth Bryan for his help in bringing the parts of this book together.

Paris, Dublin
October 2007

Matthieu Cord
Pádraig Cunningham

Contents

List of Contributors

Muhammet Baştan
Bilkent University, Ankara, Turkey, e-mail: bastan@cs.bilkent.edu.tr

Horst Bischof
Institute for Computer Graphics and Vision, Graz University of Technology, Graz, Austria, e-mail: bischof@icg.tugraz.at

Nozha Boujemaa
Projet IMEDIA, INRIA, Le Chesnay Cedex, France, e-mail: Nozha.Boujemaa@inria.fr

Ira Cohen
HP Labs, Palo Alto, CA, USA, e-mail: iracohen@hp.com

Matthieu Cord
LIP6, UPMC, Paris, France, e-mail: matthieu.cord@lip6.fr

Pádraig Cunningham
University College Dublin, Dublin, Ireland, e-mail: padraig.cunningham@ucd.ie

Rozenn Dahyot
Trinity College Dublin, Dublin, Ireland, e-mail: rozen.dahyot@tcd.ie

Sarah Jane Delany
Dublin Institute of Technology, Dublin, Ireland, e-mail: sarahjane.delany@comp.dit.ie

Ludovic Denoyer
LIP6, UPMC, Paris, France, e-mail: ludovic.denoyer@lip6.fr

Pınar Duygulu
Bilkent University, Ankara, Turkey, e-mail: duygulu@cs.bilkent.edu.tr

Julien Fauqueur
University of Cambridge, Cambridge, UK, e-mail: jf330@cam.ac.uk

Patrick Gallinari
LIP6, UPMC, Paris, France, e-mail: patrick.gallinari@lip6.fr

Theo Gevers
Faculty of Science, University of Amsterdam, Amsterdam, The Netherlands, e-mail: gevers@science.uva.nl

Philippe-Henri Gosselin
ETIS, UCP, Paris, France, e-mail: gosselin@ensea.fr

Derek Greene
University College Dublin, Dublin, Ireland, e-mail: padraig.cunningham@ucd.ie

Thomas Lidy
Vienna University of Technology, Vienna, Austria, e-mail: lidy@ifs.tuwien.ac.at

Rudolf Mayer
Vienna University of Technology, Vienna, Austria, e-mail: mayer@ifs.tuwien.ac.at

Derya Ozkan
Bilkent University, Ankara, Turkey, e-mail: deryao@cs.bilkent.edu.tr

Andreas Rauber
Vienna University of Technology, Vienna, Austria, e-mail: rauber@ifs.tuwien.ac.at

Peter M. Roth
Institute for Computer Graphics and Vision, Graz University of Technology, Graz, Austria, e-mail: roth@icg.tugraz.at

Nicu Sebe
Faculty of Science, University of Amsterdam, Amsterdam, The Netherlands, e-mail: nicu@science.uva.nl

Roberto Valenti,
Faculty of Science, University of Amsterdam, Amsterdam, The Netherlands, e-mail: rvalenti@science.uva.nl

Simon P. Wilson
Trinity College Dublin, Dublin, Ireland, e-mail: simon.wilson@tcd.ie

Part I
Introduction to Learning Principles for Multimedia Data

Chapter 1
Introduction to Bayesian Methods and Decision Theory

Simon P. Wilson, Rozenn Dahyot, and Pádraig Cunningham

Abstract Bayesian methods are a class of statistical methods that have some appealing properties for solving problems in machine learning, particularly when the process being modelled has uncertain or random aspects. In this chapter we look at the mathematical and philosophical basis for Bayesian methods and how they relate to machine learning problems in multimedia. We also discuss the notion of decision theory, for making decisions under uncertainty, that is closely related to Bayesian methods. The numerical methods needed to implement Bayesian solutions are also discussed. Two specific applications of the Bayesian approach that are often used in machine learning – naïve Bayes and Bayesian networks – are then described in more detail.

1.1 Introduction

Bayesian methods and decision theory provide a coherent framework for learning and problem solving under conditions of uncertainty. Bayesian methods in particular are a standard tool in machine learning and signal processing. For multimedia data they have been applied to statistical learning problems in image restoration and segmentation, to speech recognition, object recognition and also to content-based retrieval from multimedia databases, amongst others. We argue in this chapter that Bayesian methods, rather than any other set of statistical learning methods, are a natural tool for machine learning in multimedia. The principal argument is philosophical; the solution provided by Bayesian methods is the most easily interpretable,

Simon P. Wilson
Trinity College Dublin, Dublin, Ireland, e-mail: simon.wilson@tcd.ie

Rozenn Dahyot
Trinity College Dublin, Dublin, Ireland, e-mail: rozen.dahyot@tcd.ie

Pádraig Cunningham
University College Dublin, Dublin, Ireland, e-mail: padraig.cunningham@ucd.ie

and to demonstrate this we devote some space to justifying the laws of probability and how they should be applied to learning. We also show the other strengths of Bayesian methods as well as their strong mathematical and philosophical foundation: an easy to understand prescriptive approach, their ability to coherently incorporate data from many sources, implementable with complex models and that, as a probabilistic approach, they not only produce estimates of quantities of interest from data but also quantify the error in that estimate.

Decision theory is perhaps less well known and used in machine learning, but is a natural partner to the Bayesian approach to learning and is deeply connected to it. As a mathematical framework for decision making, its most common application in signal processing and machine learning is when one must make a point estimate of a quantity of interest. The output of a Bayesian analysis is a probability distribution on the quantity of interest; decision theory provides the method to go from this distribution to its "best" point value, e.g. predict a class in a classification problem or output a particular segmentation from a Bayesian segmentation algorithm. As we will see, most Bayesian solutions in signal processing and machine learning are implicitly using decision theory; we argue that thinking about these solutions in terms of decision theory opens up possibilities for other solutions.

One can view these two methods as a breakdown of a problem into two parts. The first concerns what and how much we know about the problem, through models that we specify and data; this is the domain of Bayesian statistical inference. The second part concerns our preferences about what makes a good solution to the problem; this is the domain of decision theory.

1.2 Uncertainty and Probability

Uncertainty is a common phenomenon, arising in many aspects of machine learning and signal processing. How is uncertainty dealt with generally? Two branches of mathematics have arisen to handle uncertainty: probability theory (for quantifying and manipulating uncertainties) and statistics (for learning in circumstances of uncertainty). In this section we will review some of the basic ideas and methods of both these fields with an emphasis on how Bayesian statistics deals with learning.

1.2.1 Quantifying Uncertainty

In many machine learning and signal processing problems the goal is to learn about some quantity of interest whose value is unknown. Examples in multimedia include whether an e-mail is spam, what the restored version of an image looks like, whether this image contains a particular object or the location of a person in a video sequence. We will call such unknowns *random quantities*. The conventional view of a random quantity is that it is the outcome of a random experiment or process; that is still true in this definition since such outcomes are also unknown, and here we extend the definition to unknown values more generally. In thinking about these random quantities, we make use of any information at our disposal; this includes our own experiences, knowledge, history and data, collectively called *background information* and denoted $\mathscr{H}$. Background information varies from individual to individual.

For a particular random quantity X, if X is ever observed or becomes known then we denote that known value with a small letter x. One specific type of random quantity is the *random event*. This is a proposition that either occurs or does not. A random quantity that takes numerical values is called a *random variable*.

The question naturally arises: given that the exact value of X is uncertain but that $\mathscr{H}$ tells us something about it, how do we quantify uncertainty in X? It is generally agreed that *probability is the only satisfactory way to quantify uncertainty*. In making this statement, we take sides in a long philosophical debate, but for the moment probability is the dominant method; see [14] for an extensive justification. This is the assumption that underlies Bayesian learning. Given any unknown quantity of interest, the goal is to define a probability distribution for it in light of any relevant information that is available. This distribution quantifies our state of knowledge about the quantity.

1.2.2 The Laws of Probability

For a discrete random quantity X, we define $P(X = x \mid \mathscr{H})$ to be the probability that X takes the value x in light of our background information $\mathscr{H}$. This probability is also denoted $P_X(x \mid \mathscr{H})$ or just $P(X \mid \mathscr{H})$. If X is a random variable, we can talk about the cumulative distribution function $F_X(x \mid \mathscr{H}) = P(X \leq x \mid \mathscr{H})$. For continuous random quantities, we assume that X is a random variable so $F_X(x \mid \mathscr{H})$ is also defined. If F_X is differentiable with respect to x then its derivative is denoted $f_X(x \mid \mathscr{H})$ (or just $f(x \mid \mathscr{H})$) and is called the *probability density function* of X.

The meaning of the vertical bar in these definitions is to separate the unknown quantity of interest on the left of the bar, from the known quantities $\mathscr{H}$ on the right. In particular, suppose we had two unknown quantities X_1 and X_2 and that the value of X_2 became known to us. Then our background knowledge is extended to include X_2 and we write our uncertainty about X_1 as $P(X_1 \mid X_2, \mathscr{H})$. These are termed *conditional probabilities* and we talk about the probability of X_1 conditional on X_2 and $\mathscr{H}$.

Probabilities must obey three laws; these laws govern the values that a probability can take and say how probabilities can be combined:

1. Convexity: $0 \leq P(X \mid \mathscr{H}) \leq 1$.
2. Additivity: If both X_1 and X_2 cannot occur together (that is, they are *mutually exclusive*) then

$$P(X_1 \text{ or } X_2 \mid \mathscr{H}) = P(X_1 \mid \mathscr{H}) + P(X_2 \mid \mathscr{H}).$$

3. Multiplicativity: $P(X_1 \text{ and } X_2 \mid \mathscr{H}) = P(X_1 \mid \mathscr{H}) \times P(X_2 \mid X_1, \mathscr{H})$.

If we assign a set of probabilities to a random quantity that obeys these laws then we call this assignment a *probability distribution*. These three rules can be justified on several grounds. There are axiomatic justifications, which argue for these laws from the perspective of pure mathematics, due to Cox [3]. A more pragmatic justification comes from de Finetti and his idea of *coherence* and *scoring rules* [4], which are related to the subjective interpretation of probability that we discuss next.

1.2.3 Interpreting Probability

What does a probability mean? For example, what does it mean to say that the probability that an e-mail is spam is 0.9? It is remarkable that there is no agreed upon answer to this question, and that it is still the subject of heated argument among statisticians, probabilists and philosophers. Again, a detailed discussion of this issue is beyond the scope of this chapter and so we will confine ourselves to describing the two main interpretations of probability that prevail today: the frequentist and the subjective.

1.2.3.1 Frequentist Probability

Under the frequentist interpretation, a probability is something physical and objective and is a property of the real world, rather like mass or volume. The probability of an event is said to be the proportion of times that the event occurs in a sequence of trials under almost identical conditions. For example, we may say that the probability of a coin landing on heads is 0.5. This would be interpreted as saying that if the coin were to be flipped many times, it would land heads on half of the tosses.

If we talk about frequentist probability then the relationship with an individual's background information $\mathscr{H}$ is lost; frequentist probabilities are independent of one's personal beliefs and history and are not conditional on $\mathscr{H}$.

1.2.3.2 Subjective Probability

A subjective or personal probability is your degree of belief in an event occurring. Subjective probabilities are always conditional on the individual's background information $\mathscr{H}$. Thus a probability is not an objective quantity and can vary legitimately between individuals, provided the three laws of probability are not compromised.

Taking the example of a spam e-mail, the probability of a message being spam is interpreted as quantifying our personal belief about that event. If our background information $\mathscr{H}$ changes, perhaps by observing some data on properties of spam messages, we are allowed to change this value as long as we are still in agreement with the laws of probability.

In his book on subjective probability, de Finetti motivated the laws of probability by considering them as subjective through the idea of coherence and scoring rules.

He argued that in order to avoid spurious probability statements, an individual must be willing to back up a subjective probability statement with a bet of that amount on the event occurring, where one unit of currency is won if it does occur and the stake is lost if not [4]. In this case, it can be shown that one should always place bets that are consistent with the laws of probability; such behaviour is called coherent. If one is not coherent then one can enter into a series of bets where one is guaranteed to lose money; such a sequence of bets is called a Dutch book. More generally de Finetti showed that under several different betting scenarios with sensible wins and losses that are a function of the probability of the event (called scoring rules), only by being coherent does one avoid a Dutch book.

1.2.3.3 Which Interpretation for Machine Learning with Multimedia Data?

Here we argue that the subjective interpretation is the most appropriate for machine learning problems with multimedia data. This is because most of the multimedia problems that we try to solve with a statistical learning method are "one-off" situations, making the frequentist interpretation invalid. For example, to interpret our statement about the probability of a message being spam, the frequentist interpretation forces us to think about a sequence of "essentially identical" messages, of which 90% are spam. However, there is only one e-mail, not a sequence of essentially identical ones; furthermore what exactly constitutes "essentially identical" in this situation is not clear. The subjective interpretation does not suffer from this requirement; the probability is simply a statement about our degree of belief in the message being spam, based on whatever knowledge we have at our disposal and made in accordance with the laws of probability.

Bayesian statistical methods, by interpreting probability as quantifying uncertainty and assuming that it depends on $\mathscr{H}$, are adopting the subjective interpretation.

1.2.4 The Partition Law and Bayes' Law

Suppose we have two random quantities X_1 and X_2 and that we have assessed their probability jointly to obtain $P(X_1, X_2 \mid \mathscr{H})$. An application of the addition rule of probability gives us the distribution of X_1 alone (the *marginal* distribution of X_1):

$$P(X_1 \mid \mathscr{H}) = \sum_{X_2} P(X_1, X_2 \mid \mathscr{H}), \tag{1.1}$$

where the summation is over all possible values of X_2. If X_1 and X_2 were continuous, the summation sign would be replaced by an integral and $P(X_1, X_2 \mid \mathscr{H})$ by the density function $f(X_1, X_2 \mid \mathscr{H})$. It also gives us the law of total probability (also called the partition law):

$$P(X_1 \mid \mathscr{H}) = \sum_{X_2} P(X_1 \mid X_2, \mathscr{H})\, P(X_2 \mid \mathscr{H}). \tag{1.2}$$

By the law of multiplication and the marginalization formula (1.1) we can say

$$P(X_1 = x \mid X_2, \mathscr{H}) = \frac{P(X_1 = x, X_2 \mid \mathscr{H})}{\sum_{X_1} P(X_1, X_2 \mid \mathscr{H})}. \tag{1.3}$$

By applying the multiplicative law to both numerator and denominator, we obtain Bayes' law:

$$P(X_1 = x \mid X_2, \mathscr{H}) = \frac{P(X_2 \mid X_1 = x, \mathscr{H}) P(X_1 = x \mid \mathscr{H})}{\sum_{X_1} P(X_2 \mid X_1, \mathscr{H}) P(X_1 \mid \mathscr{H})}, \tag{1.4}$$

with the usual replacement of sums by integrals and probabilities by densities if X_1 is continuous. Bayes' law is attributed to the Reverend Thomas Bayes (1702–1761), although it is generally accepted that Laplace was the first to develop it.

Bayes' law shows how probabilities change in the light of new information. We have explicitly written $X_1 = x$ on the left-hand side and the numerator on the right-hand side, to distinguish it from the fact that in the denominator we sum over all possible values of X_1, of which x is only one. Indeed, it should be noted that the probability on the left-hand side is a function of x alone, as both X_2 and $\mathscr{H}$ are known. On the right-hand side, x only appears in the numerator, so the denominator is just a constant of proportionality. Therefore, we can write

$$P(X_1 = x \mid X_2, \mathscr{H}) \propto P(X_2 \mid X_1 = x, \mathscr{H}) P(X_1 = x \mid \mathscr{H}). \tag{1.5}$$

The probability on the left is called the *posterior* probability of observing $X_1 = x$, since it is the probability of observing X_1 after observing X_2. On the right-hand side, we have the *prior* probability of observing $X_1 = x$, given by $P(X_1 = x \mid \mathscr{H})$. The posterior is proportional to the prior multiplied by $P(X_2 \mid X_1 = x, \mathscr{H})$; this latter term is often called the *likelihood*.

Bayes' law is just a theorem of probability, but it has become associated with Bayesian statistical methods. As we have said, in Bayesian statistics probability is interpreted subjectively and Bayes' law is frequently used to update probabilities in light of new data. It is therefore the key to learning in the Bayesian approach. However, we emphasize that the use of Bayes' law in a statistical method does not mean that the procedure is "Bayesian" – the key tenet is the belief in subjective probability. The frequency view of probability is behind *frequentist* or *classical statistics* which includes such procedures as confidence limits, significance levels and hypothesis tests with Type I and Type II errors.

1.3 Probability Models, Parameters and Likelihoods

Following the subjective interpretation, our probability assessments about X are made conditional on our background knowledge $\mathscr{H}$. Usually, $\mathscr{H}$ is large, very complex, of high dimension and may be mostly irrelevant to X. What we need is some way of abridging $\mathscr{H}$ so that it is more manageable. This introduces the idea of a parameter and a parametric model. We assume that there is another random quantity θ that summarizes the information in $\mathscr{H}$ about X, and hence makes X and $\mathscr{H}$ independent. Then by the partition law,

$$P(X \mid \mathscr{H}) = \sum_{\theta} P(X \mid \theta, \mathscr{H}) P(\theta \mid \mathscr{H}) = \sum_{\theta} P(X \mid \theta) P(\theta \mid \mathscr{H}). \tag{1.6}$$

Now our probability distribution for X is a function of two probability distributions: the first, $P(X \mid \theta)$, is the *probability model* for X with *parameter* θ. The second, $P(\theta \mid \mathscr{H})$, is called the *prior distribution* of θ.

The choice of probability model and prior distribution is a subjective one. For multimedia data, where X may be something of high dimension and complexity like an image segmentation, the choice of model is often driven by a compromise between a realistic model and one that is practical to use. The statement $P(X \mid \theta)$ may itself be decomposed and defined in terms of other distributions; the scope of probability models is vast and for audio, image and video extends through autoregressive models, hidden Markov models and Markov random fields, to name but a few. The choice of prior distribution for the parameter is a contentious issue and can be thought of as the main difference between frequentist and Bayesian statistical methods; whereas Bayesian procedures require it to be specified, frequentist procedures choose to ignore it and work only with the probability model. Various methods for specifying prior distributions have been proposed: the use of "objective" priors is one and the use of expert opinion is another (see [6]). However, this is a very under-developed field for multimedia data and often the prior is chosen for mathematical convenience, e.g. assumed constant.

However, the use of a prior is another aspect of the Bayesian approach that suits itself well to multimedia data applications of machine learning. It allows one to specify domain knowledge about the particular problem that otherwise would be ignored or incorporated in an ad hoc manner. For ill-posed problems (such as image segmentation or object recognition) it also serves as a regularization.

1.4 Bayesian Statistical Learning

For a set of random quantities $X_1, X_2, \ldots, X_n$ one can still use the model and prior approach by writing

$$P(X_1, X_2, \ldots, X_n \mid \mathscr{H}) = \sum_{\theta} P(X_1, X_2, \ldots, X_n \mid \theta) P(\theta \mid \mathscr{H}). \tag{1.7}$$

In many situations, where the X_i are a random sample of a quantity, it makes sense to assume that each X_i is independent of the others conditional on θ, thus

$$P(X_1, X_2, \ldots, X_n \mid \mathscr{H}) = \sum_{\theta} P(\theta \mid \mathscr{H}) \prod_{i=1}^{n} P(X_i \mid \theta). \tag{1.8}$$

Equation 1.8 is fundamental to statistical learning about unknown quantities from data. Under the assumptions of this equation, there are two quantities that we can learn about:

1. Our beliefs about likely values of the parameter θ given the data;
2. Our beliefs about X's given observation of X_1, …, X_n. In particular, we might want to assess the probable values of the next observation X_{n+1} in light of the data.

For Bayesian learning, in the spirit of the belief that probability is the only way to describe uncertainty, Bayesian inference strives to produce a probability distribution for the unknown quantities of interest. For inference on the parameter θ, the *posterior* distribution given the data, $P(\theta \,|\, X_1,\ldots,X_n,\mathscr{H})$, is the natural expression to look at. It reflects the fact that $X_1,\ldots,X_n$ have become known and have joined $\mathscr{H}$. Bayes' law can be written in terms of the model and prior:

$$P(\theta \,|\, X_1,\ldots,X_n,\mathscr{H}) = \frac{P(\theta \,|\, \mathscr{H}) \prod_{i=1}^{n} P(X_i \,|\, \theta)}{\sum_{\theta} P(\theta \,|\, \mathscr{H}) \prod_{i=1}^{n} P(X_i \,|\, \theta)}. \tag{1.9}$$

Since the X_i are known, $\prod_i P(X_i \,|\, \theta)$ is known as the likelihood. As in (1.5), we can then write

$$P(\theta \,|\, X_1,\ldots,X_n,\mathscr{H}) \propto P(\theta \,|\, \mathscr{H}) \prod_{i=1}^{n} P(X_i \,|\, \theta) \tag{1.10}$$

or, in words,

$$\text{posterior} \propto \text{prior} \times \text{likelihood}. \tag{1.11}$$

For our belief about the next observation, we calculate the distribution of X_{n+1} conditional on the observations and $\mathscr{H}$; this is given by (1.6), but with the posterior distribution of θ replacing the prior:

$$P(X_{n+1} \,|\, X_1,\ldots,X_n,\mathscr{H}) = \sum_{\theta} P(X_{n+1} \,|\, \theta)\, P(\theta \,|\, X_1,\ldots,X_n,\mathscr{H}). \tag{1.12}$$

This is called the posterior *predictive* distribution of X.

1.5 Implementing Bayesian Statistical Learning Methods

Implementing Bayesian methods can often be computationally demanding. When θ is continuous this is because computing posterior distributions and other functions of them will often involve high-dimensional integrals of the order of the dimension of θ. In what follows we assume that $\theta = (\theta_1,\ldots,\theta_k)$ is of dimension k. We define $X = (X_1,\ldots,X_n)$ to be the data.

The first such problem arises because the denominator of Bayes' law is an integral over θ:

$$P(\theta \,|\, X,\mathscr{H}) = \frac{P(\theta \,|\, \mathscr{H}) \prod_{i=1}^{n} P(X_i \,|\, \theta)}{\int_{\forall\theta} P(\theta \,|\, \mathscr{H}) \prod_{i=1}^{n} P(X_i \,|\, \theta)\, d\theta}, \tag{1.13}$$

a k-dimensional integral. A second problem is if we want to compute the marginal posterior distribution of a component of θ:

$$P(\theta_i \,|\, X) = \int P(\theta \,|\, X)\, \mathrm{d}\theta_{-i},$$

where $\theta_{-i} = (\theta_1, \ldots, \theta_{i-1}, \theta_i, \ldots, \theta_k)$; this is a $(k-1)$ dimensional integration. Also, if we want to calculate a posterior mean, we have to compute another k-dimensional integral:

$$E(\theta_i \,|\, X) = \int \theta_i P(\theta \,|\, X)\, \mathrm{d}\theta.$$

Except for very special cases that rarely arise in multimedia applications, these integrals are not in closed form and one must resort to numerical approximations. For $k < 4$ or so, numerical quadrature methods can be used but these quickly become computationally infeasible as k becomes large. For larger k, the dominant method is Monte Carlo simulation, where one attempts to simulate values of θ from $P(\theta \,|\, X)$. Many methods for Monte Carlo sampling exist that only require the distribution to be known up to a constant, e.g. to simulate from $P(\theta \,|\, X)$ it is sufficient to know only the numerator of Bayes' law $P(\theta \,|\, \mathscr{H}) \prod_{i=1}^{n} P(X_i \,|\, \theta)$ and we do not have to evaluate the integral in the denominator.

We also mention that there are alternatives to Monte Carlo simulation that are becoming possible. They are often faster than Monte Carlo methods but so far their scope is more limited. The most widely used is the variational Bayes' approach; a recent work on its use in signal processing is [20].

1.5.1 Direct Simulation Methods

Direct simulation methods are those methods that produce a sample of values of θ from exactly the required distribution $P(\theta \,|\, X)$. The most common are the inverse transform method and the rejection method.

The inverse distribution method works by first using the decomposition of $P(\theta \,|\, X)$ by the multiplication law:

$$P(\theta \,|\, X) = \prod_{i=1}^{k} P(\theta_i \,|\, X, \theta_j; j < i). \tag{1.14}$$

This implies that a sample of θ may be drawn by first simulating θ_1 from $P(\theta_1 \,|\, X)$, then θ_2 from $P(\theta_2 \,|\, X, \theta_1)$ and so on. At the ith stage, a uniform random number u between 0 and 1 is generated and the equation $F_i(t \,|\, X, \theta_j; j < i) = u$, where $F_i(t \,|\, X, \theta_j; j < i) = P(\theta_i < t \,|\, X, \theta_j; j < i)$ is the cumulative distribution function, is solved for t. The solution is a sample of θ_i.

For many applications this method is not practical. The conditional distributions $P(\theta_i \,|\, X, \theta_j; j < i)$ are obtained from $P(\theta \,|\, X)$ by integration which may not be in closed form and unavailable numerically.

The rejection method either can be used by simulating each θ_i from its distribution in the decomposition of (1.14), as the inverse distribution method does, or works directly with $P(\theta \,|\, X)$. In the latter case, some other distribution of θ, denoted $Q(\theta)$, is selected from which it is easy to generate samples by inverse transform and where there exists a constant c such that $cQ(\theta)$ bounds $P(\theta \,|\, X)$. In fact, it is sufficient to bound the numerator of $P(\theta \,|\, X)$ from Bayes' law, $P(\theta \,|\, \mathscr{H}) \prod_{i=1}^{n} P(X_i \,|\, \theta)$, thus eliminating the need to evaluate the integral in the denominator. A value θ^* is simulated from Q but then it is only accepted as a value from $P(\theta \,|\, X)$ with probability $\min(1, P(\theta^* \,|\, X)/cQ(\theta^*))$; if the bound c is on the numerator from Bayes' law then this probability is

$$\min\left(1, \left[P(\theta^* \,|\, \mathscr{H}) \prod_{i=1}^{n} P(X_i \,|\, \theta^*)\right] / cQ(\theta^*)\right).$$

If the value is not accepted then the process is repeated until this is the case.

There are many subtleties to obtaining efficient direct simulation by inverse distribution or the rejection method and there are other approaches that are more general; an example that is often used in Bayesian methods is adaptive rejection sampling [8, 9]. A good introductory text to simulation is [17]. Open source code for many direct simulation techniques is also available.

Unfortunately it is often the case that these methods prove impossible to implement when the dimension of θ is large. Often in multimedia applications it is very large, e.g. for an image restoration problem, θ will include the unknown original pixel values and hence be the dimension of the image. In these cases the usual techniques are indirect and generate approximate samples. The dominant approximate sampling technique is Markov chain Monte Carlo (MCMC).

1.5.2 Markov Chain Monte Carlo

This set of techniques does not create a set of values of θ from $P(\theta \,|\, X)$, rather it creates a sequence $\theta^{(1)}, \theta^{(2)}, \ldots$ that converges to samples from $P(\theta \,|\, X)$ in the limit (so-called convergence in distribution, see [10, Chap. 7]). It does this by generating a Markov chain on θ with stationary distribution $P(\theta \,|\, X)$. For those not familiar with the concept of a Markov chain, we refer to [18, Chap. 4] or for a more technical treatment [10, Chap. 6].

The Metropolis algorithm is the most general Markov chain Monte Carlo (MCMC) technique. It works by defining a starting value $\theta^{(1)}$. Then at any stage i, a value θ^* is generated from a distribution (the proposal distribution) $Q(\theta)$. Typically, Q depends on the last simulated value $\theta^{(m-1)}$; for example θ^* may be a random peturbation of $\theta^{(m-1)}$ such as a Gaussian with mean $\theta^{(m-1)}$. To make this relationship clear, we write $Q(\theta^{(m-1)} \to \theta^*)$ to indicate that we propose θ^* from the current value $\theta^{(m-1)}$. The proposed value θ^* is then accepted to be $\theta^{(m)}$ with a probability

$$\min\left(1, \frac{P(\theta^*\,|\,X)\,Q(\theta^* \to \theta^{(m-1)})}{P(\theta^{(m-1)}\,|\,X)\,Q(\theta^{(m-1)} \to \theta^*)}\right)$$

$$= \min\left(1, \frac{\prod_i P(X_i\,|\,\theta^*)\,P(\theta*)\,Q(\theta^* \to \theta^{(m-1)})}{\prod_i P(X_i\,|\,\theta^{(m-1)})\,P(\theta^{(m-1)})\,Q(\theta^{(m-1)} \to \theta^*)}\right), \tag{1.15}$$

otherwise $\theta^{(m)} = \theta^{(m-1)}$.

The choice of Q is quite general as long as it is reversible, e.g. for any possible proposal θ^* from $\theta^{(m-1)}$, it should also be possible to propose $\theta^{(m-1)}$ from θ^*. This flexibility gives the method its wide applicability.

There are some specific cases of the Metropolis algorithm. One is the Gibbs sampler, in which θ is partitioned into components (either univariate or multivariate) and proposals are generated from the full conditional distribution of that component; this is the distribution of the component conditional on X and the remaining components of θ. For example, a proposal for a single component θ_i would be simulated from $P(\theta_i\,|\,X, \theta_{-i})$. This specific form for Q ensures the acceptance probability of (1.15) is always 1. For many models it turns out that such full conditional distributions are of an amenable form that can be easily simulated from.

The advantages of MCMC are that one never needs to attempt simulation from $P(\theta\,|\,X)$ directly and that it only appears in the algorithm as a ratio in which the annoying denominator term from Bayes' law cancels. Thus we avoid the need to ever evaluate this integral. The disadvantages are that the method is computationally expensive and that, since it is converging to sampling from the posterior distribution, one has to ensure that convergence has occurred and that, once it has been achieved, the method explores the whole support of $P(\theta\,|\,X)$. This makes MCMC impractical for real-time applications. There is a certain art to ensuring convergence in complex models, and many diagnostics have been proposed to check for lack of convergence [15]. This is still an area of active research.

There are many books on MCMC methods. Good introductions to their use in simulating from posterior distributions are [5] and [7], while [21] describes their use in image analysis.

1.5.3 Monte Carlo Integration

Suppose that we have obtained M samples from $P(\theta\,|\,X_1,\ldots,X_n)$. The concept of Monte Carlo integration allows us to approximate many of the quantities defined from this distribution by integration. This concept arises from the law of large numbers and states that an expectation can be approximated by a sample average, e.g. for any function $g(\theta)$ such that its expectation exists and samples $\theta^{(1)},\ldots,\theta^{(M)}$,

$$E(g(\theta)\,|\,X_1,\ldots,X_n) \approx \frac{1}{M}\sum_m g(\theta^{(m)}). \tag{1.16}$$

Most of the integrals that interest us can be expressed as expectations. For example, expectations and variances of θ_i are approximated as the sample mean and variance. The predictive distribution for X_{n+1} is approximated as the expected value of the probability model, since

$$\begin{aligned} P(X_{n+1}\,|\,X_1,\ldots,X_n) &= \int P(X_{n+1}\,|\,\theta)\,P(\theta\,|\,X_1,\ldots,X_n)\,\mathrm{d}\theta \\ &= E(P(X_{n+1}\,|\,\theta)) \ \text{(expectation with respect to } P(\theta\,|\,X_1,\ldots,X_n)) \\ &\approx \frac{1}{M}\sum_m P(X_{n+1}\,|\,\theta^{(m)}). \end{aligned} \tag{1.17}$$

Table 1.1 lists the different uses of Monte Carlo integration in Bayesian learning.

1.5.4 Optimization Methods

In many applications, instead of wishing to compute $P(\theta\,|\,X)$, one decides to simply compute the posterior mode; this is known as the MAP estimate (maximum a posteriori):

$$\theta_{\mathrm{MAP}} = \arg\max_{\theta} P(\theta\,|\,X).$$

Table 1.1 Uses of Monte Carlo integration in Bayesian learning

Use	How to implement
Density function	$P(\theta\,\|\,X) \approx \frac{1}{M\|N(\theta)\|}\sum_m I(\theta^{(m)} \in N(\theta))$, for a small volume $N(\theta)$ about θ, where $\theta^{(m)}$ are simulated from $P(\theta\,\|\,X)$
Expected value	$E(\theta\,\|\,X) \approx \frac{1}{M}\sum_m \theta^{(m)}$ where $\theta^{(m)}$ are simulated from $P(\theta\,\|\,X)$
Expected value of a function	$E(h(\theta)\,\|\,X) \approx \frac{1}{M}\sum_m h(\theta^{(m)})$ where $\theta^{(m)}$ are simulated from $P(\theta\,\|\,X)$
Normalizing constant	$\int_{\forall\theta} P(X\,\|\,\theta)\,P(\theta)\,\mathrm{d}\theta = E(P(X\,\|\,\theta)) \approx \frac{1}{M}\sum_m P(X\,\|\,\theta^{(m)})$ where $\theta^{(m)}$ are simulated from $P(\theta)$ (the prior)
Marginalization	$P(\theta\,\|\,X) = \int P(\theta,\psi\,\|\,X)\,\mathrm{d}\psi = \int P(\theta\,\|\,\psi,X)\,P(\psi\,\|\,X)\,\mathrm{d}\psi = E(P(\theta\,\|\,\psi,X)) \approx \frac{1}{M}\sum_m p(\theta\,\|\,\psi^{(m)},X)$ where we get $\psi^{(m)}$ by simulating $(\theta^{(1)},\psi^{(1)}),\ldots,(\theta^{(M)},\psi^{(M)})$ from $P(\theta,\psi\,\|\,X)$

Now the computational problem is to maximize the posterior distribution. Most numerical maximization techniques have been used in the literature. If θ is continuous then gradient ascent methods are common.

There are also methods of Monte Carlo optimization where θ_{MAP} is searched for stochastically. The most common of these is simulated annealing. This is an iterative search algorithm with some analogies to the MCMC. First, a function $T(m)$ is defined (the temperature) of the iteration m of the process; this is a decreasing function that tends to 0. At stage m of the search, having reached a value $\theta^{(m-1)}$ at the last stage, a new value $\theta^{(*)}$ is proposed. As with the Metropolis algorithm it can depend on $\theta^{(m-1)}$ and can be generated randomly. This new value is accepted as $\theta^{(m)}$ with a probability that depends on $T(m)$, $\theta^{(m-1)}$ and θ^* with the following properties: if $P(\theta^* \mid X) > P(\theta^{(m-1)} \mid X)$ then θ^* is accepted with probability 1; if $P(\theta^* \mid X) < P(\theta^{(m-1)} \mid X)$ then θ^* has some non-zero probability of being accepted; and that as $T(m) \to 0$ so this latter acceptance probability tends to 0. For example, an acceptance probability of the form $\max(1, [P(\theta^* \mid X)/P(\theta^{(m-1)} \mid X)]^{1/T(m)})$ satisfies these conditions. Under some conditions on how quickly the sequence $T(m)$ tends to zero and properties of $P(\theta \mid X)$, this method can converge to the MAP.

There are many variants of simulated annealing; it is particularly common in audio, image and video reconstructions and segmentations.

1.6 Decision Theory

The quantification of uncertainty and learning from data are not always the final goals in a statistical machine learning task. Often we must make a decision, the consequences of which are dependent on the outcome of the uncertain quantity. The making of decisions under uncertainty is the focus of decision theory. In multimedia applications, the decision is most often which value of the quantity of interest to take as the solution, given that a posterior distribution on it has been computed. While probability theory is a coherent method of quantifying and managing uncertainties, decision theory attempts to do the same for making decisions.

Most decision problems can be divided into three components:

1. *Actions:* There are a set of available actions or decisions that we can take. This set may be discrete or continuous. The decision problem is to choose the "best" action from this group. For example, it may be to retrieve a set of images from a database in a retrieval problem, decide if an e-mail is spam or not or select a segmentation for an image.
2. *States of nature:* These are the unknowns in the decision problem that will affect the outcome of whatever action is taken. As the states of nature are uncertain, it will be necessary to assign probabilities to them. Hopefully we have data that inform us about this probability distribution; usually this will be a posterior distribution. The set of states of nature may change according to the action. In multimedia applications these are usually the "true" state of the quantity of interest, and perhaps also some parameters associated with it, e.g. the true target image

in an image retrieval, the true nature of an e-mail (spam/not spam) or the true segmentation of an image.
3. *Consequences:* Connected with every action and state of nature is an outcome or consequence. As with uncertainty, it will be necessary to quantify the consequences somehow; as with probability, this is done by subjectively assigning a number to each consequence called its *utility*. This function increases with increasing preference, e.g. if consequence c is preferred to consequence c^* then the utility of c is greater than that of c^*.

In most multimedia applications of Bayesian methods, actions and states of nature are explicitly defined in the problem. Utility is not usually defined, although we will see that it is implicitly.

The general approach to a decision problem is to enumerate the possible actions and states of nature, assign probabilities and utilities where needed and use these assignments to solve the problem by producing the "best" action. We have seen how probabilities are assigned and updated. That leaves the issues of assigning utilities and the definition of the best action.

1.6.1 Utility and Choosing the Optimal Decision

Consider a decision problem where there are a finite number m possible actions a_1, $a_2,\ldots, a_m$ and n possible states of nature $\theta_1, \theta_2,\ldots, \theta_n$. If action a_i is chosen then we denote the probability that θ_j occurs by p_{ij}. After choosing action a_i, a state of nature θ_j will occur and there will be a consequence c_{ij}. We denote the utility of that consequence by $\mathscr{U}(c_{ij})$, for some real-valued utility function $\mathscr{U}$. The idea generalizes to a continuous space of actions, states of nature and consequences. Utilities, like probabilities, should be defined subjectively.

The optimal action is that which yields the highest utility to the decision maker. However, since it is not known which state of nature will occur before an action is taken, these utilities are not known. However, we can calculate the *expected utility* of a particular action, using the probabilities p_{ij}:

$$E(\mathscr{U}(a_i)) = \sum_{j=1}^{n} p_{ij}\mathscr{U}(c_{ij}), \tag{1.18}$$

for $i = 1,\ldots,m$. We choose the action a_i for which the expected utility is a maximum. This is called the *principle of maximizing expected utility* and is the decision criterion for choosing a decision under uncertainty. The principle generalizes in the usual way when there is a continuum of possible actions and states of nature; the p_{ij}s are replaced by densities and one forms expected utilities by integration.

There are strong mathematical arguments to back the use of this principle as a decision rule and they are linked to the ideas of de Finetti on subjective probability and betting. de Finetti shows that following any other rule to choose a decision leaves the decision maker vulnerable to making decisions where loss is inevitable;

this "Dutch book" is identical to that used to justify the laws of probability. Lindley discusses this in some detail, see [14, Chap. 4].

1.6.2 Where Is the Utility?

We have introduced earlier the idea of the MAP estimate of θ: the mode of the posterior distribution. This is very common in image reconstruction and segmentation. It is trivial to show that this is the solution to the decision problem where the action is to decide a value $\hat{\theta}$, the state of nature is the "true" value θ with probability distribution $P(\theta\,|\,X)$ and the utility is

$$U(\hat{\theta},\theta) = \begin{cases} 1, \text{ if } \hat{\theta}=\theta, \\ 0, \text{ otherwise.} \end{cases} \tag{1.19}$$

Another solution is the MPM (marginal posterior mode), the θ that maximizes the marginal posterior of each component $P(\theta_i\,|\,X)$. This is the solution with utility

$$U(\hat{\theta},\theta) = ||\{i\,|\,\hat{\theta}_i=\theta_i\}||. \tag{1.20}$$

This is also used in image restoration and segmentation.

Other solutions, where they make sense, are to use the posterior mean of each component of θ or perhaps the posterior median. These in turn are the solutions with utilities $U(\hat{\theta},\theta)=\Sigma_i(\hat{\theta}_i-\theta_i)^2$ and $U(\hat{\theta},\theta)=\Sigma_i\,|\hat{\theta}_i-\theta_i|$, respectively.

The point we are trying to make here is that in fact many machine learning solutions using a Bayesian method are using implicitly a utility; if the task is to make a decision on the basis of a posterior distribution then this is always the case. In the case of the MAP solution, the utility is of a very simple form. While this utility is sensible in many situations, thinking about utilities other than a 0–1 type yields a very large class of alternative solutions that specify more richly the nature of a good solution. If the MAP is being computed numerically, by simulated annealing for example, using a richer utility function may not be more expensive computationally either.

1.7 Naive Bayes

While a comprehensive Bayesian analysis of data can be very complex it is often the case that a naïve Bayesian analysis can be quite effective. In fact the naïve Bayes or simple Bayes classifier is in widespread use in data analysis. The naïve Bayes classifier is based on the key restrictive assumption that the attributes X_i are independent of each other, i.e. $P(X_i|X_k)=P(X_i)$. This will almost never be true in practice so it is perhaps surprising that the naïve Bayes classifier is often very effective on real data. One reason for this is that we do not need to know the precise values for the class probabilities $P(\theta\in\{1,...,k\})$, we simply require the classifier to be able to *rank* them correctly. Since naïve Bayes scales well to high-dimension data it is used frequently in multimedia applications, particularly in text processing where it has been shown to be quite accurate [13].

In order to explain the operation of a naïve Bayes classifier we can begin with the MAP equation presented in Sect. 1.5.4:

$$\theta_{\text{MAP}} = \arg\max_{\theta} P(\theta \,|\, X).$$

In classification terms θ is the class label for an object described by the set of attributes $X = \{X_1, ..., X_n\}$ and θ is one of the k possible values that class label can have:

$$\theta_{\text{MAP}} = \arg\max_{\theta} P(\theta \,|\, X_1, X_2, ..., X_n).$$

Using Bayes rule this can be rewritten as

$$\theta_{\text{MAP}} = \arg\max_{\theta} \frac{P(\theta)P(X_1, X_2, ..., X_n \,|\, \theta)}{P(X_1, X_2, ..., X_n).}$$

Since the prior probabilities in the denominator are constant for all class labels this can be rewritten as

$$\theta_{\text{MAP}} = \arg\max_{\theta} P(\theta)P(X_1, X_2, ..., X_n \,|\, \theta).$$

The big leap with the naïve Bayes classifier is to approximate this as

$$\theta_{\text{NB}} = \arg\max_{\theta} P(\theta) \prod_{i=1}^{n} P(X_i \,|\, \theta). \tag{1.21}$$

This assumes that the X_i terms are independent which will rarely be true as stated above; nevertheless this classifier is often very effective on multimedia data.

In text classification, the conditional probabilities can be estimated by $P(X_i|\theta = j) = n_{ij}/n_j$ where n_{ij} is the number of times that attributes X_i occurs in those documents with classification $\theta = j$ and n_j is the number of documents with classification $\theta = j$. This provides a good estimate of the probability in many situations but in situations where n_{ij} is very small or even equal to zero this probability will dominate, resulting in an overall zero probability. A solution to this is to incorporate a small-sample correction into all probabilities called the Laplace correction [16]. The corrected probability estimate is $P(X_i|\theta = j) = (n_{ij} + f)/(n_j + f \times n_{ki})$, where n_{ki} is the number of values for attribute θ_i. Kohavi et al. [11] suggest a value of $f = 1/m$ where m is equal to the number of training documents.

1.8 Further Reading

The number of books on Bayesian statistical inference is large. For Bayesian methods, a good introductory text is by Lee [12]. More technical texts are by Berger [1] and Bernardo and Smith [2]. A good guide to using and implementing Bayesian methods for real data analysis is [7]. For the philosophical background to subjective probability and Bayesian methods, the two volumes by de Finetti are the classic texts [4]. For the links between decision theory and Bayesian methods, see [14]. For a thorough review of numerical methods for Bayesian inference, see [19].

Acknowledgements This material has been written in light of notes taken by the authors from Nozer Singpurwalla at The George Washington University and under the auspices of the European Union Network of Excellence MUSCLE; see www.muscle-noe.org.

References

1. J. O. Berger. *Statistical decision theory and Bayesian analysis*. Springer-Verlag, New York, second edition, 1993.
2. J. M. Bernardo and A. F. M. Smith. *Bayesian theory*. Wiley, Chichester, 1994.
3. R. T. Cox. Probability, frequency and reasonable expectation. *Am. J. Phys.*, 14:1–13, 1946.
4. B. de Finetti. *Theory of probability*, volume 1. Wiley, New York, 1974.
5. D. Gamerman. *Markov chain Monte Carlo: stochastic simulations for Bayesian inference*. Chapman and Hall, New York, 1997.
6. P. H. Garthwaite, J. B. Kadane, and A. O'Hagan. Statistical methods for eliciting probability distributions. *J. Am. Stat. Assoc.*, 100:680–701, 2005.
7. A. Gelman, J. B. Carlin, H. S. Stern, and D. B. Rubin. *Bayesian data analysis*. Chapman and Hall, London, second edition, 2003.
8. W. R. Gilks, N. G. Best, and K. K. C. Tan. Adaptive rejection metropolis sampling within Gibb's sampling. *Appl. Stat.*, 44:455–472, 1995.
9. W. R. Gilks, G. O. Roberts, and E. I. George. Adaptive rejection sampling. *Statistician*, 43:179–189, 1994.
10. G. Grimmett and D. Stirzaker. *Probability and Random Processes*. Oxford University Press, Oxford, third edition, 2001.
11. R. Kohavi, B. Becker, and D. Sommerfield. Improving simple bayes. In *Proceedings of the European Conference on Machine Learning (ECML-87)*, pages 78–97, 1997.
12. P. M. Lee. *Bayesian statistics: an introduction*. Hodder Arnold H&S, London, third edition, 2004.
13. D. D. Lewis and M. Ringuette. A comparison of two learning algorithms for text categorization. In *Proceedings of SDAIR-94, 3rd Annual Symposium on Document Analysis and Information Retrieval*, pp. 81–93, Las Vegas, US, 1994.
14. D. V. Lindley. *Making decisions*. Wiley, London, second edition, 1982.
15. K. L. Mengersen, C. P. Robert, and C. Guihenneuc-Jouyaux. Mcmc convergence diagnostics: a "reviewww". In J. Berger, J. Bernardo, A. P. Dawid, and A.F.M. Smith, editors, *Bayesian Statistics 6*, pp. 415–440. Oxford Science Publications, 1999.
16. T. Niblett. Constructing decision trees in noisy domains. In *2nd European Working Session on Learning*, pp. 67–78, Bled, Yugoslavia, 1987.
17. S. M. Ross. *Simulation*. Academic Press, San Diego, third edition, 2001.
18. S. M. Ross. *Introduction to probability models*. Academic Press, San Diego, eighth edition, 2003.
19. M. A. Tanner. *Tools for statistical inference: methods for the exploration of posterior distributions and likelihood functions*. Springer-Verlag, New York, third edition, 1996.
20. V. uSmídl and A. Quinn. *The variational Bayes method in signal processing*. Springer, New York, 2005.
21. G. Winkler. *Image analysis, random fields and dynamic Monte Carlo methods*. Springer-Verlag, Berlin, second edition, 2006.

Chapter 2
Supervised Learning

Pádraig Cunningham, Matthieu Cord, and Sarah Jane Delany

Abstract Supervised learning accounts for a lot of research activity in machine learning and many supervised learning techniques have found application in the processing of multimedia content. The defining characteristic of supervised learning is the availability of annotated training data. The name invokes the idea of a 'supervisor' that instructs the learning system on the labels to associate with training examples. Typically these labels are class labels in classification problems. Supervised learning algorithms *induce* models from these training data and these models can be used to classify other unlabelled data. In this chapter we ground or analysis of supervised learning on the theory of risk minimization. We provide an overview of support vector machines and nearest neighbour classifiers – probably the two most popular supervised learning techniques employed in multimedia research.

2.1 Introduction

Supervised learning entails learning a mapping between a set of *input* variables $\mathscr{X}$ and an *output* variable $\mathscr{Y}$ and applying this mapping to predict the outputs for unseen data. Supervised learning is the most important methodology in machine learning and it also has a central importance in the processing of multimedia data. In this chapter we focus on kernel-based approaches to supervised learning. We review support vector machines which represent the dominant supervised learning technology these days – particularly in the processing of multimedia data. We also review nearest neighbour classifiers which can (loosely speaking) be considered

Pádraig Cunningham
University College Dublin, Dublin, Ireland, e-mail: padraig.cunningham@ucd.ie

Matthieu Cord
LIP6, UPMC, Paris, France, e-mail: matthieu.cord@lip6.fr

Sarah Jane Delany
Dublin Institute of Technology, Dublin, Ireland, e-mail: sarahjane.delany@comp.dit.ie

a kernel-based strategy. Nearest neighbour techniques are popular in multimedia because the emphasis on similarity is appropriate for multimedia data where a rich array of similarity assessment techniques is available.

To complete this review of supervised learning we also discuss the ensemble idea, an important strategy for increasing the stability and accuracy of a classifier whereby a single classifier is replaced by a *committee* of classifiers.

The chapter begins with a summary of the principles of statistical learning theory as this offers a general framework to analyze learning algorithms and provides useful tools for solving real world applications. We present basic notions and theorems of statistical learning before presenting some algorithms.

2.2 Introduction to Statistical Learning

2.2.1 Risk Minimization

In the supervised learning paradigm, the goal is to infer a function $f : \mathscr{X} \to \mathscr{Y}$, the classifier, from a sample data or training set $\mathscr{A}_n$ composed of pairs of (input, output) points, $\mathbf{x}_i$ belonging to some feature set $\mathscr{X}$, and $y_i \in \mathscr{Y}$:

$$\mathscr{A}_n = ((\mathbf{x}_1, y_1), \ldots, (\mathbf{x}_n, y_n)) \in (\mathscr{X} \times \mathscr{Y})^n.$$

Typically $\mathscr{X} \subset \mathbb{R}^d$, and $y_i \in \mathbb{R}$ for regression problems, and y_i is discrete for classification problems. We will often use examples with $y_i \in \{-1, +1\}$ for binary classification.

In the statistical learning framework, the first fundamental hypothesis is that the training data are independently and identically generated from an unknown but fixed joint probability distribution function $P(\mathbf{x}, y)$. The goal of the learning is to find a function f attempting to model the dependency encoded in $P(\mathbf{x}, y)$ between the input $\mathbf{x}$ and the output y. $\mathscr{H}$ will denote the set of functions where the solution is sought: $f \in \mathscr{H}$.

The second fundamental concept is the notion of error or *loss* to measure the agreement between the prediction $f(\mathbf{x})$ and the desired output y. A loss (or *cost*) function $L : \mathscr{Y} \times \mathscr{Y} \to \mathbb{R}^+$ is introduced to evaluate this error. The choice of the loss function $L(f(\mathbf{x}), y)$ depends on the learning problem being solved. Loss functions are classified according to their regularity or singularity properties and according to their ability to produce convex or non-convex criteria for optimization.

In the case of pattern recognition, where $\mathscr{Y} = \{-1, +1\}$, a common choice for L is the misclassification error:

$$L(f(\mathbf{x}), y)) = \frac{1}{2}|f(\mathbf{x}) - y|.$$

This cost is singular and symmetric. Practical algorithmic considerations may bias the choice of L. For instance, singular functions may be selected for their ability to provide *sparse* solutions. For unsupervised learning developed in Chap. 3.6, the problem may be expressed in a similar way using a loss function: $L_u : \mathscr{Y} \to I\!R^+$ defined by: $L_u(f(\mathbf{x})) = -\log(f(\mathbf{x}))$.

The loss function L leads to the definition of the *risk* for a function f, also called the *generalization error*:

$$R(f) = \int L(f(\mathbf{x}),y)\mathrm{d}P(\mathbf{x},y). \tag{2.1}$$

In classification, the objective could be to find the function f in $\mathscr{H}$ that minimizes $R(f)$. Unfortunately, it is not possible because the joint probability $P(\mathbf{x},y)$ is unknown.

From a probabilistic point of view, using the input and output random variable notations X and Y, the risk can be expressed as

$$R(f) = I\!E(L(f(X),Y))$$

which can be rewritten in two expectations:

$$R(f) = I\!E[I\!E(L(f(X),Y)|X)]$$

offering the possibility to separately minimize $I\!E(L(f(\mathbf{x}),Y)|X = \mathbf{x})$ with respect to the scalar value $f(\mathbf{x})$. This expression is connected to the expected utility function introduced in Sect. 1.6.2 of this book. The resulting function is called the Bayes estimator associated with the risk R.

The learning problem is expressed as a minimization of R for any classifier f. As the joint probability is unknown, the solution is inferred from the available training set $\mathscr{A}_n = ((\mathbf{x}_1,y_1),...,(\mathbf{x}_n,y_n))$.

There are two ways to address this problem. The first approach, called generative-based, tries to approximate the joint probability $P(X,Y)$, or $P(Y|X)P(X)$, and then compute the Bayes estimator with the obtained probability. The second approach, called discriminative-based, attacks the estimation of the risk $R(f)$ head on.

Some interesting developments on probability models and estimation may be found in the Chap. 1. We focus in the following on the discriminative strategies, offering nice insights into learning theory.

2.2.2 Empirical Risk Minimization

This strategy tackles the problem of risk minimization by approximating the integral given in (2.1), using a data set $\mathscr{S}_n \in (\mathscr{X} \times \mathscr{Y})^n$ that can be the training set $\mathscr{A}_n$ or any other set:

$$R_{\mathrm{emp}}(f,\mathscr{S}_n) = \frac{1}{n}\sum_{i=I}^{n} L(f(\mathbf{x}_i),y_i). \tag{2.2}$$

This approximation may be viewed as a Monte Carlo integration of the (2.1) as described in Chap. 1. The question is to know if the empirical error is a good approximation of the risk R.

According to the law of large numbers, there is a point-wise convergence of the empirical risk for f to $R(f)$ (as n goes to infinity). This is a motivation to minimize R_{emp} instead of the true risk over the training set $\mathscr{A}_n$; it is the principle of *empirical risk minimization* (ERM):

$$f_{\text{ERM}} = \text{Arg}\min_{f\in\mathscr{H}} R_{\text{emp}}(f,\mathscr{A}_n) = \text{Arg}\min_{f\in\mathscr{H}} \left(\frac{1}{n}\sum_{i=I}^{n} L(f(\mathbf{x}_i),y_i)\right).$$

However, it is not true that, for an arbitrary set of functions $\mathscr{H}$, the empirical risk minimizer will converge to the minimal risk in the class function $\mathscr{H}$ (as n goes to infinity). There are classical examples where, considering the set of all possible functions, the minimizer has a null empirical risk on the training data, but an empirical risk or a risk equal to 1 for a test data set. This shows that learning is impossible in that case. The *no free lunch* theorem [51] is related to this point.

A desirable property for the minimizers is consistency, which can be expressed in terms of probability as the data size n goes to infinity: [45, 46]:

$$\text{ERM consistent iff } \forall\varepsilon > 0 \ \ \lim_{n\to\infty} P(\sup_{f\in\mathscr{H}} (R(f) - R_{\text{emp}}(f,\mathscr{A}_n)) > \varepsilon) = 0.$$

Thus, the learning crucially depends on the set of functions $\mathscr{H}$, and this dependency may be expressed in terms of uniform convergence that is theoretically intriguing, but not so helpful in practice. For instance, characterizations of the set of functions $\mathscr{H}$ may be useful. A set of functions with smooth decision boundaries is chosen underlying the smoothness of the decision function in real world problems. It is possible to restrict $\mathscr{H}$ by imposing a constraint of regularity to the function f. This strategy belongs to regularization theory. Instead of minimizing the empirical risk, the following regularized risk is considered:

$$R_{\text{reg}}(f) = R_{\text{emp}}(f,\mathscr{A}_n) + \lambda\Omega(f),$$

where $\Omega(f)$ is a functional introducing a roughness penalty. It will be large for functions f varying too rapidly.

One can show that minimizing a regularized risk is equivalent to ERM on a restricted set $\mathscr{F}$ of functions f. Another way to deal with the tradeoff between ERM and constraints on $f \in \mathscr{H}$ is to investigate the characterization of $\mathscr{H}$ in terms of strength or complexity for learning.

2.2.3 *Risk Bounds*

The idea is to find a bound depending on $\mathscr{H}$, $\mathscr{A}_n$ and δ, such that, for any $f \in \mathscr{H}$, with a probability at least $1-\delta$:

$$R(f) \leq R_{\text{emp}}(f, \mathscr{A}_n) + B(\mathscr{H}, \mathscr{A}_n, \delta).$$

First, we consider the case of a finite class of functions, $|\mathscr{H}| = N$. Using the Hoeffding inequality (1963), by summing the probability over the whole set, one can show the following result:

$$\forall \delta \in]0,1],\ P\left(\forall f \in \mathscr{H},\ R(f) \leq R_{\text{emp}}(f, \mathscr{A}_n) + \sqrt{\frac{\log(N/\delta)}{2n}}\right) \geq 1 - \delta.$$

This inequality shows how the risk is limited by a sum of two terms, the empirical error and a bound depending on the size of the set of functions and on the empirical risk.

As $\lim_{n \to \infty} \sqrt{\frac{\log(N/\delta)}{2n}} = 0$, we have the result: for any finite class of functions, the ERM principle is consistent for any data distribution. The tradeoff between these two terms is fundamental in machine learning, it is also called the bias/variance dilemma in the literature. It is easy to see that if $\mathscr{H}$ is large, then one can find an f that fits the training data well, but at the expense of undesirable behaviour at other points, such as lack of smoothness, that will give poor performance on test data. This scenario where there is no generalization is termed overfitting. On the other hand, if $\mathscr{H}$ is too small, there is no way to find an f function that correctly fits the training data.

To go one step further, there is an extension to infinite sets of functions. Instead of working on the size of the set, a notion of complexity, the Vapnik–Chervonenkis (VC) dimension, provides a measure of the capacity of the functions to differently label the data (in classification context) [45]: The VC dimension h of a class of functions $\mathscr{F}$ is defined as the maximum number of points that can be learnt exactly by a function of $\mathscr{F}$:

$$\text{h} = \max\left\{|\mathbf{X}|, \mathbf{X} \subset \mathscr{X}, \forall b \in \{-1,+1\}^{|\mathbf{X}|}, \exists f \in \mathscr{F} \mid \forall \mathbf{x}_i \in \mathbf{X}, f(\mathbf{x}_i) = b_i\right\}.$$

A strategy of bounding the risk has been developed in order to produce a new bound, depending on h, n and δ) [45]:

$$B(h,n,\delta) = \sqrt{\frac{h(\log(2n/h)+1) + \log(4/\delta)}{n}},$$

for which we have, for any $f \in \mathscr{H}$, with a probability at least $1 - \delta$:

$$R(f) \leq R_{\text{emp}}(f, \mathscr{A}_n) + B(h,n,\delta).$$

Moreover, a nice result due to Vapnik [45] stipulates: *ERM is consistent (for any data distribution) iff the VC dimension h is finite.*

The tradeoff is now between controlling $B(h,n,\delta)$, which increases monotonically with the VC dimension h, and having a small empirical error on training data. The structural risk minimization (SRM) principle introduced by Vapnik exploits this last bound by considering classes of functions embedded by increasing h values.

We illustrate an aspect of the SRM over the family c_γ of linear functions to classify data $\mathscr{X} \subset I\!R^d$ with a γ margin:

$$\begin{aligned} c_{w,b,\gamma}(\mathbf{x}) &= 1 \text{ if } <\mathbf{x},\mathbf{w}> +b \geq \gamma \\ c_{w,b,\gamma}(\mathbf{x}) &= -1 \text{ if } <\mathbf{x},\mathbf{w}> +b \leq -\gamma \end{aligned}$$

for $\mathbf{w} \in I\!R^d, ||\mathbf{w}|| = 1$. Let $\mathscr{X}$ be included in a ball of radius R, then we have the following bound on the VC dimension h_γ:

$$h_\gamma \leq \left\lceil \frac{R^2}{\gamma^2} \right\rceil + 1.$$

If we can choose between two linear functions that perfectly classify the training data, it makes sense from the last bound to select the one that maximizes γ. Indeed, such a classifier is ensured to have a null empirical error $R_{\text{emp}}(f,\mathscr{A}_n)$ on the training set and to belong to a class of classifiers c_γ with the hardest bounding on h_γ (hence on $B(h_\gamma,n,\delta)$). This rule is used to build the famous support vector machines classifiers.

2.3 Support Vector Machines and Kernels

Support vector machines (SVM) are a type of learning algorithm developed in the 1990s. They are based on results from statistical learning theory introduced by Vapnik [45] described previously. These learning machines are also closely connected to kernel functions [37], which are a central concept for a number of learning tasks.

The kernel framework and SVM are used now in a variety of fields, including multimedia information retrieval (see for instance [42, 48] for CBIR applications), bioinformatics and pattern recognition.

We focus here on the introduction of SVM as linear discriminant functions for binary classification. A complete introduction to SVM and kernel theory can be found in [10] and [36].

2.3.1 Linear Classification: SVM Principle

To introduce the basic concepts of these learning machines, we start with the linear support vector approach for binary classification. We assume here that both classes are linearly separable. Let $(\mathbf{x}_i)_{i\in[1,N]}$, $\mathbf{x}_i \in \mathbb{R}^p$ be the feature vectors representing the training data and $(y_i)_{i\in[1,N]}$, $y_i \in \{-1,1\}$ be their respective class labels. We can define a hyperplane by $<\mathbf{w},\mathbf{x}> +b = 0$ where $\mathbf{w} \in I\!R^p$ and $b \in I\!R$. Since the classes are linearly separable, we can find a function f, $f(\mathbf{x}) =<\mathbf{w},\mathbf{x}> +b$ with

$$y_i f(\mathbf{x}_i) = y_i(<\mathbf{w},\mathbf{x}_i> +b) > 0,\ \forall i \in [1,N]. \tag{2.3}$$

The decision function may be expressed as $f_d(\mathbf{x}) = \text{sign}(< \mathbf{w}, \mathbf{x} > +b)$ with

$$f_d(\mathbf{x}_i) = \text{sign}(y_i),\ \forall i \in [1,N].$$

Since many functions realize the correct separation between training data, additional constraints are used. One way described in Sect. 2.2 aims to investigate the generalization properties for any of the candidates. In that context, the larger the margin, the smaller the VC dimension. This is the basic rule to express the SVM optimization: the SVM classification method aims at finding the *optimal* hyperplane based on the maximization of the *margin*[1] between the training data for both classes.

Because the distance between a point $\mathbf{x}$ and the hyperplane is $\frac{yf(\mathbf{x})}{||\mathbf{w}||}$, it is easy to show [10] that the optimization problem may be expressed as the following minimization:

$$\min \frac{1}{2}||\mathbf{w}||^2 \text{ subject to } y_i(<\mathbf{w}, \mathbf{x}_i> +b) \geq 1,\ \forall i \in [1,N]. \tag{2.4}$$

The *support vectors* are the training points for which we have an equality in (2.4). They are all equally close to the optimal hyperplane. One can prove that they are enough to compute the separating hyperplane (hence their name).

This is a convex optimization problem (quadratic criterion, linear inequality constraints). Usually, the dual formulation is favoured for its easy solution with standard techniques. Using Lagrange multipliers, the problem may be re-expressed in the equivalent maximization on α (dual form):

$$\alpha^\star = \underset{\alpha}{\text{argmax}} \sum_{i=1}^{N} \alpha_i - \frac{1}{2} \sum_{i,j=1}^{N} \alpha_i \alpha_j y_i y_j <\mathbf{x}_i, \mathbf{x}_j> \tag{2.5}$$

$$\text{s.t.} \sum_{i=1}^{N} \alpha_i y_i = 0 \quad \text{and} \quad \forall i \in [1,N] \quad \alpha_i \geq 0.$$

The hyperplane decision function can be written as

$$f_d(\mathbf{x}) = \text{sign}(\sum_{i=1}^{N} y_i \alpha_i^\star <\mathbf{x}, \mathbf{x}_i> +b). \tag{2.6}$$

2.3.2 *Soft Margin*

The previous method is applicable to linearly separable data. When data cannot be linearly separated, the linear SVM method may be adapted. A soft margin may be used in order to get better efficiency in a noisy situation.

[1] The margin is defined as the distance from the hyperplane of the closest points, on either side.

In order to carry out the optimization, the constraints of (2.3) are relaxed using slack variables ξ_i:

$$y_i f(\mathbf{x}_i) = y_i(<\mathbf{w}, \mathbf{x}_i> + b) > 1 - \xi_i, \text{ with } \xi_i \geq 0 \ \forall i \in [1, N]. \tag{2.7}$$

The minimization may be then expressed as follows:

$$\min \frac{1}{2}||\mathbf{w}||^2 + C\sum_{i=1}^{N} \xi_i \tag{2.8}$$

with a constant $C > 0$, subject to the constraints (2.7).

The dual representation is obtained in a similar way:

$$\alpha^\star = \underset{\alpha}{\text{argmax}} \sum_{i=1}^{N} \alpha_i - \frac{1}{2} \sum_{i,j=1}^{N} \alpha_i \alpha_j y_i y_j < \mathbf{x}_i, \mathbf{x}_j > \tag{2.9}$$

subject to

$$\sum_{i=1}^{N} \alpha_i y_i = 0 \quad \text{and} \quad \forall i \in [1, N] \quad 0 \leq \alpha_i \leq C.$$

It is a very simple adaptation of the original algorithm by introducing a bound C [47]. The constant C is used to tune the tradeoff between having a large margin and few classification errors.

2.3.3 Kernel-Based Classification

The linear SVM classifier previously described finds linear boundaries in the input feature space. To obtain more general decision surfaces, the feature space may be mapped into a larger space before undertaking linear classification. Linear boundaries in the enlarged space translate to non-linear boundaries in the original space.

Let us denote the induced space $\mathscr{H}$ via a map Φ:

$$\begin{aligned} \Phi : \mathbb{R}^p &\rightarrow \mathscr{H} \\ \mathbf{x} &\mapsto \Phi(\mathbf{x}). \end{aligned}$$

In the SVM framework, there is no assumption on the dimensionality of $\mathscr{H}$, which could be very large, and sometimes infinite. We just suppose that $\mathscr{H}$ is equipped with a dot product. Maximizing (2.9) now requires the computation of dot products $< \Phi(\mathbf{x}_i), \Phi(\mathbf{x}_j) >$. In practice, $\Phi(\mathbf{x})$ is never computed due to the *kernel trick*: kernel functions $k(.,.)$, respecting some conditions (positive definite, Mercer's conditions[2]), are introduced such that

$$k(\mathbf{x}, \mathbf{y}) = < \Phi(\mathbf{x}), \Phi(\mathbf{y}) > .$$

[2] We are dealing with the class of kernels k that correspond to dot product in the enlarged space.

Using these functions allows us to avoid to explicitly compute $\Phi(\mathbf{x})$. Everything about the linear case also applies to non-linear cases, using a suitable kernel k instead of the Euclidean dot product. The resulting optimization problem may be expressed as follows:

$$\alpha^\star = \underset{\alpha}{\operatorname{argmax}} \sum_{i=1}^{N} \alpha_i - \frac{1}{2} \sum_{i,j=1}^{N} \alpha_i \alpha_j y_i y_j k(\mathbf{x}_i, \mathbf{x}_j)$$
$$\text{with} \begin{cases} \sum_{i=1}^{N} \alpha_i y_i = 0 \\ \forall i \in [1,N] \quad 0 \leq \alpha_i \leq C. \end{cases} \tag{2.10}$$

Thanks to the optimal $\alpha^\star$ value, the decision function is

$$f_d(\mathbf{x}) = \operatorname{sign}(\sum_{i=1}^{N} y_i \alpha_i^\star k(\mathbf{x}, \mathbf{x}_i) + b). \tag{2.11}$$

Some popular choices for k in the SVM literature are, for $\mathscr{X} \subset \mathbb{R}^d$:

- Gaussian radial basis function kernel:

$$k(\mathbf{x}, \mathbf{y}) = \mathrm{e}^{-\frac{1}{2}\left(\frac{\|\mathbf{x}-\mathbf{y}\|}{\sigma}\right)^2} \tag{2.12}$$

- Sigmoid kernel: $k(\mathbf{x}, \mathbf{y}) = \tanh(\eta_1 < \mathbf{x}, \mathbf{y} > + \eta_2)$
- Polynomial kernel: $k(\mathbf{x}, \mathbf{y}) = < \mathbf{x}, \mathbf{y} >^d$
- Rational kernel: $k(\mathbf{x}, \mathbf{y}) = 1 - \frac{\|\mathbf{x}-\mathbf{y}\|^2}{\|\mathbf{x}-\mathbf{y}\|^2 + b}$

For multicategory problems, the linear SVM approach may be extended using many linear machines to create boundaries consisting of sections of hyperplanes. When linear discrimination is not effective, an appropriate nonlinear mapping can be found.

For density estimation problems, adaptation of binary SVM machines is also proposed. The one-class SVM method estimates the density support of a vector set $(\mathbf{x}_i)_{i \in [1,N]}$.

2.4 Nearest Neighbour Classification

The principle underlying nearest neighbour classification is quite straightforward, examples are classified based on the class of their nearest neighbours. It is often useful to take more than one neighbour into account so the technique is more commonly referred to as k-nearest neighbour (k-NN) Classification where k nearest neighbours are used in determining the class. Since the training examples are needed at run-time, i.e. they need to be in memory at run-time, it is sometimes also called memory-based classification. Because induction is delayed to run-time, it is considered a *lazy*

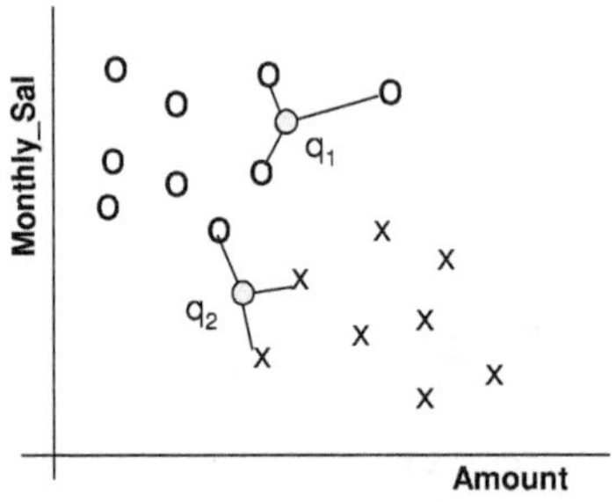

Fig. 2.1 A simple example of 3-nearest neighbour classification

learning technique. Because classification is based directly on the training examples it is also called example-based classification or case-based classification.

The basic idea is as shown in Fig. 2.1 which depicts a 3-nearest neighbour classifier on a two-class problem in a two-dimensional feature space. In this example the decision for q_1 is straightforward – all three of its nearest neighbours are of class O so it is classified as an O. The situation for q_2 is a bit more complicated at it has two neighbours of class X and one of class O. This can be resolved by simple majority voting or by distance-weighted voting (see below).

So k-NN classification has two stages; the first is the determination of the nearest neighbours and the second is the determination of the class using those neighbours.

Let us assume that we have a training data set D made up of $(\mathbf{x}_i)_{i\in[1,|D|]}$ training samples. The examples are described by a set of features F and any numeric features have been normalized to the range [0,1]. Each training example is labelled with a class label $y_j \in Y$. Our objective is to classify an unknown example $\mathbf{q}$. For each $\mathbf{x}_i \in D$ we can calculate the distance between $\mathbf{q}$ and $\mathbf{x}_i$ as follows:

$$d(\mathbf{q},\mathbf{x}_i) = \sum_{f\in F} w_f \delta(\mathbf{q}_f, \mathbf{x}_{if}). \tag{2.13}$$

There are a large range of possibilities for this distance metric; a basic version for continuous and discrete attributes would be

$$\delta(\mathbf{q}_f, \mathbf{x}_{if}) = \begin{cases} 0 & f \text{ discrete and } \mathbf{q}_f = \mathbf{x}_{if} \\ 1 & f \text{ discrete and } \mathbf{q}_f \neq \mathbf{x}_{if} \\ |\mathbf{q}_f - \mathbf{x}_{if}| & f \text{ continuous} \end{cases}. \tag{2.14}$$

The k nearest neighbours are selected based on this distance metric. Then there are a variety of ways in which the k nearest neighbours can be used to determine the class of $\mathbf{q}$. The most straightforward approach is to assign the majority class among the nearest neighbours to the query.

It will often make sense to assign more weight to the nearer neighbours in deciding the class of the query. A fairly general technique to achieve this is distance-weighted voting where the neighbours get to vote on the class of the query case with votes weighted by the inverse of their distance to the query:

$$\text{Vote}(y_j) = \sum_{c=1}^{k} \frac{1}{d(\mathbf{q},\mathbf{x}_c)^n} 1(y_j, y_c). \tag{2.15}$$

Thus the vote assigned to class y_j by neighbour $\mathbf{x}_c$ is 1 divided by the distance to that neighbour, i.e. $1(y_j, y_c)$ returns 1 if the class labels match and 0 otherwise. In (2.15) n would normally be 1 but values greater than 1 can be used to further reduce the influence of more distant neighbours.

Another approach to voting is based on Shepard's work [38] and uses an exponential function rather than inverse distance, i.e.:

$$\text{Vote}(y_j) = \sum_{c=1}^{k} \mathrm{e}^{-\frac{d(\mathbf{q},\mathbf{x}_c)}{h}} 1(y_j, y_c). \tag{2.16}$$

This simplicity of k-NN classifiers makes them an attractive option for use in analysing multimedia content. In this section we consider three important issues that arise with the use of k-NN classifiers. In the next section we look at the core issue of similarity and distance measures and explore some *exotic* (dis)similarity measures to illustrate the generality of the k-NN idea. In Sect. 2.4.3 we look at computational complexity issues and review some speed-up techniques for k-NN. In Sect. 4.4 we look at dimension reduction – both feature selection and sample selection. Dimension reduction is of particular importance with k-NN as it has a big impact on computational performance and accuracy. The section concludes with a summary of the advantages and disadvantages of k-NN.

2.4.1 Similarity and Distance Metrics

While the terms *similarity metric* and *distance metric* are often used colloquially to refer to any measure of affinity between two objects, the term *metric* has a formal meaning in mathematics. A metric must conform to the following four criteria (where $d(x,y)$ refers to the distance between two objects x and y):

1. $d(x,y) \geq 0$; non-negativity
2. $d(x,y) = 0$ only if $x = y$; identity
3. $d(x,y) = d(y,x)$; symmetry
4. $d(x,z) \leq d(x,y) + d(y,z)$; triangle inequality

It is possible to build a k-NN classifier that incorporates an affinity measure that is not a proper metric, however, there are some performance optimizations to the basic k-NN algorithm that require the use of a proper metric [3, 34]. In brief, these techniques can identify the nearest neighbour of an object without comparing that object to every other object but the affinity measure must be a metric, in particular it must satisfy the triangle inequality.

The basic distance metric described in (2.13) and (2.14) is a special case of the Minkowski distance metric – in fact it is the 1-norm (L_1) Minkowski distance. The general formula for the Minkowski distance is

$$MD_p(\mathbf{q},\mathbf{x}_i) = \left(\sum_{f \in F} |\mathbf{q}_f - \mathbf{x}_{if}|^p \right)^{\frac{1}{p}}. \tag{2.17}$$

The L_1 Minkowski distance is the Manhattan distance and the L_2 distance is the Euclidean distance. It is unusual but not unheard of to use p values greater than 2. Larger values of p have the effect of giving greater weight to the attributes on which the objects differ most. To illustrate this we can consider three points in 2D space; $A = (1,1), B = (5,1)$ and $C = (4,4)$. Since A and B differ on one attribute only the $MD_p(A,B)$ is 4 for all p, whereas $MD_p(A,C)$ is 6, 4.24 and 3.78 for p values of 1, 2 and 3, respectively. So C becomes the nearer neighbour to A for p values of 3 and greater.

The other important Minkowski distance is the L_∞ or Chebyshev distance:

$$MD_\infty(\mathbf{q},\mathbf{x}_i) = \max_{f \in F} |\mathbf{q}_f - \mathbf{x}_{if}|.$$

This is simply the distance in the dimension in which the two examples are most different; it is sometimes referred to as the chessboard distance as it is the number of moves it takes a chess king to reach any square on the board.

In the remainder of this section we will review a selection of other metric distances that are important in multimedia analysis.

2.4.2 Other Distance Metrics for Multimedia Data

The Minkowski distance defined in (2.17) is a very general metric that can be used in a k-NN classifier for any data that is represented as a feature vector. When working with image data a convenient representation for the purpose of calculating distances is a colour histogram. An image can be considered as a grey-scale histogram H of N levels or bins where h_i is the number of pixels that fall into the interval represented by bin i (this vector h is the feature vector). The Minkowski distance formula (2.17) can be used to compare two images described as histograms. L_1, L_2 and less often L_∞ norms are used.

Other popular measures for comparing histograms are the Kullback–Leibler divergence (2.18) [24] and the χ^2 statistic (2.19) [33]:

$$d_{KL}(H,K) = \sum_{i=1}^{N} h_i \log\left(\frac{h_i}{k_i}\right), \tag{2.18}$$

$$d_{\chi^2}(H,K) = \sum_{i=1}^{N} \frac{h_i - m_i}{h_i}, \tag{2.19}$$

where H and K are two histograms, h and k are the corresponding vectors of bin values and $m_i = \frac{h_i + k_i}{2}$.

While these measures have sound theoretical support in information theory and in statistics they have some significant drawbacks. The first drawback is that they are not metrics in that they do not satisfy the symmetry requirement. However, this problem can easily be overcome by defining a modified distance between x and y that is in some way an average of $d(x,y)$ and $d(y,x)$ – see [33] for the Jeffrey divergence which is a symmetric version of the Kullback–Leibler divergence.

A more significant drawback is that these measures are prone to errors due to bin boundaries. The distance between an image and a slightly darker version of itself can be great if pixels fall into an adjacent bin as there is no consideration of adjacency of bins in these measures.

2.4.2.1 Earth Mover Distance

The earth mover distance (EMD) is a distance measure that overcomes many of these problems that arise from the arbitrariness of binning. As the name implies, the distance is based on the notion of the amount of effort required to convert one image to another based on the analogy of transporting *mass* from one distribution to another. If we think of two images as distributions and view one distribution as a mass of earth in space and the other distribution as a hole (or set of holes) in the same space then the EMD is the minimum amount of work involved in filling the holes with the earth.

In their analysis of the EMD Rubner et al. argue that a measure based on the notion of a *signature* is better than one based on a histogram. A signature $\{\mathbf{s}_j = \mathbf{m}_j, w_{\mathbf{m}_j}\}$ is a set of j clusters where $\mathbf{m}_j$ is a vector describing the mode of cluster j and $w_{\mathbf{m}_j}$ is the fraction of pixels falling into that cluster. Thus a signature is a generalization of the notion of a histogram where boundaries and the number of partitions are not set in advance; instead j should be 'appropriate' to the complexity of the image [33].

The example in Fig. 2.2 illustrates this idea. We can think of the clustering as a quantization of the image in some colour space so that the image is represented by a set of cluster modes and their weights. In the figure the source image is represented in a 2D space as two points of weights 0.6 and 0.4; the target image is represented by three points with weights 0.5, 0.3 and 0.2. In this example the EMD is calculated to be the sum of the amounts moved (0.2, 0.2, 0.1 and 0.5) multiplied by the distances they are moved. Calculating the EMD involves discovering an assignment that minimizes this amount.

For two images described by signatures $S = \{\mathbf{m}_j, w_{\mathbf{m}_j}\}_{j=1}^{n}$ and $Q = \{\mathbf{p}_k, w_{\mathbf{p}_k}\}_{k=1}^{r}$ we are interested in the work required to transfer from one to the other for a given flow pattern $\mathbf{F}$:

$$\text{WORK}(S, Q, \mathbf{F}) = \sum_{j=1}^{n} \sum_{k=1}^{r} d_{jk} f_{jk}, \tag{2.20}$$

where d_{jk} is the distance between clusters $\mathbf{m}_j$ and $\mathbf{p}_k$ and f_{jk} is the flow between $\mathbf{m}_j$ and $\mathbf{p}_k$ that minimizes overall cost. An example of this in a 2D colour space is shown in Fig. 2.2. Once the transportation problem of identifying the flow that

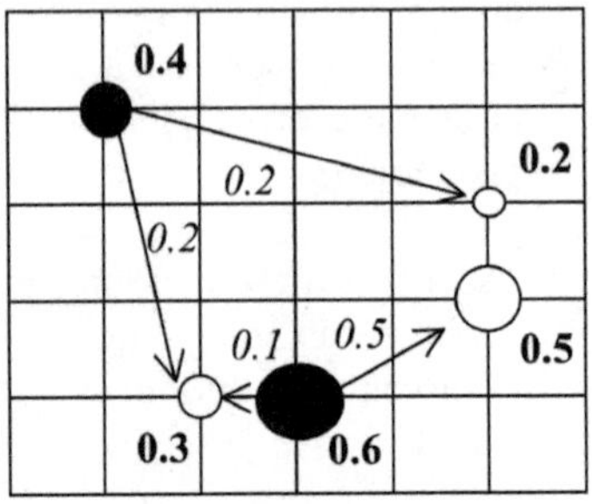

Fig. 2.2 An example of the EMD between two 2D signatures with two points (clusters) in one signature and three in the other (based on example in [32])

minimizes effort is solved (using dynamic programming) the EMD is defined to be

$$\text{EMD}(S,Q) = \frac{\sum_{j=1}^{n}\sum_{k=1}^{r} d_{jk} f_{jk}}{\sum_{j=1}^{n}\sum_{k=1}^{r} f_{jk}}. \tag{2.21}$$

Efficient algorithms for the EMD are described in [33]; however, this measure is expensive to compute with cost increasing more than linearly with the number of clusters. Nevertheless it is an effective measure for capturing similarity between images.

2.4.2.2 Compression-Based Dissimilarity

In recent years the idea of basing a similarity metric on compression has received a lot of attention [21, 28]. Indeed Li et al. [28] refer to this as *The* similarity metric. The basic idea is quite straightforward; if two documents are very similar then the compressed size of the two documents concatenated together will not be much greater than the compressed size of a single document. This will not be true for two documents that are very different. Slightly more formally, the difference between two documents A and B is related to the compressed size of document B when compressed using the codebook produced when compressing document A.

The theoretical basis of this metric is in the field of Kolmogorov complexity, specifically in conditional Kolmogorov complexity:

$$d_{Kv}(x,y) = \frac{Kv(x|y) + Kv(y|x)}{Kv(xy)}, \tag{2.22}$$

where $Kv(x|y)$ is the length of the shortest program that computes x when y is given as an auxiliary input to the program and $Kv(xy)$ is the length of the shortest program that outputs y concatenated to x. While this is an abstract idea it can be approximated using compression

$$d_C(x,y) = \frac{C(x|y) + C(y|x)}{C(xy)}. \quad (2.23)$$

$C(x)$ is the size of data x after compression, and $C(x|y)$ is the size of x after compressing it with the compression model built for y. If we assume that $Kv(x|y) \approx Kv(xy) - Kv(y)$ then we can define a normalized compression distance

$$d_{NC}(x,y) = \frac{C(xy) - \min(C(x),C(y))}{\max(C(x),C(y))}. \quad (2.24)$$

It is important that $C(.)$ should be an appropriate compression metric for the data. Delany and Bridge [12] show that compression using Lempel–Ziv (GZip) is effective for text. They show that this compression-based metric is more accurate in k-NN classification than distance-based metrics on a bag-of-words representation of the text.

2.4.3 Computational Complexity

Computationally expensive metrics such as the earth mover's distance and compression-based (dis)similarity metrics focus attention on the computational issues associated with k-NN classifiers. Basic k-NN classifiers that use a simple Minkowski distance will have a time behaviour that is $O(|D||F|)$ where D is the training set and F is the set of features that describe the data, i.e. the distance metric is linear in the number of features and the comparison process increases linearly with the amount of data. The computational complexity of the EMD and compression metrics is more difficult to characterize but a k-NN classifier that incorporates an EMD metric is likely to be $O(|D|n^3 \log n)$ where n is the number of clusters [33].

For these reasons there has been considerable research on editing down the training data and on reducing the number of features used to describe the data (see Sect. 4.4). There has also been considerable research on alternatives to the exhaustive search strategy that is used in the standard k-NN algorithm. Here is a summary of four of the strategies for speeding up nearest neighbour retrieval:

- **Case-retrieval nets:** k-NN retrieval is widely used in case-based reasoning and case-retrieval nets (CRNs) are perhaps the most popular technique for speeding up the retrieval process. Again, the cases are pre-processed, but this time to form a network structure that is used at retrieval time. The retrieval process is done by *spreading activation* in this network structure. CRNs can be configured to return exactly the same cases as k-NN [26, 27].
- **Footprint-based retrieval:** As with all strategies for speeding up nearest neighbour retrieval, footprint-based retrieval involves a preprocessing stage to organize the training data into a two-level hierarchy on which a two-stage retrieval process operates. The preprocessing constructs a competence model which identifies 'footprint' cases which are landmark cases in the data. This process is not

guaranteed to retrieve the same cases as k-NN but the results of the evaluation of speed-up and retrieval quality are nevertheless impressive [40].

- **Fish & Shrink:** This technique requires the distance to be a true metric as it exploits the triangle inequality property to produce an organization of the case-base into candidate neighbours and cases excluded from consideration. Cases that are remote from the query can be bounded out so that they need not be considered in the retrieval process. Fish & Shrink can be guaranteed to be equivalent to k-NN [34].
- **Cover trees for nearest neighbour:** This technique might be considered the state-of-the-art in nearest neighbour speed-up. It uses a data structure called a cover tree to organize the cases for efficient retrieval. The use of cover trees requires that the distance measure is a true metric; however, they have attractive characteristics in terms of space requirements and speed-up performance. The space requirement is $O(n)$ where n is the number of cases; the construction time is $O(c^6 n \log n)$ and the retrieval time is $O(c^{12} \log n)$ where c is a measure of the inherent dimensionality of the data [3].

These techniques involve additional preprocessing to construct data structures that are used to speed-up retrieval. Consequently they are more difficult to implement than the standard k-NN algorithm. As emphasized at the beginning of this section, the alternative speed-up strategy is to reduce the dimension of the data – this is covered in the next section.

2.4.4 Instance Selection and Noise Reduction

An area of instance-based learning that has prompted much recent research is case-base editing, which involves reducing the number of cases in the training set while maintaining or even improving performance.

2.4.4.1 Early Techniques

Case-base editing techniques have been categorized by Brighton and Mellish [6] as competence preservation or competence enhancement techniques. Competence preservation corresponds to redundancy reduction, removing superfluous cases that do not contribute to classification competence. Competence enhancement is effectively noise reduction, removing noisy or corrupt cases from the training set. Figure 2.3 illustrates both of these, where cases of one class are represented by stars and cases of the other class are represented by circles. Competence preservation techniques aim to remove internal cases in a cluster of cases of the same class and can predispose towards preserving noisy cases as exceptions or border cases. Noise reduction on the other hand aims to remove noisy or corrupt cases but can remove exceptional or border cases which may not be distinguishable from true noise, so a balance of both can be useful.

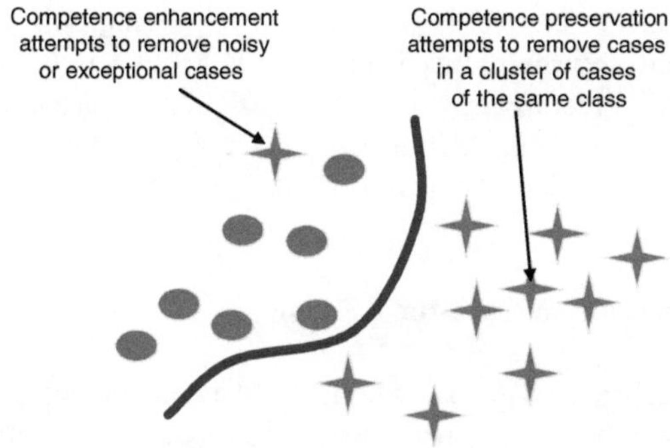

Fig. 2.3 Case-base editing techniques demonstrating competence preservation and competence enhancement

Editing strategies normally operate in one of two ways; *incremental* which involves adding selected cases from the training set to an initially empty edited set, and *decremental* which involves contracting the training set by removing selected cases.

An early competence preservation technique is Hart's condensed nearest neighbour (CNN) [18]. CNN is an incremental technique which adds to an initially empty edited set any case from the training set that cannot be classified correctly by the edited set. This technique is very sensitive to noise and to the order of presentation of the training set cases, in fact CNN by definition will tend to preserve noisy cases. Ritter et al. [31] reported improvements on the CNN with their selective nearest neighbour (SNN) which imposes the rule that every case in the training set must be closer to a case of the same class in the edited set than to any other training case of a different class. Gates [16] introduced a decremental technique which starts with the edited set equal to the training set and removes a case from the edited set where its removal does not cause any other training case to be misclassified. This technique will allow for the removal of noisy cases but is sensitive to the order of presentation of cases.

Competence enhancement or noise reduction techniques start with Wilson's edited nearest neighbour (ENN) algorithm [50], a decremental strategy, which removes cases from the training set which do not agree with their k nearest neighbours. These cases are considered to be noise and appear as exceptional cases in a group of cases of the same class.

Tomek [41] extended this with his repeated ENN (RENN) and his *all k-NN* algorithms. Both make multiple passes over the training set, the former repeating the ENN algorithm until no further eliminations can be made from the training set and the latter using incrementing values of k. These techniques focus on noisy or exceptional cases and do not result in the same storage reduction gains as the competence preservation approaches.

Later editing techniques can be classified as hybrid techniques incorporating both competence preservation and competence enhancement stages. Aha et al. [1] presented a series of instance-based learning algorithms to reduce storage requirements and tolerate noisy instances. IB2 is similar to CNN adding only cases that cannot be classified correctly by the reduced training set. IB2's susceptibility to noise is handled by IB3 which records how well cases are classified and only keeps those that classify correctly to a statistically significant degree. Other researchers have provided variations on the IBn algorithms [7, 8, 52].

2.4.4.2 Competence-Based Case-Base Editing

More recent approaches to case-base editing build a competence model of the training data and use the competence properties of the cases to determine which cases to include in the edited set. Measuring and using case competence to guide case-base maintenance was first introduced by Smyth and Keane [39] and developed by Zu and Yang [53]. Smyth and Keane [39] introduce two important competence properties, the *reachability* and *coverage* sets for a case in a case-base. The *reachability set* of a case *c* is the set of all cases that can successfully classify *c*, and the *coverage set* of a case *c* is the set of all cases that *c* can successfully classify. The coverage and reachability sets represent the local competence characteristics of a case and are used as the basis of a number of editing techniques.

McKenna and Smyth [29] presented a family of competence-guided editing methods for case-bases which combine both incremental and decremental strategies. The family of algorithms is based on four features;

(i) An *ordering policy* for the presentation of the cases that is based on the competence characteristics of the cases,
(ii) An *addition rule* to determine the cases to be added to the edited set,
(iii) A *deletion rule* to determine the cases to be removed from the training set and
(iv) An *update policy* which indicates whether the competence model is updated after each editing step.

The different combinations of ordering policy, addition rule, deletion rule and update policy produce the family of algorithms.

Brighton and Mellish [6] also use the coverage and reachability properties of cases in their iterative case filtering (ICF) algorithm. ICF is a decremental strategy contracting the training set by removing those cases *c*, where the number of other cases that can correctly classify *c* is higher that the number of cases that *c* can correctly classify. This strategy focuses on removing cases far from class borders. After each pass over the training set, the competence model is updated and the process repeated until no more cases can be removed. ICF includes a preprocessing noise reduction stage, effectively RENN, to remove noisy cases. McKenna and Smyth compared their family of algorithms to ICF and concluded that the overall best algorithm of the family delivered improved accuracy (albeit marginal, 0.22%) with less than 50% of the cases needed by the ICF edited set [29].

Wilson and Martinez [49] present a series of reduction technique (RT) algorithms, RT1, RT2 and RT3 which, although published before the definitions of coverage and reachability, could also be considered to use a competence model. They define the set of associates of a case *c* which is comparable to the coverage set of

McKenna and Smyth except that the associates set will include cases of a different class from case c whereas the coverage set will only include cases of the same class as c. The RTn algorithms use a decremental strategy. RT1, the basic algorithm, removes a case c if at least as many of its associates would still be classified correctly without c. This algorithm focuses on removing noisy cases and cases at the centre of clusters of cases of the same class as their associates which will most probably still be classified correctly without them. RT2 fixes the order of presentation of cases as those furthest from their nearest unlike neighbour (i.e. nearest case of a different class) to remove cases furthest from the class borders first. RT2 also uses the original set of associates when making the deletion decision, which effectively means that the associate's competence model is not rebuilt after each editing step which is done in RT1. RT3 adds a noise reduction preprocessing pass based on Wilson's noise reduction algorithm.

Wilson and Martinez [49] concluded from their evaluation of the RTn algorithms against IB3 that RT3 had a higher average generalization accuracy and lower storage requirements overall but that certain data sets seem well suited to the techniques while others were unsuited. Brighton and Mellish [6] evaluated their ICF against RT3 and found that neither algorithm consistently outperformed the other and both represented the 'cutting edge in instance set reduction techniques'.

2.4.5 *k*-NN: Advantages and Disadvantages

k-NN is very simple to understand and easy to implement. So it should be considered in seeking a solution to any classification problem. Some advantages of k-NN are as follows (many of these derive from its simplicity and interpretability):

- Because the process is transparent, it is easy to implement and debug.
- In situations where an explanation of the output of the classifier is useful, k-NN can be very effective if an analysis of the neighbours is useful as explanation.
- There are some noise reduction techniques that work only for k-NN that can be effective in improving the accuracy of the classifier [13].
- Case-retrieval nets [26] are an elaboration of the memory-based classifier idea that can greatly improve run-time performance on large case-bases.

These advantages of k-NN, particularly those that derive from its interpretability, should not be underestimated. On the other hand, some significant disadvantages are as follows:

- Because all the work is done at run-time, k-NN can have poor run-time performance if the training set is large.
- k-NN is very sensitive to irrelevant or redundant features because all features contribute to the similarity (see (2.13)) and thus to the classification. This can be ameliorated by careful feature selection or feature weighting.
- On very difficult classification tasks, k-NN may be outperformed by more *exotic* techniques such as support vector machines or neural networks.

2.5 Ensemble Techniques

2.5.1 Introduction

The key idea in ensemble research is in many situations a committee of classifiers will produce better results than a single classifier – better in terms of stability and accuracy. This is particularly the case when the component classifiers are unstable as is the case with neural networks and decision trees. While the use of ensembles in machine learning research is fairly new, the idea that aggregating the opinions of a committee of experts will increase accuracy is not new. The Condorcet jury theorem states that:

If each voter has a probability p of being correct and the probability of a majority of voters being correct is P, then $p > 0.5$ *implies* $P > p$. *In the limit, P approaches* 1, *for all* $p > 0.5$, *as the number of voters approaches infinity.*

This theorem was proposed by the Marquis of Condorcet in 1784 [9] – a more accessible reference is by Nitzan and Paroush [30]. We know now that P will be greater than p only if there is diversity in the pool of voters. And we know that the probability of the ensemble being correct will only increase as the ensemble grows if the diversity in the ensemble continues to grow as well. Typically the diversity of the ensemble will plateau as will the accuracy of the ensemble at some size between 10 and 50 members.

In ML research it is well known that ensembling will improve the performance of unstable learners. Unstable learners are learners where small changes in the training data can produce quite different models and thus different predictions. Thus, a ready source of diversity is to train models on different subsets of the training data. This approach has been applied with great success in eager learning systems such as neural networks [17] or decision trees [4]. This research shows that, for difficult classification and regression tasks, ensembling will improve the performance of unstable learning techniques such as neural networks and decision trees. Ensembling will also improve the accuracy of more stable learners such as k-NN or Naïve Bayes classifiers; however, these techniques are relatively stable in the face of changes in training data so other sources of diversity must be employed. Perhaps the most popular choice for stable classifiers is to achieve diversity by training different classifiers on using different feature subsets [11, 19, 20].

Krogh and Vedelsby [23] have shown that the reduction in error due to an ensemble is directly proportionate to the diversity or ambiguity in the predictions of the components of the ensemble as measured by variance. It is difficult to show such a direct relationship for classification tasks but it is clear that the uplift due to the ensemble depends on the diversity in the ensemble members.

Colloquially, we can say that if the ensemble members are more likely on average to be right, and when they are wrong they are wrong at different points, then their decisions by majority voting are more likely to be right than that of individual members. But they must be more likely on average to be right and when they are wrong they must be wrong in different ways.

2.5.2 Bias–Variance Analysis of Error

In their seminal paper on the error analysis of zero–one loss functions Kohavi and Wolpert [22] develop the idea that the error of a classifier can be divided into three components.

- **Intrinsic 'target noise':** This noise is inherent in the learning problem and is effectively a lower bound on the error that can be achieved by the classifier. It reflects shortcomings in the potential of the available features to capture the phenomenon.
- **Bias:** This captures how the average guess of the learning algorithm (overall possible training sets of the given training set size) matches the target.
- **Variance:** This quantifies how much the learning algorithm 'bounces around' for the different training sets of a given size.

In this analysis the intrinsic error is something we cannot do anything about. However, in adjusting a classifier to reduce error, it is useful to be mindful of what aspect of error we are trying to reduce. It is well known that while neural nets can fit a model to training data very well they are unstable learners. In terms of the bias–variance decomposition of error a neural net has low bias but high variance. By contrast a logistic regression model will have low variance error but a high bias. This is because logistic regression models are simpler than neural networks. It is normally the case that attempts to reduce bias will increase variance and vice versa; thus there is a *tension* between these two types of error.

It should also be clear from the definition of the variance component of error given above that a process that averages the output of several models should have the effect of reducing error due to model variance. This is what happens in bagging as explained in Sect. 2.5.3. The boosting approach to building ensembles (Sect. 2.5.5) is even better in that it can reduce *both* the variance and bias components of error.

2.5.3 Bagging

The simplest way to generate an ensemble of unstable classifiers such as neural nets or decision trees is to use bootstrap aggregation, more commonly known as bagging [4]. The basic idea for a bagging ensemble is shown in Fig. 2.4; given a set of training data D and a query sample $\mathbf{q}$ the key steps are as follows:

1. For an ensemble of S members, generate S training sets T_i $(i = 1, S)$ from D by bootstrap sampling, i.e. sampling with replacement. Often $|D_i| = |D|$.
2. For each D_i; let Dv_i be the set of training examples not selected in D_i (this set can be used as a validation set to control the overfitting of the ensemble member trained with D_i). Dv_i is often called the 'out-of-bag' (OOB) data.
3. Train S classifiers $f_i(,D_i)$ using the D_i training sets. The validation sets Dv_i can be used to control overfitting.

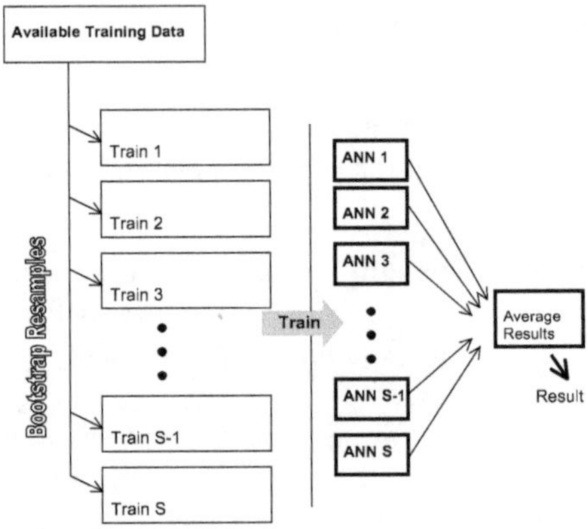

Fig. 2.4 An overview of a bagging ensemble

4. Generate S predictions for $\mathbf{q}$ using the S classifiers $f_i(,D_i)$.
5. Aggregate these S predictions $f_i(\mathbf{q},D_i)$ to get a single prediction for $\mathbf{q}$ using some aggregation function.

The formula for this ensemble prediction would be

$$f_E(\mathbf{q},D) = F(f_1(\mathbf{q},D_1), f_2(\mathbf{q},D_2), ..., f_S(\mathbf{q},D_S)). \tag{2.25}$$

The simplest approach to aggregate a regression ensemble is by averaging and the simplest for a classification ensemble is by averaging or by *weighted* averaging:

$$f_E(\mathbf{q},D) = \sum_{i=1}^{S} w_i \times f_i(\mathbf{q},D_i), \tag{2.26}$$

where $\sum_{i=1}^{S} w_i = 1$.

Provided there is diversity in the ensemble (and the bootstrap resampling should deliver this), the predictions of the ensemble $f_E(\mathbf{q},D)$ will be more accurate than the predictions from the ensemble members $f_i(\mathbf{q},D_i)$.

It is important to note that 'bagging' (i.e. sub-sampling the training data) is only a source of diversity if the classifier is unstable. For stable classifiers such as k-NN or Naïve Bayes sub-sampling the training data will not produce diverse ensemble members. In that situation an alternative strategy is to sub-sample the features rather than the examples. This can be quite effective when the data is described by a large number of features and there is redundancy in this representation.

2.5.3.1 Quantifying Diversity

Krogh and Vedelsby [23] have shown that the following very simple relationship holds for regression ensembles:

$$E = \bar{E} - \bar{A}. \tag{2.27}$$

This says that the reduction in error due to an ensemble is directly proportionate to the diversity or ambiguity in the predictions of the components of the ensemble as measured by variance of the predictions. Unfortunately there appears to be no intuitive means of quantifying diversity in classification that has the same direct relationship to the reduction in error due to the ensemble. Nevertheless several measures of diversity have been evaluated to assess their ability to quantify the reduction in error in an ensemble [25, 43].

The evaluation presented by Tsymbal et al. [43] found the following four measures to be effective:

- **Plain disagreement:** This is a pair-wise measure that quantifies the proportion of instances on which a pair of classifiers i and j disagree:

$$\text{div\_plain}_{i,j} = \frac{1}{|D|} \sum_{k=1}^{|D|} \text{Diff}(C_i(\mathbf{x}_k), C_j(\mathbf{x}_k)), \tag{2.28}$$

  where D is the data set, $C_i(\mathbf{x}_k)$ is the class assigned to instance k by classifier i and Diff() returns 1 if its two arguments are different.
- **Fail/non-fail disagreement:** This is the percentage of test instances for which the classifiers make different predictions but for which one of them is correct:

$$\text{div\_dis}_{i,j} = \frac{N^{01} + N^{10}}{|D|}, \tag{2.29}$$

  where N^{01} is the number of instances correctly classified by classifier j and not correctly classified by classifier i.
- **Entropy:** A non-pairwise measure based on entropy was proposed by Cunningham and Carney [11]:

$$\text{div\_ent} = \frac{1}{|D|} \sum_{i=1}^{|D|} \sum_{k=1}^{l} -\frac{N_k^i}{S} \log \frac{N_k^i}{S}, \tag{2.30}$$

  where S is the number of classifiers in the ensemble, l the number of classes and N_k^i the number of classifiers that assign class k to instance i.
- **Ambiguity:** This measure adapts the variance-based diversity from regression problems [23] for use in classification. An l class classification task is considered as l pseudo-regression problems. The diversity of the classification ensemble can then be calculated as the average ambiguity over these pseudo-regression tasks for each of the instances:

$$\begin{aligned} \text{div\_amb} &= \frac{1}{l|D|} \sum_{i=1}^{l} \sum_{j=1}^{|D|} \text{Ambiguity}_{i,j} \\ &= \frac{1}{l|D|} \sum_{i=1}^{l} \sum_{j=1}^{|D|} \sum_{k=1}^{S} \left(\text{Is}(C_k(\mathbf{x}_j) = i) - \frac{N_i^j}{S} \right)^2, \end{aligned} \tag{2.31}$$

where Is() is a truth predicate.

The entropy and ambiguity measures quantify the diversity of the ensemble as a whole. The two 'disagreement' measures are pair-wise measures that quantify the disagreement between a pair of base-classifiers; for these, the total ensemble diversity is the average of the disagreement of all the pairs of classifiers in the ensemble.

The evaluation presented by Tsymbal et al. [43] showed that these four diversity measures were well correlated with the uplift in accuracy due to the ensemble with *div_dis* showing slightly better correlation than the other three.

2.5.4 Random Forests

In Sect. 2.5.3 we mentioned that the two most popular strategies for introducing diversity into an ensemble are sub-sampling the examples (bagging) and sub-sampling the features (feature selection). Breiman has used both of these ideas in the random forest strategy for developing ensembles [5]. It is worth looking at random frorests in some detail because some important additional benefits emerge from the random forest strategy – these are described at the end of this section.

As the name suggests, a random forest is an ensemble of decision trees. The two sources of diversity are manifest in the algorithm as follows:

1. As in bagging, for each ensemble member the training set D is sub-sampled with replacement to produce a training set of size $|D|$.
2. Where F is the set of features that describes the data, $m << |F|$ is selected as the number of features to be used in the feature selection process. At each stage (i.e. node) in the building of a tree m features are selected at random to be the candidates for splitting at that node.

In order to ensure diversity among the component trees no pruning is employed as would be normal in building decision trees. The OOB data can be used to assess the generalization accuracy of the ensemble members. It is normal when building a random forest to generate many more ensemble members that would be used in bagging – 100 or even 1000 trees might be built. The effort expended on building these trees has the added benefit of providing an analysis of the data. It is worth highlighting three of these benefits here:

- **Estimation of generalization error:** The OOB data directly offers an estimate of the generalization error of the component trees. However, it can also be used to get an estimate of the generalization error of the complete ensemble that is unbiased [5]. Since each case is out-of-bag in roughly 1/3 of trees, the majority vote from those trees can be taken as the class the ensemble would predict for that case. The error on these aggregated out-of-bag predictions is an unbiased estimate of the error of the random forest as a whole.
- **Case proximities:** When many trees are being built, an interesting statistic to track is the frequency with which cases (both training and OOB) are located at the same leaf node. Every leaf node in every tree is examined and an $|D| \times |D|$ matrix is maintained where cell (i, j) is incremented each time cases i and j share

the same leaf node. If the matrix entries are divided by the number of trees we have a proximity measure that is *in tune* with the classification algorithm (the random forest).

- **Variable importance:** The basic ideas in assessing the importance of a variable using a random forest is to look at the impact of randomly permuting values of that variable in OOB cases and reclassifying these permuted cases. If the error increases significantly when a variable is *noised* in this way then that variable is important. If the error does not increase then that variable is not useful for the classification.

In recent years there has been a shift of emphasis in ML research as increases in computing power remove concerns about computational aspects of algorithms. Instead we are interested in useful ways to *spend* the significant computational resources available [15]. The random forest idea is a useful innovation in this regard.

2.5.5 Boosting

Boosting is a more deliberate approach to ensemble building. Instead of building the component classifiers all at once as in bagging, in boosting, the classifiers are built sequentially. The principle can be explained with reference to Fig. 2.5. This figure shows a true decision surface and a decision surface that has been learned by the classifier. The errors due to the discrepancy between the true decision surface and the learned one are highlighted. The idea with boosting is to focus on these errors in building subsequent classifiers. It is because of this that boosting reduces the bias component of error as well as the variance component. In boosting, classifiers are built sequentially with subsequent classifiers focusing on training examples that have not been learned well by earlier classifiers. This is achieved by adjusting the sampling distribution of the training data. The details as outlined by Schapire [35] are as follows (where $(x_1,y_1),...,(x_{|D|},y_{|D|})$ with $x_i \in \mathbf{X}, y_i \in Y = \{-1,+1\}$ is the available training data):

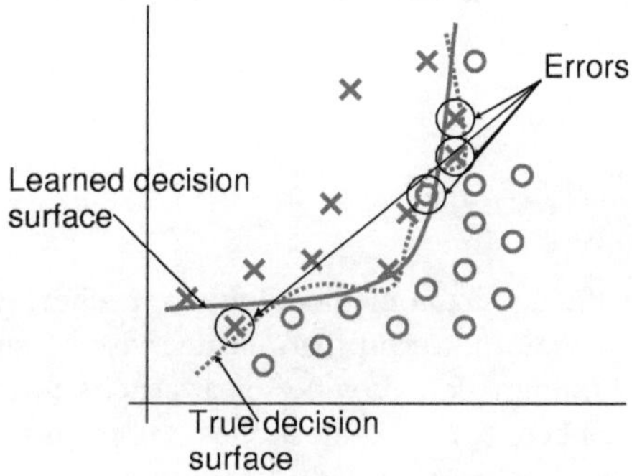

Fig. 2.5 Difficult to classify examples near the decision surface

1. Initialize the sampling distribution $P_1(i) = 1/|D|$
2. For $t = 1, ..., T$:
   a. Train classifier using distribution P_t.
   b. Hypothesis this classifier represents is $h_t : \mathbf{X} \rightarrow \{-1, +1\}$.
   c. Estimate error of this classifier as $\varepsilon_t = \sum_{i:h_t(x_i) \neq y_i} P_t(i)$.
   d. Let $\alpha_t = \frac{1}{2} \ln \frac{1-\varepsilon_t}{\varepsilon_t}$.
   e. Update

$$P_{t+1}(i) = \frac{P_t(i)}{Z_t} \times \begin{cases} e^{-\alpha t} & \text{if } h_t(x_i) = y_i \\ e^{\alpha t} & \text{if } h_t(x_i) \neq y_i \end{cases}, \tag{2.32}$$

   where Z_t is a normalization factor set to ensure that P_{t+1} will be a distribution.
   f. Continue training new classifiers while $\varepsilon_t < 0.5$.
3. This ensemble of classifiers can be used to produce a classification as follows:

$$H(x) = \text{sign}(\sum_{t=1}^{T} \alpha_t h_t(x)). \tag{2.33}$$

Clearly, the role that classification error plays in this algorithm means that this formulation only works for classification problems. However, extensions to the boosting idea for regression problems exist; two representative examples are AD-ABoost.R2 by Drucker [14] and BEM by Avnimelech and Intrator [2].

ssec: It is generally the case that the generalization error of a classifier will be improved by building an ensemble of such classifiers and aggregating the results from these component classifiers. Even if the reduction in error is modest (of the order of a couple of per cent) and the computational cost increases by an order of magnitude it will probably still be worth it as the computational resources are available.

The simplest strategy for building an ensemble is bagging where diverse component classifiers are built by sub-sampling the training cases or sub-sampling the features. Boosting will often produce even better improvements on error than bagging as it has the potential to reduce the bias component of error in addition to the variance component. Finally, random forests are an interesting strategy for building ensembles that can provide some useful insights into the data in addition to providing a very effective classifier.

2.6 Summary

Supervised learning is the dominant methodology in machine learning. Supervised learning techniques are more powerful than unsupervised techniques because the availability of labelled training data provides clear criteria for model optimization. In the analysis presented here risk minimization is presented as the appropriate criteria to optimize in supervised learning and SVMs are presented as a learning model that implements risk minimization. This overview of supervised learning techniques for multimedia data analysis is completed with descriptions of nearest neighbour classifiers and ensembles, two other techniques that are widely used in multimedia data analysis.

Readers interested in exploring the theoretical underpinnings of these learning techniques should explore PAC learning theory (probably approximately correct) as presented by Valiant [44] and the work of Vapnik on risk minimization and SVMs [45, 46]. The theoretical foundations of boosting as presented by Schapire [35] are also worth exploration.

References

1. D. W. Aha, D. Kibler, and M. K. Albert. Instance-based learning algorithms. *Machine Learning*, 6:37–66, 1991.
2. R. Avnimelech and N. Intrator. Boosted mixture of experts: An ensemble learning scheme. *Neural Computation*, 11(2):483–497, 1999.
3. A. Beygelzimer, S. Kakade, and J. Langford. Cover trees for nearest neighbor. In *Proceedings of 23rd International Conference on Machine Learning (ICML 2006)*, 2006.
4. L. Breiman. Bagging predictors. *Machine Learning*, 24(2):123–140, 1996.
5. L. Breiman. Random forests. *Machine Learning*, 45(1):5–32, 2001.
6. H. Brighton and C. Mellish. Advances in instance selection for instance-based learning algorithms. *Data Mining and Knowledge Discovery*, 6(2):153–172, 2002.
7. C. Brodley. Addressing the selective superiority problem: Automatic algorithm/mode class selection. In *Proceedings of the 10th International Conference on Machine Learning (ICML 93)*, pages 17–24. Morgan Kaufmann Publishers Inc., San Francisco, CA, USA, 1993.
8. R. M. Cameron-Jones. Minimum description length instance-based learning. In *Proceedings of the 5th Australian Joint Conference on Artificial Intelligence*, pages 368–373. Morgan Kaufmann Publishers Inc., San Francisco, CA, USA, 1992.
9. Marquis J. A. Condorcet. Sur les elections par scrutiny. *Histoire de l'Academie Royale des Sciences*, 31–34, 1781.
10. N. Cristianini and J. Shawe-Taylor. An introduction to support vector machines. Cambridge University Press, Cambridge, 2000.
11. P. Cunningham and J. Carney. Diversity versus quality in classification ensembles based on feature selection. In Ramon Lṕez de Mántaras and Enric Plaza, editors, *Machine Learning: ECML 2000, 11th European Conference on Machine Learning, Barcelona, Catalonia, Spain, May 31–June 2, 2000, Proceedings*, pages 109–116. Springer, New York, 2000.
12. S.J. Delany and D. Bridge. Feature-based and feature-free textual cbr: A comparison in spam filtering. In D. Bell, P. Milligan, and P. Sage, editors, *Proceedings of the 17th Irish Conference on Artificial Intelligence and Cognitive Science (AICS'06)*, pages 244–253, 2006.
13. S.J. Delany and P. Cunningham. An analysis of case-base editing in a spam filtering system. In *7th European Conference on Case-Based Reasoning*. Springer Verlag, New York, 2004.
14. H. Drucker. Improving regressors using boosting techniques. In D. H. Fisher, editor, *Proceedings of the Fourteenth International Conference on Machine Learning (ICML 1997), Nashville, Tennessee, USA, July 8–12, 1997*, pages 107–115. Morgan Kaufmann, San Francisco, CA, USA, 1997.
15. S. Esmeir and S. Markovitch. Anytime induction of decision trees: An iterative improvement approach. In *AAAI*. AAAI Press, Menlo Park, CA, USA, 2006.
16. G. W. Gates. The reduced nearest neighbor rule. *IEEE Transactions on Information Theory*, 18(3):431–433, 1972.

17. L. K. Hansen and P. Salamon. Neural network ensembles. *IEEE Transactions on Pattern Analysis and Machine Intelligence*, 12(10):993–1001, 1990.
18. P. E. Hart. The condensed nearest neighbor rule. *IEEE Transactions on Information Theory*, 14(3):515–516, 1968.
19. T. K. Ho. Nearest neighbors in random subspaces. In Adnan Amin, Dov Dori, Pavel Pudil, and Herbert Freeman, editors, *Advances in Pattern Recognition, Joint IAPR International Workshops SSPR '98 and SPR '98, Sydney, NSW, Australia, August 11–13, 1998, Proceedings*, pages 640–648. Springer, New York, 1998.
20. T. K. Ho. The random subspace method for constructing decision forests. *IEEE Transactions on Pattern Analysis and Machine Intelligence*, 20(8):832–844, 1998.
21. E. J. Keogh, S. Lonardi, and C. Ratanamahatana. Towards parameter-free data mining. In W. Kim, R. Kohavi, J. Gehrke, and W. DuMouchel, editors, *KDD*, pages 206–215. ACM, New York, Ny, USA, 2004.
22. R. Kohavi and D. Wolpert. Bias plus variance decomposition for zero–one loss functions. In *ICML*, pages 275–283. Morgan Kaufmann, 1996.
23. A. Krogh and J. Vedelsby. Neural network ensembles, cross validation, and active learning. In Gerald Tesauro, David S. Touretzky, and Todd K. Leen, editors, *Advances in Neural Information Processing Systems 7, [NIPS Conference, Denver, Colorado, USA, 1994]*, pages 231–238. MIT Press, Cambridge, MA, USA, 1994.
24. S. Kullback and R. A. Leibler. On information and sufficiency. *Annals of Mathematical Statistics*, 22:79–86, 1951.
25. L. I. Kuncheva and C. J. Whitaker. Measures of diversity in classifier ensembles and their relationship with the ensemble accuracy. *Machine Learning*, 51(2):181–207, 2003.
26. M. Lenz and H-D. Burkhard. Case retrieval nets: Basic ideas and extensions. In *KI - Kunstliche Intelligenz*, pages 227–239, 1996.
27. M. Lenz, H.-D.Burkhard, and S. Brückner. Applying case retrieval nets to diagnostic tasks in technical domains. In Ian F. C. Smith and Boi Faltings, editors, *EWCBR*, volume 1168 of *Lecture Notes in Computer Science*, pages 219–233. Springer, New York, 1996.
28. M. Li, X. Chen, X. Li, B. Ma, and P. M. B. Vitányi. The similarity metric. *IEEE Transactions on Information Theory*, 50(12):3250–3264, 2004.
29. E. McKenna and B. Smyth. Competence-guided editing methods for lazy learning. In W. Horn, editor, *ECAI 2000, Proceedings of the 14th European Conference on Artificial Intelligence*, pages 60–64. IOS Press, The Netherlands 2000.
30. S.I. Nitzan and J. Paroush. *Collective Decision Making*. Cambridge University Press, Cambridge, 1985.
31. G. L. Ritter, H. B. Woodruff, S. R. Lowry, and T. L. Isenhour. An algorithm for a selective nearest neighbor decision rule. *IEEE Transactions on Information Theory*, 21(6):665–669, 1975.
32. Y. Rubner, L. J. Guibas, and C. Tomasi. The earth mover's distance, multi-dimensional scaling, and color-based image retrieval. In *Proceedings of the ARPA Image Understanding Workshop*, pages 661–668, 1997.
33. Y. Rubner, C. Tomasi, and L. J. Guibas. The earth mover's distance as a metric for image retrieval. *International Journal of Computer Vision*, 40(2):99–121, 2000.
34. J.W. Schaaf. Fish and Shrink. A next step towards efficient case retrieval in large-scale case bases. In I. Smith and B. Faltings, editors, *European Conference on Case-Based Reasoning (EWCBR'96*, pages 362–376. Springer, New York, 1996.
35. R. E. Schapire. A brief introduction to boosting. In T. Dean, editor, *Proceedings of the Sixteenth International Joint Conference on Artificial Intelligence, IJCAI 99, Stockholm, Sweden, July 31–August 6, 1999. 2 Volumes, 1450 pages*, pages 1401–1406. Morgan Kaufmann, San Francisco, CA, USA, 1999.
36. B. Schölkopf and A. Smola. *Learning with Kernels*. MIT Press, Cambridge, MA, 2002.
37. J. Shawe-Taylor and N. Cristianini. *Kernel methods for Pattern Analysis*. Cambridge University Press, Cambridge ISBN 0-521-81397-2, 2004.
38. R. N. Shepard. Toward a universal law of generalization for psychological science. *Science*, 237:1317–1228, 1987.

39. B. Smyth and M. Keane. Remembering to forget: A competence preserving case deletion policy for cbr system. In C. Mellish, editor, *Proceedings of the Fourteenth International Joint Conference on Artificial Intelligence, IJCAI (1995)*, pages 337–382. Morgan Kaufmann, San Francisco, CA, USA, 1995.
40. B. Smyth and E. McKenna. Footprint-based retrieval. In Klaus-Dieter Althoff, Ralph Bergmann, and Karl Branting, editors, *ICCBR*, volume 1650 of *Lecture Notes in Computer Science*, pages 343–357. Springer, New York, 1999.
41. I. Tomek. An experiment with the nearest neighbor rule. *IEEE Transactions on Information Theory*, 6(6):448–452, 1976.
42. S. Tong. *Active Learning: Theory and Applications*. PhD thesis, Stanford University, 2001.
43. A. Tsymbal, M. Pechenizkiy, and P. Cunningham. Diversity in random subspacing ensembles. In Yahiko Kambayashi, Mukesh K. Mohania, and Wolfram Wöß, editors, *DaWaK*, volume 3181 of *Lecture Notes in Computer Science*, pages 309–319. Springer, New York, 2004.
44. L. G. Valiant. A theory of the learnable. *Communications of the ACM*, 27(11):1134–42, 1984.
45. V. Vapnik. *Statistical Learning Theory*. John Wiley, New York, 1998.
46. V. N. Vapnik and A.Y. Chervonenkis. On the uniform convergence of relative frequencies of events to their probabilities. *Theory of Probability and its Applications*, 16(2):264–280, 1971.
47. K. Veropoulos. Controlling the sensivity of support vector machines. In *International Joint Conference on Artificial Intelligence (IJCAI99)*, Stockholm, Sweden, 1999.
48. L. Wang. Image retrieval with svm active learning embedding euclidean search. In *IEEE International Conference on Image Processing*, Barcelona, September 2003.
49. D. Wilson and T. Martinez. Instance pruning techniques. In *ICML '97: Proceedings of the Fourteenth International Conference on Machine Learning*, pages 403–411. Morgan Kaufmann Publishers Inc., San Francisco, CA, USA, 1997.
50. D. L. Wilson. Asymptotic properties of nearest neighbor rules using edited data. *IEEE Transactions on Systems, Man and Cybernetics*, 2(3):408–421, 1972.
51. D. H. Wolpert The lack of a priori distinctions between learning algorithms. In *Neural Computation*, 7, pages 1341–1390, 1996.
52. J. Zhang. Selecting typical instances in instance-based learning. In *Proceedings of the 9th International Conference on Machine Learning (ICML 92)*, pages 470–479. Morgan Kaufmann Publishers Inc., San Francisco, CA, USA, 1992.
53. J. Zu and Q. Yang. Remembering to add: competence preserving case-addition policies for case-base maintenance. In *Proceedings of the 16th International Joint Conference on Artificial Intelligence (IJCAI 97)*, pages 234–239. Morgan Kaufmann Publishers Inc., San Francisco, CA, USA, 1997.

Chapter 3
Unsupervised Learning and Clustering

Derek Greene, Pádraig Cunningham, and Rudolf Mayer

Abstract Unsupervised learning is very important in the processing of multimedia content as clustering or partitioning of data in the absence of class labels is often a requirement. This chapter begins with a review of the classic clustering techniques of k-means clustering and hierarchical clustering. Modern advances in clustering are covered with an analysis of kernel-based clustering and spectral clustering. One of the most popular unsupervised learning techniques for processing multimedia content is the self-organizing map, so a review of self-organizing maps and variants is presented in this chapter. The absence of class labels in unsupervised learning makes the question of evaluation and cluster quality assessment more complicated than in supervised learning. So this chapter also includes a comprehensive analysis of cluster validity assessment techniques.

3.1 Introduction

The most fundamental distinction in machine learning is that between supervised and unsupervised techniques. Supervision invokes the idea of a teacher who guides the learning process. Typically this guidance comes in the form of labelled training examples that can be used to build a classification model. This external guidance is absent in unsupervised learning; thus the process of building a model from the data is more difficult. Often all that can be done is to cluster or organize the data in some way.

Derek Greene
University College Dublin, Dublin, Ireland, e-mail: padraig.cunningham@ucd.ie

Pádraig Cunningham
University College Dublin, Dublin, Ireland, e-mail: padraig.cunningham@ucd.ie

Rudolf Mayer
Vienna University of Technology, Vienna, Austria, e-mail: mayer@ifs.tuwien.ac.at

Unsupervised learning is very important in the processing of multimedia content, as clustering or partitioning of data in the absence of class labels is often a requirement. This chapter will present an overview of classic clustering techniques (k-means, EM and hierarchical) and will introduce the modern clustering techniques such as kernel k-means and spectral clustering. Given the popularity of self-organizing maps in the processing of multimedia content, this topic will also be covered in detail. Because of the unsupervised nature of clustering, the validation of the resulting partition is a key issue: a comprehensive overview of cluster validation techniques will also be presented. Dimension reduction techniques for unsupervised learning are not discussed in this chapter but are covered in detail in the next chapter on dimension reduction.

3.2 Basic Clustering Techniques

3.2.1 k-Means Clustering

Partitional clustering methods involve directly decomposing a data set into a flat partition consisting of k disjoint clusters, denoted $\mathscr{C} = \{C_1, \ldots, C_k\}$. These methods generally seek to produce a local approximation to a global objective function, which is identified by iteratively refining an initial solution.

Standard k-means is the most widely used partitional clustering algorithm. It employs an iterative relocation scheme to produce a k-way hard clustering that locally minimizes the distortion between the data objects and a set of k cluster representatives. Each representative, referred to as a *centroid*, is computed as the mean vector of all objects assigned to a given cluster. In the classical version of the algorithm, distortion is measured using Euclidean distance, so that the goal of the clustering process becomes the minimization of the *sum-of-squared error* (SSE) between the objects and cluster centroids $\{\mu_1, \ldots, \mu_k\}$:

$$\mathrm{SSE}(\mathscr{C}) = \sum_{c=1}^{k} \sum_{x_i \in C_c} ||x_i - \mu_c||^2 \quad \text{where} \quad \mu_c = \frac{\sum_{x_i \in C_c} x_i}{|C_c|}. \tag{3.1}$$

While many variations of the basic algorithm exist, the most frequently applied version for offline clustering is the *batch k-means* algorithm, generally attributed to Forgy [20], which involves a two-step process as shown in Fig. 3.1. In the first step, each object is reassigned to the closest cluster centroid. Once all objects have been processed, the centroid vectors are updated to reflect the new cluster assignments. The iterative refinement process is repeated until a given termination criterion is satisfied. Typically this occurs when the assignment of objects to clusters no longer changes from one iteration to another. Alternatively, the procedure may be terminated if the change in the evaluation of (3.1) between two successive iterations is less than a user-defined threshold.

1. Create an arbitrary initial clustering with centroids $\{\mu_1, \ldots, \mu_k\}$.
2. For each object $x_i \in \mathscr{X}$:
    1. Compute $||x_i - \mu_c||$ for $1 \leq c \leq k$.
    2. Reassign x_i to the cluster corresponding to the nearest centroid.
3. Update cluster centroids.
4. Repeat from Step 2 until a termination criterion is satisfied.

Fig. 3.1 Standard batch k-means algorithm

The SSE function (3.1) implicitly assumes that the clusters approximate a mixture of Gaussians such that each cluster is spherical in shape and data objects are largely concentrated near its centroid. Consequently, k-means will often fail to identify a useful partition in cases where the clusters are non-spherical or differ significantly in size. The traditional objective for k-means can also give undue influence to outlying objects. Their effect in centroid construction can lead to vectors that are not representative of the underlying groups in the data, resulting in highly skewed clusters. Some authors have proposed the introduction of an "outlier cluster", which is used to hold objects that do not fit well in any other cluster [19]. Others have suggested repeatedly applying the clustering algorithm and removing poorly clustered data after each run [28]. However, both approaches require the introduction of an arbitrary threshold to determine whether an object is far enough from its current centroid to be deemed an outlier. Another problem occurs when the iterative refinement process results in the formation of empty clusters. A common strategy to deal with this is to assign the most outlying object (i.e. furthest from its current centroid) to the empty cluster. However, if the problem persists, it is more likely that the fault may lie with the choice of clustering model, such as the use of an unsuitable value for k.

3.2.2 *Fuzzy Clustering*

Dunn [16] proposed a generalization of standard k-means, the Fuzzy c-means (FCM) algorithm, which allows objects to belong to different clusters to certain degrees as expressed by probabilistic weights. These weights may be represented in the form of a $n \times k$ matrix $\mathbf{V}$, where $V_{ij} \in [0,1]$ denotes the degree of membership of the object x_i in cluster C_j, and $\sum_j V_{ij} = 1$. Once again, the task of clustering is to minimize the distortion between objects and centroids, which is now measured by the fuzzy criterion function

$$F(\mathscr{C}, \mathbf{V}) = \sum_{i=1}^{n} \sum_{j=1}^{k} V_{ij}{}^{m} \left|\left|x_i - \mu_j\right|\right|^2, \tag{3.2}$$

where the exponent $m > 1$ controls the fuzziness of object memberships. In this algorithm, centroids are computed using

$$\mu_j = \frac{\sum_{i=1}^{n} V_{ij}{}^m x_i}{\sum_{i=1}^{n} V_{ij}{}^m}. \tag{3.3}$$

Another well-known soft partitional clustering technique is the expectation maximization (EM) algorithm [10]. Unlike the other techniques described here, this algorithm takes a model-based approach to identifying groups in data. Formally, EM clustering is based on the assumption that the data objects are generated using a model θ which consists of a mixture of k underlying probability distributions $\{\theta_1, \ldots, \theta_k\}$. The task of clustering can then be viewed as the problem of determining the most likely parameters for the model, where each component in the mixture represents a cluster. The likelihood of an object x_i is given by

$$P(x_i|\theta) = \sum_{c=1}^{k} P(C_c)P(x_i|C_c).$$

In the standard formulation of the algorithm, the k distributions are assumed to be Gaussians, so that the problem becomes the approximation of the mean and covariance of each component. In practice, the algorithm begins with an initial estimate for the model parameters and subsequently applies an iterative optimization approach that alternates between two steps: first identify the expected value of the log likelihood with respect to the current parameter estimates, then find new parameter values to maximize this likelihood. Once the algorithm has converged to a local solution, each data object is probabilistically assigned to each cluster based on the estimated distributions. As with standard k-means, the choice of initial clusters can have a considerable effect on the accuracy of the final solution.

3.2.3 Hierarchical Clustering

Instead of generating a flat partition of data, it may often be useful to construct a hierarchy of concepts by producing a set of nested clusters that may be arranged to form a tree structure. While partitional clustering methods have received more attention in recent literature, hierarchical clustering algorithms represent the traditional choice for performing document clustering, since text collections often contain broad themes that may be naturally sub-divided into more specific topics. Hierarchical algorithms are generally organized into two distinct categories:

1. *Agglomerative*: Begin with each object assigned to a singleton cluster. Apply a bottom-up strategy where, at each step, the most similar pair of clusters are merged.
2. *Divisive*: Begin with a single cluster containing all n objects. Apply a top-down strategy where, at each step, a chosen cluster is split into two sub-clusters.

In either case, the resulting hierarchy may be presented visually using a tree-like structure referred to as a *dendrogram*, which contains nodes for each cluster constructed by the clustering algorithm, together with cluster relations illustrating the merge or split operations that were performed during the clustering process. Figure 3.2 provides a simple example of an agglomerative clustering process applied to a set of five data objects, together with the corresponding cluster assignments. It is worth noting that, as each merge operation is performed, the similarity between the chosen pair of cluster decreases.

Unlike the requirement in most partitional algorithms to specify a value for the number of clusters k in advance, hierarchical algorithms support the construction of a tree from which a user may manually select k by examining the resulting dendrogram and identifying an appropriate cut-off point [44]. For instance, by cutting the tree in Fig. 3.2 at the level indicated, we can derive a clustering of the data for $k = 2$ from the two leaf nodes at that level.

3.2.3.1 Agglomerative Algorithms

Agglomerative hierarchical clustering (AHC) involves the construction of a tree of clusters from the bottom upwards. A variety of agglomerative algorithms have been proposed, such as BIRCH [61] and CURE [26], which are suitable for specific types of data. However, we focus on the standard formulation that has widely been used in a range of domains, which proceeds as follows:

1. Assign each object to singleton clusters.
2. Update the pairwise inter-cluster similarity matrix.
3. Identify and merge the most similar pair of clusters.
4. Repeat from Step 2 until a single cluster remains or a given termination criterion has been satisfied.

When an estimation for the number of clusters k is given in advance, the algorithm may be terminated when the required number of leaf nodes remains in the dendrogram.

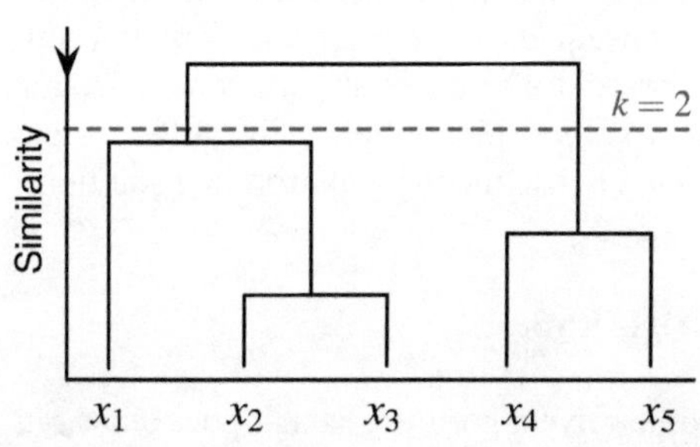

$$C = \{\{x_1, x_2, x_3, x_4, x_5\}\}$$
$$C = \{\{x_1, x_2, x_3\}, \{x_4, x_5\}\}$$
$$C = \{\{x_1\}, \{x_2, x_3\}, \{x_4, x_5\}\}$$
$$C = \{\{x_1\}, \{x_2, x_3\}, \{x_4\}, \{x_5\}\}$$
$$C = \{\{x_1\}, \{x_2\}, \{x_3\}, \{x_4\}, \{x_5\}\}$$

Fig. 3.2 Example dendrogram representing an agglomerative clustering of five data objects, together with the corresponding cluster memberships

A variety of *linkage* strategies exist for determining which pair of clusters should be merged from among all possible pairs. While these strategies are typically expressed in terms of distances, they may be easily adapted to use similarity values such as those produced by the cosine measure. Given a symmetric matrix $\mathbf{S} \in \mathbb{R}^{n \times n}$, where S_{ij} denotes the similarity between a pair of objects x_i and x_j, the most popular linkage strategies are defined as follows:

Single linkage: The most common strategy, also known as the nearest neighbour technique, defines the similarity between two clusters (C_a, C_b) as the maximum similarity between an object assigned to C_a and an object assigned to C_b:

$$\text{sim}(C_a, C_b) = \max_{x_i \in C_a, x_j \in C_b} S_{ij}.$$

While this approach is widely used, it can often produce clusters of poor quality as it is subject to the phenomenon of "chaining", where singletons are repeatedly merged with an existing cluster, resulting in one large, elongated cluster with highly dissimilar objects at either end.

Complete linkage: The similarity between two clusters (C_a, C_b) is defined as the minimum similarity between an object assigned to C_a and an object assigned to C_b:

$$\text{sim}(C_a, C_b) = \min_{x_i \in C_a, x_j \in C_b} S_{ij}.$$

This strategy tends to favour strongly compact, tightly coupled clusters and is often highly sensitive to the presence of outliers.

Average linkage: The similarity between a pair of clusters (C_a, C_b) is calculated as the mean similarity between objects assigned to C_a and objects assigned to C_b:

$$\text{sim}(C_a, C_b) = \frac{\sum_{x_i \in C_a} \sum_{x_j \in C_b} S_{ij}}{|C_a|\,|C_b|}.$$

This strategy is often referred to as *unweighted pair group method using arithmetic averages* (UPGMA), since normalizing by cluster size has the effect of giving equal weights to objects that are assigned to clusters of different sizes.

Clearly, the choice of linkage strategy can significantly affect the structure of the clusters that are generated by AHC. As a consequence, the prior selection of a suitable strategy for a given data set may represent a non-trivial parameter selection problem. In practice, a user may generate several hierarchies using different approaches and manually inspect the results to choose the most appropriate solution.

3.2.3.2 Limitations of Agglomerative Algorithms

A substantial drawback of standard agglomerative algorithms is that poor decisions made early in the clustering process can greatly influence the accuracy of the final solution. Without the use of a global objective function, many potential mergers at these stages may appear to be equally valid. Once a merging decision has been made, there exists no facility to rectify an erroneous choice at a later stage. On the contrary, the adverse effects of these decisions are often exaggerated as the clustering process continues. In addition to deficiencies in clustering accuracy, hierarchical clustering algorithms are generally considerably more computationally costly than their partitional counterparts, typically having time complexity $O(n^3)$.

3.2.3.3 Divisive Algorithms

In contrast to agglomerative methods, divisive hierarchical clustering involves building a cluster tree from the root node downwards:

1. Assign all data objects to a single cluster.
2. Select a cluster to split.
3. Replace the selected cluster with two new subclusters.
4. Repeat from Step 2 until k leaf clusters have been generated or a given termination criterion has been satisfied.

Several authors have empirically shown divisive algorithms to be superior to agglomerative techniques on text data [13]. In addition, these algorithms are often less time consuming than traditional bottom-up clustering. However, in general, they have been employed less frequently due to the non-trivial problems of selecting a cluster to split and finding the optimal sub-division of the chosen cluster.

3.2.3.4 Bisecting *k*-Means

As a representative example of divisive clustering, we consider the algorithm proposed by Steinbach et al. [56], which combines aspects of hierarchical and partitional clustering. Initially, all objects are assigned to a single root cluster. The algorithm involves repeatedly selecting an existing cluster and splitting the cluster into two subclusters using the generalized k-means algorithm with cosine similarity. The process is repeated until k clusters have been obtained. To split a cluster, a fixed number of randomly initialized bisections τ may be performed, from which the best candidate is selected. This choice is determined by a cluster evaluation criterion, such as mean object–centroid distance:

$$\mathrm{Cen}(C_c) = \frac{\sum_{x_i \in C_c} d(x_i, \mu_c)}{|C_c|}. \tag{3.4}$$

A larger value for τ renders the algorithm less sensitive to the choice of initial clusters than the partitional algorithms described previously, although it does increase the computational cost of applying the algorithm. A summary of the complete procedure is given in Fig. 3.3.

Several strategies have been proposed to identify the most appropriate cluster to split. A naïve approach is to divide the largest cluster into two sub-clusters at each stage [56]. However, this may be inappropriate when working with text corpora,

1. Assign all n objects to a single cluster.
2. Select a cluster C_c to split according to a chosen splitting criterion.
3. Generate τ 2-way partitions of the cluster C_c using randomly initialized k-means.
4. Replace C_c with the best pair of clusters as determined by a given clustering criterion.
5. Repeat from Step 2 until k leaf clusters have been generated.

Fig. 3.3 Bisecting k-means (BKM) algorithm

which frequently contain "unbalanced" clusters that differ in their relative proportions. An alternative strategy is to split the cluster with maximal distortion [13]. Given k potential candidates for splitting, Zhao and Karypis [63] proposed evaluating all possible candidates and selecting the split that leads to the best subsequent partition containing $k+1$ clusters. However, this approach requires significantly more computational time than the standard formulation of the algorithm.

3.3 Modern Clustering Techniques

3.3.1 Kernel Clustering

Kernel methods involve the transformation of a data set to a new, possibly high-dimensional space where non-linear relationships between objects may be more easily identified. Rather than explicitly computing the transformed representation $\phi(x)$ of each data object x, the application of the "kernel trick" [1] allows us to consider the affinity between a pair of objects (x_i,x_j) using a kernel function κ, which is defined in terms of the dot product:

$$\kappa(x_i,x_j) = \langle \phi(x_i), \phi(x_j) \rangle . \tag{3.5}$$

In practice, the function κ is represented by an $n \times n$ symmetric, positive semi-definite *kernel matrix* (or Gram matrix) $\mathbf{K}$, such that $K_{ij} = \kappa(x_i,x_j)$. By re- formulating algorithms using only dot products and subsequently replacing these with affinity values from $\mathbf{K}$, we can efficiently apply learning algorithms in the new kernel space.

Another significant advantage of kernel methods is their modularity, where every method is composed of two decoupled components: a generic learning algorithm and a problem-specific kernel function. Consequently, it is possible to develop algorithms that can readily be deployed in a wide range of domains without requiring any customization. Novel kernels can also be constructed in a modular fashion by chaining together multiple existing functions.

The main focus of research in this area has been on the development of techniques for supervised tasks, notably the well-known *support vector machine* (SVM) classifier [8]. However, kernel methods have also been shown to be effective in unsupervised problems [12]. We now describe several algorithms and kernel functions that are relevant in this context.

3.3.1.1 Kernel *k*-Means

A variety of popular clustering techniques have been re-formulated for use in a kernel-induced space, including the standard *k*-means algorithm. To describe the algorithm, we first observe that, using the notation given above, the squared Euclidean distance between a pair of objects in the kernel space represented by a matrix $\mathbf{K}$ can be expressed as

$$||\phi(x_i) - \phi(x_j)||^2 = K_{ii} + K_{jj} - 2K_{ij}. \tag{3.6}$$

This may be used as a starting point for the identification of cluster structures. Formally, given a set of objects $\{x_1, \ldots, x_n\}$, the kernel *k*-means algorithm [53] seeks to minimize the distortion between the objects and the "pseudo-centroids" $\{\mu_1, \ldots, \mu_k\}$ in the new space (Fig. 3.4):

$$\sum_{c=1}^{k} \sum_{x_i \in C_c} ||\phi(x_i) - \mu_c||^2 \text{ where } \mu_c = \frac{\sum_{x_i \in C_c} \phi(x_i)}{|C_c|}. \tag{3.7}$$

Note that this expression is analogous to the SSE objective (3.1) used in standard *k*-means. However, rather than explicitly constructing centroid vectors in the kernel space, distances are computed using dot products only. From (3.6), we can formulate squared object–centroid distance by the expression

$$||\phi(x_i) - \mu_c||^2 = K_{ii} + \frac{\sum_{x_j, x_l \in C_c} K_{jl}}{|C_c|^2} - \frac{2\sum_{x_j \in C_c} K_{ij}}{|C_c|}. \tag{3.8}$$

The first term in (3.8) may be excluded as it remains constant; the second is a common term representing the self-similarity of the centroid μ_c, which only needs to be calculated once for each cluster; the third term represents the affinity between x_i and the centroid of the cluster C_c.

This kernelized algorithm has a significant advantage over standard *k*-means in the sense that, given an appropriate kernel function, it can be used to identify

---

1. Select k arbitrary initial clusters $\{C_1, \ldots, C_k\}$.
2. For each object $x_i \in \mathscr{X}$ and centroid μ_c, compute the distance:

$$d(x_i, \mu_c) = K_{ii} + \frac{\sum_{x_j, x_l \in C_c} K_{jl}}{|C_c|^2} - \frac{2\sum_{x_j \in C_c} K_{ij}}{|C_c|}$$

3. Assign each x_i to cluster corresponding to nearest centroid.
4. Repeat from Step 2 until termination criterion is satisfied.

Fig. 3.4 Kernel *k*-means (KKM) algorithm

structures that are not necessarily spherical or convex. In addition, once we have constructed a single matrix **K**, multiple partitions may be subsequently generated without referring back to the original feature space.

3.3.2 Spectral Clustering

Motivated by work in graph theory, unsupervised feature extraction methods have been developed that employ well-known techniques from linear algebra to analyse the spectral properties of a graph representing a data set. In practice, this involves constructing a reduced-dimensional space from the eigenvalue decomposition (EVD) of a matrix form of the graph. Existing clustering algorithms may subsequently be applied in the reduced space to uncover the underlying classes in the data. Spectral clustering methods have been widely used due to their efficiency and applicability in a variety of tasks, including image segmentation [55], gene expression analysis [34] and document clustering [11].

3.3.2.1 Graph Partitioning

A common way of expressing the relations between pairs of data objects is to use a symmetric similarity or affinity matrix **S**, where S_{ij} denotes the association between the objects x_i and x_j. The task of producing a disjoint clustering may then be modelled as a graph partitioning problem, where **S** becomes the adjacency matrix for a weighted undirected graph $G(\mathscr{V},\mathscr{E})$. In this model, the set of vertices $\mathscr{V}$ represents the data objects and the set of edges $\mathscr{E}$ represents pairwise similarities between objects. Using this graph-theoretic formulation, clustering becomes the problem of finding the partition $\{V_1,\ldots,V_k\}$ of G that optimizes a given cost criterion.

A variety of criteria have been used in graph partitioning, which are also relevant to the theoretical foundations of spectral clustering. The simplest of these, the *minimum cut* criterion, measures the weight of the edges crossing the partition. Formally, the optimization of the criterion involves locating a bi-partition (C_1,C_2) of the graph vertices such that $C_1 \cup C_2 = \mathscr{V}$, which minimizes the sum of the weights of edges connecting the two clusters, as denoted by

$$s(C_1,C_2) = \sum_{i \in C_1, j \in C_2} S_{ij}. \tag{3.9}$$

This expression shows that the weight of the cut is directly proportional to the number of edges that join the two sub-graphs. Consequently, the criterion favours small groups of isolated vertices. This makes it sensitive to outliers and often leads to highly unbalanced clusterings.

A more robust measure for assessing bi-partitions, the *normalized cut* criterion, was proposed in [54]. This measures the degree of association between a cluster and the remaining vertices, relative to the total association within that cluster:

$$N\text{cut}(C_1,C_2) = \frac{s(C_1,C_2)}{s(C_1,\mathscr{V})} + \frac{s(C_1,C_2)}{s(C_2,\mathscr{V})}. \tag{3.10}$$

The normalization given in the denominator makes the criterion less sensitive to the presence of outlying objects. Yu and Shi [60] subsequently generalized this objective for multi-class partitioning:

$$KN\text{cut}(\mathscr{C}) = \sum_{i=1}^{k} \frac{s(C_i,\mathscr{V}\backslash C_i)}{s(C_i,\mathscr{V})}. \tag{3.11}$$

3.3.2.2 Spectral Bi-partitioning

Unfortunately, the problem of finding an optimal partition according to the criteria described in the previous section is NP-complete. While traditional techniques such as the Kernighan–Lin algorithm [33] have been used to produce local approximations, such methods often have drawbacks in terms of partition accuracy. Rather than directly attempting to optimize a given criterion, many authors have sought to transform the optimization task into a generalized eigenvalue problem. Given a symmetric adjacency matrix **S**, this involves computing its eigenvalue decomposition:

$$\mathbf{S} = \mathbf{V}\boldsymbol{\Lambda}\mathbf{V}^{\mathsf{T}},$$

where the diagonal entries of $\boldsymbol{\Lambda}$ represent the set of eigenvalues and the columns of **V** are a corresponding set of orthogonal eigenvectors. Unlike local partitioning methods, analysing the spectrum of **S** allows grouping to be performed based on global information describing the structure of the corresponding graph.

Early work in this area [14, 18] indicated the existence of a connection between the problem of finding vertex separators for a graph and the eigenvalue decomposition of its corresponding *Laplacian matrix* $\mathbf{L} = \mathbf{D} - \mathbf{S}$, where **D** denotes a diagonal degree matrix such that $D_{ii} = \sum_{j=1}^{n} S_{ij}$. The Laplacian of a graph G containing n vertices is symmetric positive semi-definite, with non-negative eigenvalues $0 = \lambda_1 < \lambda_2 < \cdots \leq \lambda_n$ and corresponding eigenvectors $\{v_1,\ldots,v_n\}$. The most important common observation made by these authors was that the spectrum of a graph provides useful structural information that may indicate how best to partition its vertices. The use of spectral partitioning was popularized by the proposal of a formal technique by Pothen et al. [47]. Following the discussion given in [27], a bi-partition (C_1,C_2) of G may be represented by a membership indicator vector $q = \{q_1,\ldots,q_n\}$ such that

$$q_i = \begin{cases} +1 & \text{if } i \in C_1 \\ -1 & \text{if } i \in C_2 \end{cases}.$$

If the adjacency matrix **S** has a block-diagonal structure (i.e. the rows can be reorganized by cluster membership to form a checker-board pattern), we can optimize the minimum cut by finding a clustering that minimizes the sum of the weights in the off-diagonal blocks. The problem can be formulated as the search for a vector q that minimizes

$$\underset{q}{\operatorname{argmin}} \sum_{i,j=1}^{n} S_{ij}(q_i - q_j)^2 \quad \text{such that} \quad \sum_{i=1} q_i^2 = 1.$$

This objective can also be expressed in quadratic form using the Laplacian **L**:

$$\underset{q}{\operatorname{argmin}} \; \mathbf{q}^{\mathsf{T}} \mathbf{L} q.$$

Rather than solving this as a complex combinatorial problem, a solution may be found by relaxing the requirement on q to contain discrete values, so that the assignment of each vertex is continuous, with membership weights taking real values in the range $[-1,1]$. The partition approximating the minimal cut can then be found by examining the eigenvectors of **L**. Specifically, the relaxed membership weights are calculated as the components of the eigenvector v_2 corresponding to the second smallest eigenvalue λ_2 of the Laplacian (i.e. the first non-trivial eigenvector), which is often referred to as the *Fiedler vector*.

A variety of justifications for partitioning based on v_2 are given in the literature. Fiedler [18] showed the association between its corresponding eigenvalue λ_2 and the edge connectivity of a graph, while Pothen et al. [47] demonstrated a relationship between the edge separator induced by v_2 and the *isoperimetric number* of a graph, which represents the value of the smallest possible edge cut over all candidate separators. The latter has motivated several popular spectral bi-partitioning algorithms. In these, the vertices are sorted according to the values in v_2, and those vertices with values below a chosen threshold, such as 0, the mean value or the median value, are assigned to one cluster, with the remaining vertices assigned to the second cluster.

Spectral methods have also been developed to optimize other, more robust graph partitioning criteria. Notably, in [54] a spectral approach was proposed to find a bisection that minimizes the normalized cut criterion (3.10). A good approximation may be identified in a manner similar to that described previously, but rather using the spectrum of the *normalized Laplacian matrix* of the graph, which is defined as $\mathbf{L_n} = \mathbf{D}^{-\frac{1}{2}}(\mathbf{D}-\mathbf{S})\mathbf{D}^{-\frac{1}{2}}$. Two clusters are formed from the normalized Fiedler vector by sorting the entries and choosing a splitting point along the vector which results in the minimal value for (3.10).

3.3.2.3 *k*-Way Spectral Clustering

In most cases, we will typically want to partition a data set into more than two clusters. Two general approaches have been proposed in the literature to extend spectral bi-partitioning to the problem of k-way clustering. The first involves recursively applying spectral bi-partitioning to hierarchically divide each resulting sub-graph until k clusters have been recovered [54]. However, if k is not a power of 2, it is unclear as to how to choose which segments should be sub-divided.

A more effective approach involves directly producing a k-way partition by constructing an embedding from multiple eigenvectors of the affinity matrix. However, rather than using those vectors corresponding to the smallest eigenvalues, clustering may be performed using the eigenvectors associated with the largest eigenvalues, which also contain structural information. A formal justification for the benefits of clustering in the reduced space formed from these vectors was given in the *polarization theorem* proposed in [6]. This theorem asserts that, as an affinity matrix $\mathbf{S}$ is projected onto smaller subsets of successive leading eigenvectors, the angles between highly similar objects are least distorted, while the angles between dissimilar objects tend to greatly increase. Consequently, by magnifying the similarities between objects that belong to the same natural class and attenuating the associations between objects belonging to different classes, the clustering problem will often become easier to solve.

k-Way spectral clustering techniques generally consist of three principal phases: preprocessing, spectral mapping and post-processing [59]. We now describe each of these phases individually and summarize the most popular approaches that have been used to implement them:

Preprocessing: Initially, an affinity matrix $\mathbf{S}$ is constructed from the original data using an appropriate metric, such as the Gaussian kernel function for image data or cosine similarity for text documents. As with bi-partitioning, various normalization techniques may be applied to $\mathbf{S}$ to support the optimization of different partitioning criteria. Most commonly, an approximation to the k-way normalized cut (3.11) has been used, which is found by computing the truncated EVD of the normalized affinity matrix given by

$$\mathbf{S_n} = \mathbf{D}^{-\frac{1}{2}}\mathbf{S}\mathbf{D}^{-\frac{1}{2}}. \tag{3.12}$$

When using this objective, some authors have observed that removing the influence of the diagonal values by setting $S_{ii} = 0$ prior to decomposition results in improved accuracy [45].

Spectral mapping: The second phase of the spectral clustering process involves computing the eigenvalue decomposition of the normalized affinity matrix. In [45] it was shown that, when partitioning data into k clusters, the use of eigenvectors corresponding to the k largest eigenvalues affords the best discriminating power. By stacking these vectors in columns to form $\mathbf{Y} \in \mathbb{R}^{n\times k}$, a reduced-dimensional representation is produced for the original n objects.

Post-processing: The columns of a k-dimensional spectral embedding $\mathbf{Y}$ can be viewed as a set of k semantic variables. However, since these variables may take negative values, they are not immediately interpretable as clusters. In simple cases where the affinity matrix is approximately block diagonal, it may be possible to identify a partition by inspecting the values in $\mathbf{Y}$. However, for real-world data such as text corpora some form of post-processing will be required to extract the final cluster assignments.

A popular approach is to treat the rows $\{y_1,\ldots,y_k\}$ as points in a geometric space $\mathbf{R}^k$ and apply a partitional algorithm, such as standard k-means, to cluster these points. A final clustering of the original data set may be derived by simply assigning the object x_i to the cluster C_j which contains the corresponding embedded point y_j. It has been shown that the quality of this partition may often be improved by normalizing the rows of $\mathbf{Y}$ to L2 unit length prior to clustering [45]. Several other authors have focused on directly decomposing the selected eigenvectors into a set of k clusters without the need for the subsequent application of a clustering algorithm [60].

To illustrate how the three phases fit together, a summary of a popular representative algorithm, Ng–Jordan–Weiss (NJW) clustering [45], is given in Fig. 3.5.

3.3.2.4 Bipartite Spectral Co-clustering

The techniques described previously in this section focus solely on the problem of grouping the objects in a data set. However, in certain situations it may be useful to perform *co-clustering*, where both objects and features are assigned to groups simultaneously. Such techniques are related to the *principle of the duality of clustering objects and features*, which states that a clustering of objects induces a clustering of features while a clustering of features also induces a clustering of objects [11]. The co-clustering problem may be viewed as the task of partitioning a weighted bipartite graph. Formally, we build a graph $G(\mathscr{V},\mathscr{E})$ such that $\mathscr{V} = \mathscr{V}_X \cup \mathscr{V}_T$, where $\mathscr{V}_X$ is a set of vertices representing the n objects and $\mathscr{V}_T$ is a set of vertices representing the m features. Feature values are given by the weights on the edges (i,j) in $\mathscr{E}$. We can conveniently represent such a graph using a feature-object matrix $\mathbf{A}$.

While the methods in the previous section involve analysing the eigendecomposition of an affinity matrix, for the bipartite case several authors have suggested the use of the related *singular value decomposition* (SVD), which may be applied to rectangular matrices. Formally, this involves decomposing a matrix $\mathbf{A} \in \mathbb{R}^{m\times n}$ into the product of three factors:

$$\mathbf{A} = \mathbf{U}\boldsymbol{\Sigma}\mathbf{V}^{\mathrm{T}}. \tag{3.13}$$

The columns of the matrix $\mathbf{U} \in \mathbb{R}^{m\times m}$ are referred to as the left singular vectors, the rows of $\mathbf{V} \in \mathbb{R}^{n\times n}$ are the right singular vectors and the diagonal entries of $\boldsymbol{\Sigma} \in \mathbb{R}^{m\times n}$ are called the singular values of $\mathbf{A}$. Note that the left singular vectors are equivalent to the eigenvectors of $\mathbf{A}\mathbf{A}^{\mathrm{T}}$, the right singular vectors are the eigenvectors of $\mathbf{A}^{\mathrm{T}}\mathbf{A}$ and their identical sets of eigenvalues are given by the diagonal of $\boldsymbol{\Sigma}^2$.

---

1. Construct an affinity matrix $\mathbf{S} \in \mathbb{R}^{n\times n}$ on the original data $\mathscr{X}$, and set $S_{ii} = 0$.
2. Form the normalized affinity matrix:
$$\mathbf{S_n} = \mathbf{D}^{-\frac{1}{2}}\mathbf{S}\mathbf{D}^{-\frac{1}{2}}$$
3. Decompose $\mathbf{S_n}$ and construct an embedding $\mathbf{Y} \in \mathbb{R}^{n\times k}$, such that the columns are given by the eigenvectors corresponding to the k largest eigenvalues.
4. Normalize the rows of $\mathbf{Y}$ to L2 unit length.
5. Apply standard k-means to the rows of $\mathbf{Y}$ to generate a k-way clustering $\mathscr{C}$.
6. Produce a clustering of $\mathscr{X}$ by assigning each object x_i to the j-th cluster if $y_i \in C_j$.

Fig. 3.5 Ng–Jordan–Weiss (NJW) spectral clustering algorithm

Dhillon [11] suggested that an approximation for the optimal normalized cut of a bipartite graph represented by a matrix **A** may be obtained by analysing the $l = \log_2 k$ leading singular vectors of the degree-normalized matrix given by

$$\mathbf{A_n} = \mathbf{D_1}^{-\frac{1}{2}} \mathbf{A} \mathbf{D_2}^{-\frac{1}{2}}, \tag{3.14}$$

where $\mathbf{D_1}$ and $\mathbf{D_2}$ are diagonal matrices such that

$$[D_1]_{ii} = \sum_{j=1}^{n} A_{ij}, \quad [D_2]_{jj} = \sum_{i=1}^{m} A_{ij}. \tag{3.15}$$

If **A** is a feature-object matrix, the rows of the left truncated vectors $\mathbf{U_l}$ represent a l-dimensional embedding of the features, while the columns of the right truncated vectors $\mathbf{V_l}$ represent an embedding of the data objects. By selecting the leading vectors of the spectral decomposition, we can produce a reduced-dimensional space that amplifies the natural structures in the data. In this case, a unified embedding $\mathbf{Z} \in \mathrm{IR}^{(m+n)\times l}$ is constructed by normalizing and arranging the truncated factors as follows:

$$\mathbf{Z} = \begin{bmatrix} \mathbf{D_1}^{-1/2}\mathbf{U_l} \\ \mathbf{D_2}^{-1/2}\mathbf{V_l} \end{bmatrix}.$$

A partitional clustering algorithm, such as k-means, is then applied in the geometric space **Z** to produce a simultaneous k-way partitioning of both objects and features.

3.4 Self-organizing Maps

The self-organising map (SOM) [35] is a well-known and widely used neural network model. It provides a topology-preserving mapping from a high-dimensional input space to a lower-dimensional output space. In many applications, this output space is a rectangular or hexagonal grid in two-dimensional space, which is a simple, "human readable" representation. Different architectures, e.g. a three-dimensional output space, are possible as well. In its mapping, the SOM performs both a vector projection (i.e. reduction of dimensionality) and vector quantization (i.e. finding prototypical representatives).

The SOM is often applied for visualization of high-dimensional data. The lower-dimensional mapping it generates allows an easier understanding and interpretation of the underlying complex inherent structures and correlations in the data and is appealing to users due to its analogy to two-dimensional geographical maps. Self-organizing maps have been applied to a wide range of very different tasks, spanning from control interfaces for industrial processing plants and other engineering problems [36] to document organization in digital libraries [49]. A detailed list of publications and applications related to the SOM can be found in the *SOM Bibliography* [32, 46].

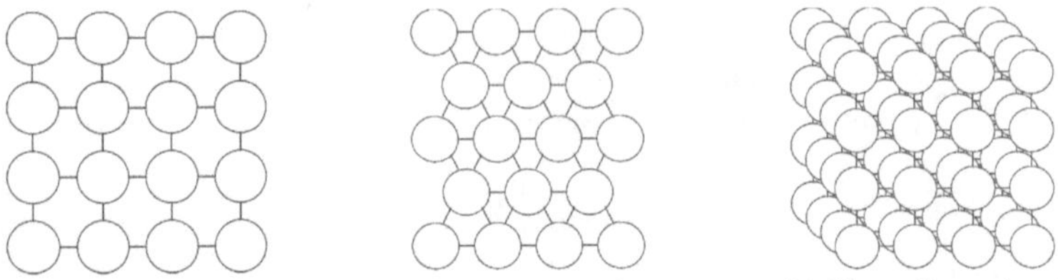

(**a**) Rectangular (**b**) Hexagonal (**c**) Three-dimensional

Fig. 3.6 Various SOM architectures

3.4.1 SOM Architecture

The SOM consists of a grid of nodes, with each node of the grid being associated with a model of some observation. In the literature, nodes are sometimes also referred to as units, cells, prototype or codebook vectors. The models are computed by the SOM algorithm so that they optimally describe the domain of observations. Mainly rectangular or hexagonal structured grids are employed, and most commonly, the grid is in two-dimensional space. However, other grid structures and three- or higher-dimensional architectures are possible. Some common architectures are illustrated in Fig. 3.6.

The models m_i assigned to each node form a so-called model vector $m_i = [m_{i1}, m_{i2}, ..., m_{in}]^{\mathsf{T}} \in \mathbb{R}^n$, which is of the same dimension as the input vectors $x_i = [x_{i1}, x_{i2}, ..., x_{in}]^{\mathsf{T}} \in \mathbb{R}^n$. To each of the nodes of the SOM, a number of input vectors x_i are assigned during the training process, with similar vectors being mapped to the same node.

3.4.2 SOM Algorithm

During the SOM training algorithm, the map is adjusted in a way that it optimally describes the domain of observations. The training process consists of the following basic steps:

- Initialization of the map
- A number of iterations of
  - Presenting input patterns and finding the best matching node
  - Adapting the model vectors of the best matching node and a certain number of neighbouring nodes.
- Fine-tuning

These steps will now be described in detail.

3.4.2.1 Initialization of the Map

Several approaches for initializing the nodes in the grid have been proposed. A basic and common approach is to initialize each node with a randomly generated model vector. Alternatively, randomly chosen vectors from the input data set could be taken as initial values. More sophisticated methods include principal component analysis, where first the eigenvalues and eigenvectors of the training data are calculated. Then the map is linearly initialized along the n greatest eigenvectors, where n is the number of dimensions of the map.

3.4.2.2 Training Iterations

After initialization, the actual learning phase is carried out. The number of learning steps (iterations) depends mainly on the number of items in the input space. A common approach therefore is to set the number of iterations as a multiple of the number of input items. One training iteration consists of two steps – finding the winning node and adapting model vectors.

Finding the Best Matching Node

A vector of the collection of input patterns is randomly selected. It is then presented to the SOM, and the model (i.e. the node's model vector) which is most similar to a presented input vector x is computed. This node is referred to as the winner, winning node, best matching node or best matching unit (BMU) c. For measuring the distance $d(x,m)$ between the model vector m_i and the input vector x several different metrics have been proposed and studied. Most commonly, the L2 norm or Euclidean distance is employed. Other commonly used distance metrics are the L1 norm (also called city-block or Manhattan distance) or other instances of Ln norms. Another widely used function is the cosine similarity.

The node c is selected according to

$$c(x,t) = \arg\min_i \{d(x(t), m_i(t))\}. \tag{3.16}$$

Model Vector Adaptation

After the best matching node has been found, the SOM learns from the input sample to improve the mapping quality, i.e. some model vectors of the SOM are adapted towards the input vector x. The new values of the model vectors are determined by their current value and two other factors, the *learning rate* α as well as the *neighbourhood function* h_{ci}. The adapted model vector can be computed according to

$$m_i(t+1) = m_i(t) + \alpha(t) \cdot h_{ci}(t)[x(t) - m_i(t)]. \tag{3.17}$$

The learning rate α, $0 < \alpha(t) < 1$, determines how much a vector is adapted, and should be a time-decreasing function; in other words, vectors should be adapted more in the beginning of the learning process, with this adaptation decreasing towards the end.

The neighbourhood function is typically designed to be symmetric around the winning node; its task is to impose a spatial structure on the amount of model vector adaptation [42]. There are mainly two different approaches to be found in the literature. The simpler one is by defining a *neighbourhood set* N_c, centred around the best matching node c. Nodes which lie inside this neighbourhood set are, all to the same degree, adapted according to the learning rate, while nodes outside the neighbourhood set are left as they are. Therefore, the neighbourhood function can be written as follows:

$$h_{ci}(t) = \begin{cases} 1, \forall i \in N_c(t) \\ 0, \forall i \notin N_c(t) \end{cases} . \tag{3.18}$$

This neighbourhood function is illustrated in Fig. 3.7(a), with the black-coloured nodes, lying within the neighbourhood set, being adapted, i.e. for these nodes, the neighbourhood function takes value 1. It is of advantage to have a time-variable width or radius of Nc, with Nc being very wide at the beginning of the training process, shrinking monotonically with time.

The second, more sophisticated and probably more widely used approach for a neighbourhood function is the use of a Gaussian function. With r_i and r_c denoting the coordinates of the nodes i and c in the two-dimensional output space $\mathbb{R}^2$, respectively, a proper form for h_{ci} is given as

$$h_{ci}(t) = \mathrm{e}^{\frac{\|r_i - r_c\|^2}{2 \cdot \sigma(t)}} . \tag{3.19}$$

Analogous to the first approach for the neighbourhood, this function decreases with time t by the usage of a monotonically decreasing function $\sigma(t)$. Unlike the simpler approach, using the Gaussian function will adapt the nodes' model vectors differently depending on their "distance" from the winning node c in the output space A. This is expressed by the term $\|r_i - r_c\|$, specifying, e.g., the Euclidean distance.

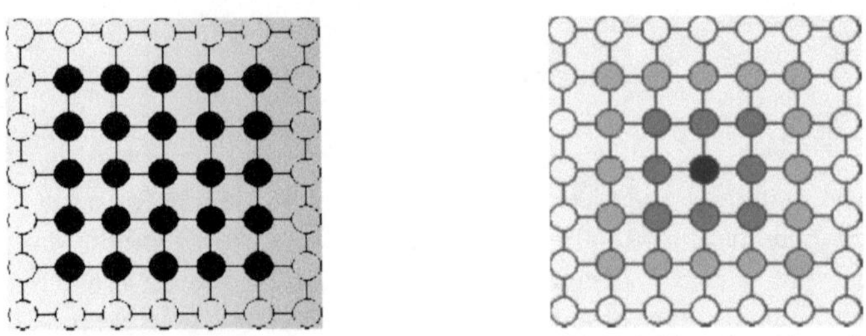

(**a**) Neighbourhood range (**b**) Neighbourhood kernel

Fig. 3.7 Two variants of neighbourhood adaptation

This behaviour is desirable as nodes close to the winning node should be adapted more than nodes further away, as closer nodes mean a closer relation. This neighbourhood function is illustrated in Fig. 3.7(b), where the black node is the winner, and the grey-shaded nodes are going to be adapted (the darker the node, the more it will be adapted).

3.4.2.3 Fine-Tuning

At the end of the training process, often a step of fine-tuning is performed. All input vectors are once more presented to the map, but in this step only the best-matching unit is adapted, employing only a very low learning rate. The rationale behind adapting only the winner node is that in this phase, the global ordering of the map is already achieved.

3.4.3 Self-organizing Map and Clustering

The mapping of the SOM is topology preserving, which is illustrated in Fig. 3.8. Topology preservation can be summarized as follows [22]:

- Input patterns that are spatially close in the input space V, i.e. are similar to each other, should also be mapped spatially close in the output space A.
- Nodes which are spatially close in A should have similar input patterns mapped on them.
- Areas of a high density in V should be represented by a corresponding high number of elements in A.

Even though the SOM is often reported to be an unsupervised clustering algorithm, the SOM itself does not provide a clustering of the input data. Rather, the SOM on the one hand performs a vector projection, i.e. a dimensionality reducing mapping, similar to, e.g., Sammon's mapping [52]. The SOM also provides a

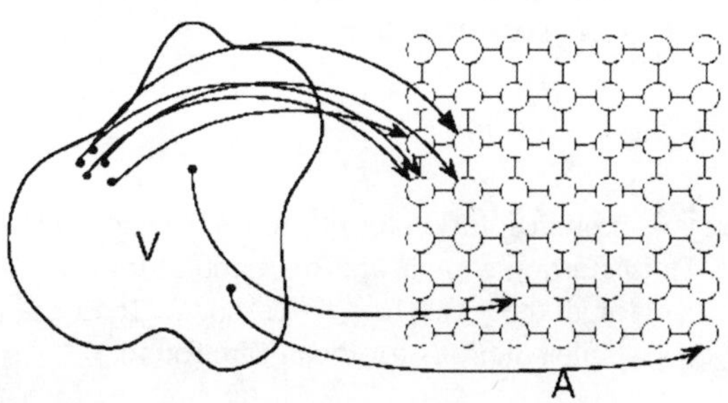

Fig. 3.8 Topology-preserving mapping: spatially close elements in V are spatially close in A as well

vector quantization, i.e. finding prototypical vectors for a reduced representation of the input data. This is similar to other clustering methods, e.g. k-means. However, in respect of the SOM, an assumption to regard each node (prototype) as a cluster cannot be done, as due to the topology-preserving mapping, clusters, especially the more densely populated ones, might easily stretch over various nodes. Cluster structures can therefore be on the one hand determined by applying any clustering algorithm (e.g. k-means or a hierarchical clustering) on the model vectors of the nodes.

3.4.4 Variations of the Self-organizing Map

Several network models based on the original SOM have been proposed. Most of them try to overcome some of the drawbacks and limitations of the initial model, e.g. by setting of learning parameters like the learning rate $\alpha(t)$ or defining the size and proportions (i.e. width and height) of the SOM in advance without a priori knowledge of the input data or the lack of hierarchical structure, albeit sometimes at the expense of the stability of the basic architecture with respect to parameter setting.

3.4.4.1 Batch Map Principle of the SOM

The incremental algorithm of the SOM, as defined by (3.16) and (3.17), can often be replaced by a significantly faster batch computation version [35], which does not require the specification of any learning rate $\alpha(t)$. This principle is based on the assumption that the SOM will converge to some ordered state, and therefore, the expectation values of $m_i(t+1)$ and $m_i(t)$ for $t \to \infty$ are required to be equal, i.e. with m_i^* denoting the model vectors in the equilibrium state,

$$\Delta m_i = m_i(t) \cdot h_{ci}(t)[x - m_i^*] = 0 \qquad (3.20)$$

holds true. For simplicity, $h_{ci}(t)$ is regarded as being time-invariant, and with a finite number (batch) of $x(t)$, (3.20) can be written as

$$m_i^* = \frac{\sum_t h_{ci} \cdot x(t)}{\sum_t h_{ci}}. \qquad (3.21)$$

However, this is not an explicit solution for m_i^*, as on the right side, h_{ci} still depends on $x(t)$ and all m_i^*. Thus, starting with an approximation for the m_i^*, (3.16) is utilized to find the indexes $c(x)$ for all the $x(t)$. Then, (3.21) and (3.16) are applied iteratively, and after a few cycles, stable solutions for the m_i^* are found.

3.4.4.2 Growing Grid

The problem of having to define the map size in advance is addressed in the *Growing Grid* [23]. The map starts small in size, e.g. 2×2 nodes, and the grid is expanded when certain conditions are fulfilled. The training process is terminated when a stopping criterion comes true.

Each node has assigned a *resource* variable τ_c, which is increased by 1 each time the node is the best matching node, thereby acting as a winner counter. After a certain number of training iterations, the node e with the maximum resource value, i.e. the node which has the highest number of mapped inputs, called the *error node*, is determined. With the aim of distributing the input patterns more evenly over all nodes, new rows or columns are inserted between e and its neighbour f, where f has the most different model vector to e, i.e. indicating a direction with high variance. This process is illustrated in Fig. 3.9.

After growing the grid, a new round of learning iterations is started, until a stopping criterion is fulfilled. Common criteria are, e.g., simply a fixed total number of nodes allowed in the grid or when the winning counter falls below a certain threshold compared to the map size.

3.4.4.3 Incremental Grid Growing

Similar to the growing grid, the *incremental grid growing* (IGG) [4] addresses the problem of specifying the grid size in advance. This is achieved by growing a grid (from an initially small size of 2×2 nodes) at the borders of the grid where the *quantization error* is the highest. The quantization error for a node i is defined as the sum of the squares of the distance between the node's model vector m_i and all the vectors $x_j \in C_i$, where C_i is the set of vectors mapped on node i:

$$qe_i = \sum_{x_j \in C_i} \|m_i - x_j\|. \tag{3.22}$$

Moreover, the IGG proposes a means of dealing with discontinuities in the input space, which in the standard SOM architecture may lead to having nodes in areas

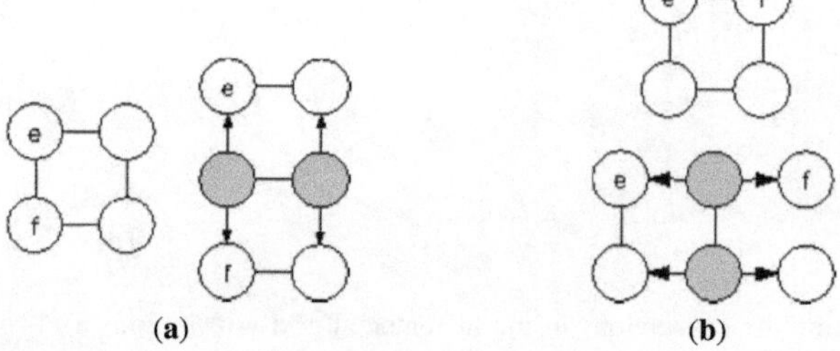

Fig. 3.9 Insertion of a row (**a**) or column (**b**) between the nodes e and f in the growing grid

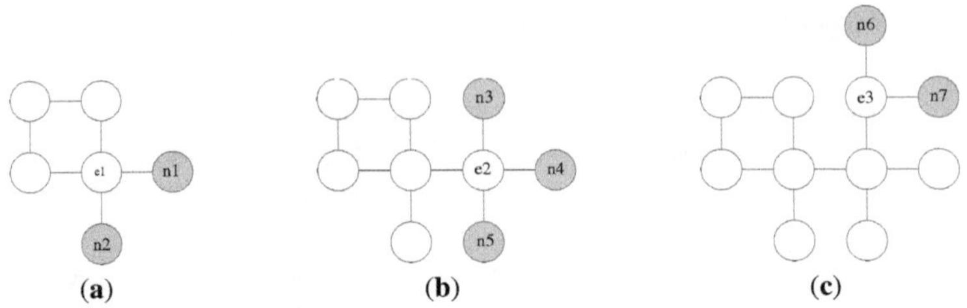

Fig. 3.10 Expanding the grid in the incremental grid growing by growing new nodes around the error node

where the input probability is 0; this is solved by having the grid not fully filled with nodes, but leaving some grid positions empty. A scenario of the growth process is illustrated in Fig. 3.10, where the nodes $e1$, $e2$ and $e3$ are the nodes with the highest error after the first, second and third training cycles, respectively, and the new nodes $n1 - n7$ are added around them.

Finally, the IGG proposes a solution to visualize clusters by using the concept of *connections* between nodes – clusters become visible by areas of connected nodes, which are separated from other areas simply by not having a connection to them. Connections are added or deleted after each step of growing the grid, depending on the pairwise distances between two nodes. This is illustrated in Fig. 3.11, where in (a) a connection is added between two close nodes $c1$ and $c2$, while in (b) a connection is deleted between two distant nodes $d1$ and $d2$.

3.4.4.4 Hierarchical Feature Map

The hierarchical feature map (HFM) [43] is a variation of the SOM that allows the creation of a hierarchical structure, thereby explicitly visualizing the hierarchical taxonomy of the input patterns. To achieve this, the hierarchical feature map uses several standard, two-dimensional self-organizing maps, arranged in hierarchical layers. For each node in one layer, a self-organizing map is added in the next layer.

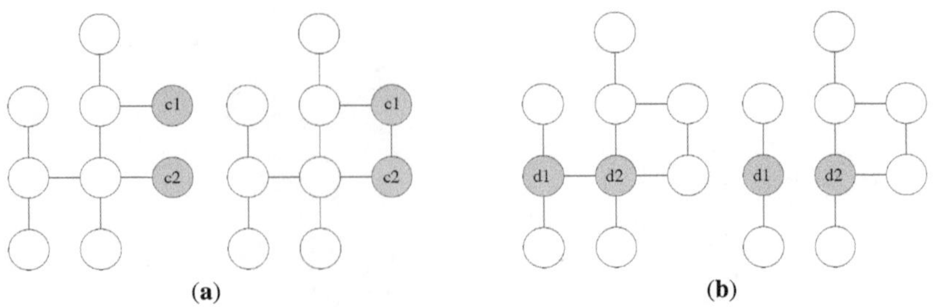

Fig. 3.11 Adapting the connections in the incremental grid growing: in (a), two similar nodes establish a connection, whereas in (b), two dissimilar nodes are disconnected

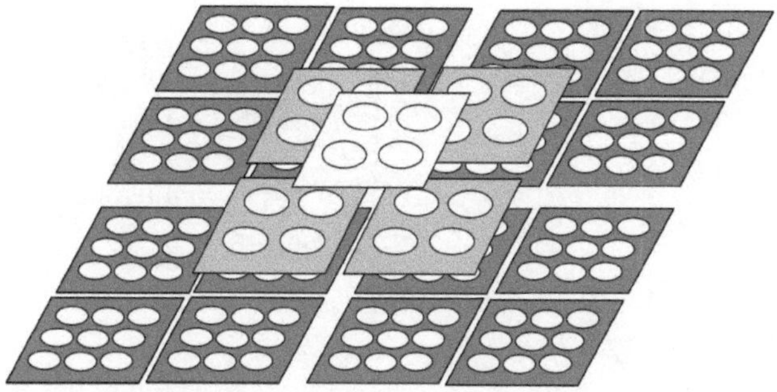

Fig. 3.12 Architecture of a hierarchical feature map with three layers

The highest level is supposed to lay out the different categories of input patterns, while on lower levels the self-organization process will form sub-categories and groups. A sample architecture with three layers is illustrated in Fig. 3.12. The size of the SOMs is predefined as is the number of hierarchy layers. Also, the hierarchies are fully balanced (i.e., all possible paths down the hierarchies have the same length).

3.4.4.5 Growing Hierarchical Self-Organizing Map

The growing hierarchical self-organizing map (GHSOM) [50] addresses the static architecture limitation of the hierarchical feature map. It combines ideas of the hierarchical feature map with the growing grid, i.e. using dynamically growing instead of standard SOMs. The hierarchical depth is chosen during run time depending on the input data, decreasing the need for an a priori knowledge about the structure of the input patterns. Further, the hierarchy is not necessarily balanced, which seems to be a more accurate representation of real-world data than in the HFM. Growth within a SOM layer and hierarchical growth depend on the quantization error of whole maps and single nodes. A sample architecture of a GHSOM with three layers is depicted in Fig. 3.13. Note that maps may have different sizes, and not all hierarchies are expanded to the same extent.

3.5 Cluster Validation

We now consider the task of assessing the validity of the output of a clustering algorithm, which represents a fundamental problem in unsupervised learning. Unlike in classification tasks, cluster analysis procedures will generally be unable to refer to predefined class labels when employed in real-world applications. Consequently, there is no clear definition of what constitutes a correct clustering for a given data set. As a result, it may be difficult to distinguish between a solution consisting of groups that accurately reflect the underlying patterns in the data and one that does

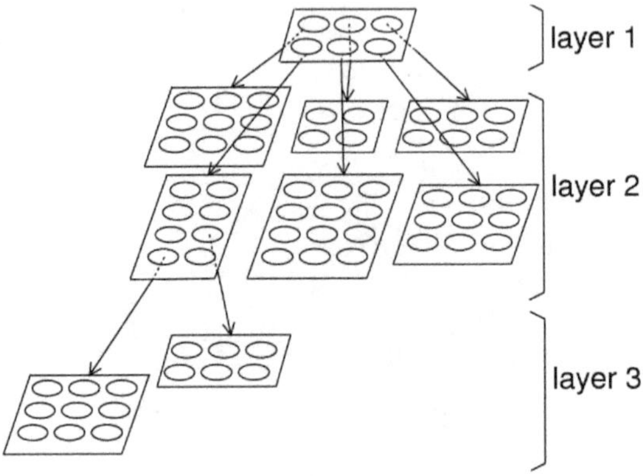

Fig. 3.13 Architecture of a growing hierarchical self-organizing map

not provide the user with any helpful insight. While it may be possible in some cases for a domain expert to manually evaluate a clustering solution, this will be unfeasible for larger data sets and may introduce an element of human bias.

In contrast, *cluster validation* methods automatically produce a quantitative evaluation, which can be highly useful in both the exploratory analysis of data and the design of new clustering algorithms. The validation problem can be viewed as comprising several different tasks:

Examining cluster tendency: In certain applications, a crucial initial step in the cluster analysis process is to determine whether any significant structures exist in a data set at all. However, in the document clustering literature it has been common to assume that text corpora will contain at least two identifiable topics.

Model selection: This task relates to the identification of an appropriate clustering algorithm and a corresponding set of parameter values. In the context of document clustering, a particularly important model selection problem is that of estimating the optimal number of clusters in a corpus, denoted by $\hat{k}$. For certain data sets, there may be several reasonable values for $\hat{k}$.

Relative comparison: It is often necessary to directly compare two or more candidate clusterings of the same data set. This comparison may be performed as part of model selection or may be used to evaluate the performance of a newly proposed clustering algorithm, relative to that afforded by existing algorithms.

Stability analysis: When a clustering solution is generated using an algorithm that contains a stochastic element or requires the selection of key parameter values, it is important to consider whether the solution represents a "definitive" solution that may easily be replicated. This can typically be determined by assessing the level of pairwise agreement between two or more clusterings of the same data.

A wide variety of validation methods have been proposed in the cluster analysis literature, which pertain to one or more of the above tasks. In the remainder of this chapter we review a range of methods, both classical and contemporary, many of which are relevant for document clustering. These methods are often organized into three distinct categories:

1. *Internal validation:* Compare clustering solutions based on the goodness-of-fit between each clustering and the raw data on which the solutions were generated.
2. *External validation:* Assess the agreement between the output of a clustering algorithm and a predefined reference partition that is unavailable during the clustering process.
3. *Stability-based validation:* Evaluate the suitability of a given clustering model by examining the consistency of solutions generated by the model over multiple trials.

3.5.1 Internal Validation

Internal validation methods are designed to provide a means of systematically assessing the quality of a clustering based on some evaluation function, which usually takes the form of an index measuring the goodness-of-fit between the clustering and the data on which it was generated. This evaluation is based solely on aspects of the features and metrics used during the clustering process, without considering any additional information or external supervision. In certain cases, these indices can also be used to provide an objective function for clustering, although many are intractable to optimize directly. Consequently, internal validation techniques are generally applied after the completion of the clustering procedure.

Model selection in areas such as bioinformatics has frequently been performed by using internal techniques [5]. Specifically, it is common to generate multiple clusterings of the data for a range of reasonable parameter values. For knowledge discovery tasks, it may be necessary to repeatedly adjust parameter values and reapply the clustering algorithm until a useful solution is obtained. To guide this process, one or more internal validation indices may be employed to assess the quality of different solutions. A set of suitable parameter values may be identified by locating a solution which optimizes these indices.

It is common for internal validation indices to measure goodness-of-fit by examining aspects of a clustering solution such as intra-cluster compactness and inter-cluster separation. To illustrate this idea, we consider the simple two-dimensional data depicted in Fig. 3.14, for which we wish to choose a suitable value for the number of clusters k. An internal index is likely to favour the first clustering in Fig. 3.14(a), which consists of three clusters that are relatively compact and well separated. However, the partition shown in Fig. 3.14(b) is clearly a poor fit for the data, as two of the clusters are not well separated. Therefore, evaluating the second partition with the same index should result in a relatively poor score. From this, we can conclude that $k = 3$ is likely to represent a more suitable choice for the data.

We now discuss a number of internal indices that have been traditionally used to evaluate hard clusterings.

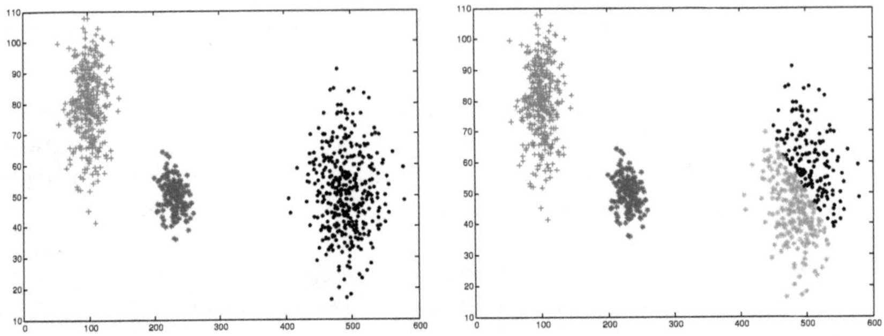

(**a**) Well-separated clusters ($k = 3$)

(**b**) Poorly-separated clusters ($k = 4$)

Fig. 3.14 Clusterings on simple data set containing three well-separated groups

3.5.1.1 Calinski–Harabasz Index

Motivated by the clustering objectives used in well-known partitional algorithms, a number of internal indices have been proposed which assess cluster quality by considering the squared distances between data objects and cluster representatives. Formally, the *within-cluster sum of squares* is the total of the squared distances between each object x_i and the centroid of the cluster C_c to which it has been assigned:

$$W(\mathscr{C}) = \sum_{c=1}^{k} \sum_{x_i \in C_c} d(x_i, \mu_c)^2.$$

When employing Euclidean distance, this is equivalent to the SSE function (3.1) used in the standard k-means algorithm. The *between-cluster sum of squares* is the total of the squares of the distances between each cluster centroid and the centroid of the entire data set, denoted $\hat{\mu}$:

$$B(\mathscr{C}) = \sum_{c=1}^{k} |C_c| \; d(\mu_c, \hat{\mu})^2 \quad \text{where} \quad \hat{\mu} = \frac{1}{n} \sum_{i=1}^{n} x_i.$$

The statistics $W(\mathscr{C})$ and $B(\mathscr{C})$ have been combined in a number ways by different authors for the purposes of validation. A representative example, the *Calinski–Harabasz* (CH) index [7], involves computing the normalized ratio of within-cluster relative to inter-cluster scatter:

$$CH(\mathscr{C}) = \frac{B(\mathscr{C})/(k-1)}{W(\mathscr{C})/(n-k)}. \tag{3.23}$$

A larger value is indicative of greater internal cohesion and a large degree of separation between the clusters in $\mathscr{C}$. This index has been frequently used as a means of automatically selecting the number of cluster in data, particularly in conjunction with agglomerative hierarchical clustering methods.

3.5.1.2 Generalized Dunn's Index

Many popular cluster validation indices are based on the assumption that a correct clustering will minimize intra-cluster dissimilarity, while simultaneously maximizing inter-cluster dissimilarity. A prototypical index is that proposed in [17], which was designed to reward "compact and well separated clusters". This index was generalized in [3] to support the use of arbitrary cluster evaluation criteria. Formally, for a disjoint k-way clustering, we let $\Delta : C \rightarrow \mathbf{R}$ denote a function that evaluates the intra-cluster dissimilarity or diameter of a cluster in $\mathscr{C}$, and let $\delta : C \times C \rightarrow \mathbf{R}$ denote a function that evaluates the inter-cluster dissimilarity between a pair of clusters. An overall evaluation for $\mathscr{C}$ is calculated using the expression

$$D(\mathscr{C}) = \min_{1\leq i\leq k} \left\{ \min_{1\leq j\leq k, i\neq j} \left\{ \frac{\delta(C_i, C_j)}{\max_{1\leq l\leq k}\{\Delta(C_l)\}} \right\} \right\}. \tag{3.24}$$

A larger value for $D(\mathscr{C})$ indicates that the clustering $\mathscr{C}$ consists of compact clusters which are well separated.

To evaluate clusters, the original formulation of Dunn's index made use of complete intra-cluster diameter and single-linkage inter-cluster distance, as defined by

$$\Delta_1(C_i) = \max_{x\in C_i, y\in C_i} \{d(x,y)\} \qquad \delta_1(C_i, C_j) = \min_{x\in C_i, y\in C_j} \{d(x,y)\}. \tag{3.25}$$

Since both functions only make use of a single distance value corresponding to the most extreme case, this formulation is highly sensitive to the presence of outliers. An alternative approach is to include the contribution of all objects in a cluster by considering object–centroid scatter and measuring inter-cluster dissimilarity in terms of the distance between centroids:

$$\Delta_2(C_i) = 2\left(\frac{\sum_{x\in C_i} d(x,\mu_i)}{|C_i|}\right) \qquad \delta_2(C_i, C_j) = d(\mu_i, \mu_j). \tag{3.26}$$

Bezdek and Pal [3] suggested that, by considering average object–centroid distance together with average-linkage inter-cluster distance, more robust cluster evaluations can be produced:

$$\Delta_3(C_i) = 2\left(\frac{\sum_{x\in C_i} d(x,\mu_i)}{|C_i|}\right) \qquad \delta_3(C_i, C_j) = \frac{\sum_{x\in C_i, y\in C_j} d(x,y)}{|C_i|\,|C_j|}. \tag{3.27}$$

It should be noted that evaluation criteria based on object–centroid distances will often exhibit an unfair bias towards spherical clusters in the same way as the standard k-means algorithm, leading to the production of misleading results on data where the underlying groups are elongated or non-convex in structure.

3.5.1.3 Davies–Bouldin Index

A related internal validation technique was proposed in [9] that considers the ratio of intra-cluster scatter to inter-cluster separability across all k groups in a clustering.

Formally, the *DB index* is defined as a function of the proximity between each cluster and its nearest neighbour:

$$\mathrm{DB}(\mathscr{C}) = \frac{1}{k}\sum_{i=1}^{k} \max_{i \neq j} \left\{ \frac{\Delta(C_i) + \Delta(C_j)}{\delta(C_i, C_j)} \right\}. \tag{3.28}$$

This value will decrease as clusters become more compact and more distinctly separated, making smaller values for this index desirable. As with Dunn's index, arbitrary cluster evaluation functions can potentially be used in (3.28). However, typically the centroid-based metrics defined in (3.26) are employed to assess scatter and inter-cluster dissimilarity. A significant disadvantage of the DB index is that it does not have a fixed range, with an output value only constrained to be non-negative, making interpretation problematic. In addition, empirical analysis has shown that, when attempting to select k, this index tends to underestimate the number of groups, particularly for weakly clustered data [15].

3.5.1.4 *C*-Index

Hubert and Levin [30] proposed a cluster validation measure that evaluates the homogeneity of a set of clusters by comparing the weight of the intra-cluster distances induced to a similar proportion of inter-cluster distances. Formally, let d_w denote the sum of all l_w intra-cluster distances induced by a clustering $\mathscr{C}$. Furthermore, let $d_{\min}$ denote the sum of the l_w smallest and $d_{\max}$ denote the sum of the l_w largest distances across all pairs of objects in the data set. Having examined all pairwise distances, the *C-index* for a clustering is calculated as the ratio

$$\mathrm{HL}(\mathscr{C}) = \frac{d_w - d_{\min}}{d_{\max} - d_{\min}}. \tag{3.29}$$

A small value for this ratio is generally indicative of a more cohesive clustering.

3.5.1.5 Silhouette Index

Rousseeuw [51] suggested computing a "silhouette value" for each object in a clustering, which measures the degree to which the object belongs to its current cluster relative to the other $k-1$ clusters. Formally, for each $x_i \in C_a$, let $a(i)$ denote the average distance between the object and all other objects in C_a, and let $b(i)$ denote the average distance between x_i and all objects in the nearest competing cluster C_b:

$$a(i) = \frac{1}{|C_a|} \sum_{j \in C_a, i \neq j} d(i,j) \qquad b(i) = \frac{1}{|C_b|} \sum_{j \in C_b} d(i,j).$$

The silhouette width for x_i is then computed using the expression

$$\mathrm{sil}(i) = \frac{b(i) - a(i)}{\max\{a(i), b(i)\}}. \tag{3.30}$$

This produces a score in the range $[-1,1]$, indicating how well the object fits in its own cluster when compared to how well it would fit if moved to another cluster. A value close to 1 indicates that x_i is likely to have been assigned to the appropriate cluster, a silhouette closer to 0 suggests that x_i could also have been assigned to the nearest alternative cluster, while a negative value suggests that x_i is likely to have been incorrectly assigned. The latter case can also be interpreted to mean that the object is an outlier. An overall evaluation for a k-way clustering, the *average silhouette width*, can be computed by taking the mean silhouette of all n participating objects:

$$\text{ASW}(\mathscr{C}) = \frac{1}{n}\sum_{i=1}^{n} \text{sil}(i). \tag{3.31}$$

A higher value for this expression signifies a superior clustering of the data.

3.5.2 External Validation

A significant disadvantage of internal techniques is that useful comparisons may only be made between clusterings that are generated using the same data model and similarity metric [24]. We have also seen that many well-known internal indices make assumptions about the structure of the clusters in data, so that they favour clusters with certain geometric properties.

An alternative approach to validation is to apply the algorithm to a data set for which a reference partition or "ground truth" is available, typically in the form of predefined class labels. External validation indices make use of this information, unavailable to the clustering algorithm itself, to quantify the level of agreement between the algorithm's output and the set of k' natural classes $\mathscr{C}' = \{C'_1, \ldots, C'_{k'}\}$ in a reference partition. Since these indices generally only consider the final partition of the data, they are independent of the representation and metrics used during the clustering procedure. In this section we provide a comprehensive review of external validation indices that are suitable for evaluating disjoint clusterings. When describing these indices, we let n'_i denote the number of objects in class C'_i, let n_j denote the number of objects in cluster C_j and let n_{ij} denote the number of objects common to both the class C'_i and cluster C_j.

It should be acknowledged that the main role for external measures in machine learning has been in the development and comparative evaluation of clustering algorithms. In the literature it is common for authors to select a fixed value for the number of clusters k, with one or more external indices being subsequently used to gauge the relative merit of the clustering techniques under consideration. In contrast to the common usage of internal indices, it is generally inappropriate to use external criteria to directly select parameters such as k, as this form of a priori information is generally inaccessible to the learning algorithm in real applications.

We now provide a review of common external validation indices that may be used to evaluate disjoint clusterings.

3.5.2.1 Set Matching Measures

A simple external validation approach is to identify a match between each cluster and a corresponding natural class in the reference partition. Once a mapping has been found, evaluations can be readily computed based on the $k' \times k$ *confusion matrix* $\mathbf{N}$, where the entry N_{ij} denotes the size of the intersection $|C'_i \cap C_j|$ between the class C'_i and cluster C_j.

Motivated by conventional evaluation techniques in supervised learning, several authors have suggested assessing the quality of a partition by assigning a unique dominant natural class to each cluster and counting the number of objects that have been assigned to the correct cluster [41]. To do this, a heuristic correspondence procedure is applied, which first identifies the largest intersection N_{ij}, resulting in a match between C'_i and C_j. The next match is chosen based on the highest value N_{ij} from the remaining pairs, with the procedure continuing until $\min(k',k)$ matches have been found. Note that no class may be matched to more than one cluster. The *classification accuracy* for the clustering $\mathscr{C}$ is then calculated using the expression

$$H(\mathscr{C}',\mathscr{C}) = \frac{1}{n} \sum_{j'=\text{match}(j)} N_{jj'}, \tag{3.32}$$

where $\text{match}(j)$ denotes the index of the class selected as a match for the cluster C_j.

Zhao and Karypis [62] suggested measuring the extent to which each cluster contains objects from a single dominant natural class. The *purity* of a cluster C_j is defined as the fraction of objects in the cluster that belong to the dominant class contained within that cluster:

$$P(C'_i,C_j) = \frac{1}{n_j} \max_i \{N_{ij}\}. \tag{3.33}$$

Unlike the classification accuracy measure, purity allows multiple clusters to be matched to the same dominant class. The overall purity of a clustering is defined as the sum of the individual cluster purities, weighted by the size of each cluster:

$$P(\mathscr{C}',\mathscr{C}) = \sum_{j=1}^{k} \frac{n_j}{n} P(C',C_j). \tag{3.34}$$

This measure provides a naïve estimate of partition quality, where larger purity values are intended to indicate a better clustering. However, the index favours small clusters, with the degenerate case of a singleton cluster resulting in a maximal cluster purity score [57].

The *F-measure* [38] is based on the *recall* and *precision* criteria that are commonly used in information retrieval tasks. Each cluster is viewed as the result of a query operation, and each natural class is viewed as the target set of documents for the query. In the ideal case, each cluster will directly correspond to a natural class. Using our notation, precision and recall for a class C'_i and cluster C_j are defined, respectively, as

$$p(C_j, C_i') = \frac{N_{ij}}{n_j} \qquad r(C_j, C_i') = \frac{N_{ij}}{n_i'}.$$

High precision implies that most objects in a given cluster belong to the same class, while high recall suggests that most objects from a single class were assigned to the same cluster. The F-measure for a pair (C_i', C_j) is given by the harmonic mean of their precision and recall, calculated as

$$F_{ij} = \frac{2 \cdot r_{ij} \cdot p_{ij}}{p_{ij} + r_{ij}}. \tag{3.35}$$

For each class C_i', a unique matching cluster C_j is selected so as to maximize the value F_{ij}. An overall score for a clustering $\mathscr{C}$ is obtained by taking the weighted average of the maximum F-values across all k' classes:

$$F(\mathscr{C}', \mathscr{C}) = \sum_{i=1}^{k'} \frac{n_i'}{n} \max_j \{F_{ij}\}. \tag{3.36}$$

Ghosh [24] notes that this measure tends to favour lower values of k, resulting in coarser clusterings.

3.5.2.2 Pairwise Co-assignment Measures

An alternative approach to external validation is to count the pairs of objects for which the clusters and natural classes agree on their co-assignment. By considering all pairs, we can calculate statistics for each of four possible cases:

- a = number of pairs in the same class in $\mathscr{C}'$ and assigned to the same cluster in $\mathscr{C}$.
- b = number of pairs in the same class in $\mathscr{C}'$, but in different clusters in $\mathscr{C}$.
- c = number of pairs assigned to the same cluster in $\mathscr{C}$, but in different classes in $\mathscr{C}'$.
- d = number of pairs belonging to different classes in $\mathscr{C}'$ and assigned to different clusters in $\mathscr{C}$.

Note that $a+d$ corresponds to the number of agreements between $\mathscr{C}'$ and $\mathscr{C}$, $b+c$ corresponds to the disagreements and $M = a+b+c+d = \frac{n(n-1)}{2}$ is the total number of unique pairs.

The *Jaccard coefficient* [31] has been commonly applied to assess the similarity between binary sets. It is also possible for this measure to be used in the context of external validation, where the level of agreement between the disjoint partitions $\mathscr{C}'$ and $\mathscr{C}$ is given by normalizing the number of positive agreements:

$$J(\mathscr{C}', \mathscr{C}) = \frac{a}{a+b+c}. \tag{3.37}$$

This index produces a result in the range $[0,1]$, where a value of 1 indicates that $\mathscr{C}'$ and $\mathscr{C}$ are identical. It was observed in [15] that (3.37) tends to produce high values for random clusterings and favours lower values of k.

The *Rand index* [48] not only is similar to the above measure, but also considers cases where both partitions assign a pair of objects to different groups. This results in an evaluation in the range $[0,1]$ based on the fraction of pairs for which there is an agreement

$$R(\mathscr{C}',\mathscr{C}) = \frac{a+d}{a+b+c+d}. \tag{3.38}$$

To eliminate biases related to different cluster size distributions and the number of clusters, Hubert and Arabie [29] proposed the *corrected Rand index*, which is computed as follows:

$$CR(\mathscr{C}',\mathscr{C}) = \frac{2(ad-bc)}{(a+b)(b+d)+(a+c)(c+d)}. \tag{3.39}$$

After applying this correction, a value of 1 indicates a perfect agreement between the two groupings, while a value of 0 indicates that a clustering is no better than a random partitioning of the data.

Another popular index for assessing the similarity between partitions was proposed in [21], which is based on the calculation of two probability scores: the probability that a pair of objects are assigned to the same cluster given that they belong to the same class and the probability that a pair objects belong to the same class given that they were assigned to the same cluster. A value for the *Fowlkes–Mallows index* (FM) is found by taking the geometric mean of these probabilities:

$$\mathrm{FM}(\mathscr{C}',\mathscr{C}) = \sqrt{\left(\frac{a}{a+b}\right)\left(\frac{a}{a+c}\right)}. \tag{3.40}$$

A value close to 1 indicates that the clusters in $\mathscr{C}$ provide a good estimate for the reference partition.

3.5.2.3 Information Theoretic Measures

Recent research relating to cluster validation has focused on concepts from information theory, which consider the uncertainty of predicting a set of natural classes based on the information provided by a clustering of the same data. We now describe two indices, based on these concepts, which have frequently been applied to evaluate clusterings of text data.

Steinbach et al. [56] suggested an entropy-based measure for assessing the agreement between two partitions. By considering the probability $\frac{N_{ij}}{n_j}$ that an object assigned to cluster C_j belongs to a class C_i', we can compute the entropy for the assignments in C_j:

$$E(C_j) = -\sum_{i=1}^{k'} \frac{N_{ij}}{n_j} . \log \frac{N_{ij}}{n_j} \tag{3.41}$$

An overall score for a clustering $\mathscr{C}$ is given by the sum of the entropy values for each cluster weighted by the fraction of objects assigned to that cluster:

$$E(\mathscr{C}', \mathscr{C}) = \sum_{j=1}^{k} \frac{n_j}{n} E(C_j). \tag{3.42}$$

Smaller values for this measure are desirable, with a value of 0 indicating that each cluster contains instances from a single class. To eliminate the strong bias of (4.15) with respect to k, a variant of this index was proposed in [62], where the normalized entropy for a cluster C_j is calculated as

$$\mathrm{NE}(C_j) = -\frac{1}{\log k'} \sum_{i=1}^{k'} \frac{N_{ij}}{n_j} . \log \frac{N_{ij}}{n_j} \tag{3.43}$$

Unlike purity and classification accuracy, entropy considers the distribution of all classes in a cluster, rather than a single dominant class. However, this index still exhibits a bias in favour of smaller clusters.

Strehl and Ghosh [57] observed that external measures such as purity and entropy are biased with respect to the number of clusters k, since the probability of each cluster solely containing objects from a single natural class increases as k increases. To address this problem, an alternative index was proposed, based on *mutual information*, which quantifies the amount of information shared between the random variables describing a pair of disjoint partitions.

Formally, let $p'(i)$ and $p(j)$ denote the probabilities that an object belongs to class C'_i and cluster C_j, respectively. Furthermore, let $p(i,j)$ denote the joint probability that an object belongs to both C'_i and C_j. For each data object assigned to a class in $\mathscr{C}'$, mutual information evaluates the degree to which knowledge of this assignment reduces the uncertainty regarding the assignment of the object in $\mathscr{C}$. The mean reduction in uncertainty across all objects can be expressed as

$$I(\mathscr{C}', \mathscr{C}) = \sum_{i=1}^{k'} \sum_{j=1}^{k} p(i,j) \log \frac{p(i,j)}{p'(i)p(j)}. \tag{3.44}$$

$I(\mathscr{C}', \mathscr{C})$ takes values between zero and $\min(E(\mathscr{C}'), E(\mathscr{C}))$, where the upper bound is the minimum of the entropy values for the two clusterings. To produce values in the range $[0,1]$, the authors in [57] proposed *normalized mutual information* (NMI), where the degree of information shared between the two clusterings is normalized with respect to the geometric mean of their entropies:

$$\mathrm{NI}(\mathscr{C}', \mathscr{C}) = \frac{I(\mathscr{C}', \mathscr{C})}{\sqrt{E(\mathscr{C}')E(\mathscr{C})}}. \tag{3.45}$$

In practice, an approximation for this quantity, based on cluster assignments, can be calculated using

$$\mathrm{NMI}(\mathscr{C}',\mathscr{C}) = \frac{\sum_{i=1}^{k'}\sum_{j=1}^{k} n_{ij}\log\left(\frac{n\cdot n_{ij}}{n'_i n_j}\right)}{\sqrt{\left(\sum_{i=1}^{k'} n'_i \log\frac{n'_i}{n}\right)\left(\sum_{j=1}^{k} n_j \log\frac{n_j}{n}\right)}}. \tag{3.46}$$

An accurate clustering should maximize this score, where a value of 1 indicates an exact correspondence between the assignment of objects in $\mathscr{C}'$ and $\mathscr{C}$, while a value of 0 indicates that knowledge of $\mathscr{C}$ provides no information about the true classes $\mathscr{C}'$. Equation (3.46) does have a slight tendency to favour clusterings for larger values of k, although it exhibits no bias against unbalanced cluster sizes.

3.5.3 Stability-Based Techniques

Recently, a number of methods based on the concept of *stability analysis* have been proposed for the task of model selection. The *stability* of a clustering algorithm refers to its ability to consistently produce similar solutions on data originating from the same source [37]. Since only a single set of data objects will be generally available in unsupervised learning tasks, clusterings are generated on perturbations of the original data set. A key advantage of stability analysis methods lies in their ability to evaluate a model independently of any specific clustering algorithm or similarity measure. Thus, they represent a robust approach for selecting key algorithm parameters [39].

In this section, we focus on stability-based methods that are relevant when estimating the optimal number of clusters $\hat{k}$ in a data set. These methods are motivated by the observation that, if the number of clusters in a model is too large, repeated clusterings will lead to arbitrary partitions of the data, resulting in unstable solutions. On the other hand, if the number of clusters is too small, the clustering algorithm will be constrained to merge subsets of objects which should remain separated, also leading to unstable solutions. In contrast, repeated clusterings generated using the optimal number of clusters $\hat{k}$ will generally be consistent, even when the data are perturbed or distorted.

3.5.3.1 Stability Analysis Based on Resampling

The most common approach to stability analysis involves perturbing the data by randomly sampling the original objects to produce a set of τ non-disjoint subsets. For each potential value of k in a reasonable range $[k_{\min}, k_{\max}]$, a corresponding set of τ clusterings are generated on the data subsets. The stability of the clustering model for each candidate value of k is evaluated using indices operating on pairs of hard clusterings, such as the external validation indices described previously. A higher overall stability score suggests that k is a better estimate for the optimal value $\hat{k}$.

A representative example of this approach is the algorithm proposed in [40]. For each value of k, an initial partition $\mathscr{C}_0$ is generated on the entire data set using a partitional clustering algorithm, which represents a "gold standard" for analysing the stability afforded by using k clusters. Subsequently, τ samples of the data are constructed by randomly selecting a subset of βn data objects without replacement, where $0 \leq \beta \leq 1$ denotes the sampling ratio controlling the number of objects in each sample. A set of clusterings $\{\mathscr{C}_1, \ldots, \mathscr{C}_\tau\}$ is then generated by applying the clustering algorithm to each sample. For each clustering $\mathscr{C}_i$, the fraction of co-assignments preserved from $\mathscr{C}_0$ is calculated, which is equivalent to the Rand index (3.38). An overall evaluation for the stability afforded by k is found by averaging the agreement scores across all τ runs. This process is repeated for each potential $k \in [k_{\min}, k_{\max}]$. A final estimation for $\hat{k}$ is chosen by identifying the value k leading to the highest average agreement.

Law and Jain [39] proposed an alternative stability analysis approach for model selection where the data are perturbed by bootstrapping. This involves generating τ samples of size n by randomly sampling with replacement. Rather than comparing each clustering to a single gold standard solution, stability is evaluated by considering the level of agreement between each pair of clusterings. A number of indices were considered for assessing agreement, including the Jaccard index (3.37) and the Fowlkes–Mallows index (3.40). The authors note that scores produced by these indices should be corrected for chance to eliminate biases towards smaller values of k. After computing the variance of the corrected agreement scores for each potential value k, the model resulting in the lowest variance is selected as the best estimate for $\hat{k}$.

Ben-Hur et al. [2] described a similar approach based on pairwise stability analysis, where agglomerative hierarchical clustering is applied to each sample. By using different cut-off levels from the same hierarchy, the output of a single clustering procedure may be used in the evaluation of all potential values of k. In [25] this approach was extended further to encompass the problem of cluster tendency. This is achieved by setting a threshold value θ for the minimum average pairwise stability that is sufficient to indicate a consistent clustering model. If no stability evaluation exceeds this threshold for any candidate $k \in [k_{\min}, k_{\max}]$, the data are assumed to have no significant underlying structure. The choice of θ largely depends on the index used to measure the agreement between clusterings.

3.5.3.2 Prediction-Based Validation

In supervised learning problems, model selection is typically performed by identifying a learning model whose estimated prediction accuracy is highest. A number of authors have suggested that the concept of prediction accuracy can be adapted to the problem of evaluating models in clustering tasks. Recent work by Tibshirani et al. [58] has provided a theoretical basis for *prediction-based validation* methods, which assess the stability of a clustering model by measuring the degree to which it allows us to consistently construct a classifier on a training set that will predict the assignment of objects in a clustering of a corresponding test set.

Formally, the validation process involves applying twofold cross-validation to randomly split a data set $\mathscr{X}$ into disjoint training and test sets, denoted by $\mathscr{X}_a$ and $\mathscr{X}_b$, respectively. Both sets are then clustered to produce partitions $\mathscr{C}_a$ and $\mathscr{C}_b$, typically using the standard k-means clustering algorithm. Subsequently, a prediction $\mathscr{P}_b$ for the assignment of objects in the test set is produced by assigning each $x_i \in \mathscr{X}_b$ to the nearest centroid in $\mathscr{C}_a$. Prediction accuracy is measured by evaluating the degree to which the class memberships in $\mathscr{P}_b$ correspond to the cluster assignments in $\mathscr{C}_b$.

To numerically evaluate prediction accuracy, a new pairwise measure for comparing partitions was proposed in [58], referred to as *prediction strength*. For each cluster in the test clustering $\mathscr{C}_b = \{C_1, \ldots, C_k\}$, we identify the number of pairs of objects assigned to the same cluster that also belong to the same class in the prediction $\mathscr{P}_b$. These associations can be represented as an $\frac{n}{2} \times \frac{n}{2}$ binary matrix $\mathbf{M}$, where $M_{ij} = 1$ only if the pair (x_i, x_j) is co-assigned in both $\mathscr{C}_b$ and $\mathscr{P}_b$. From this matrix, an evaluation is computed based on the cluster containing the smallest fraction of correctly predicted pairs:

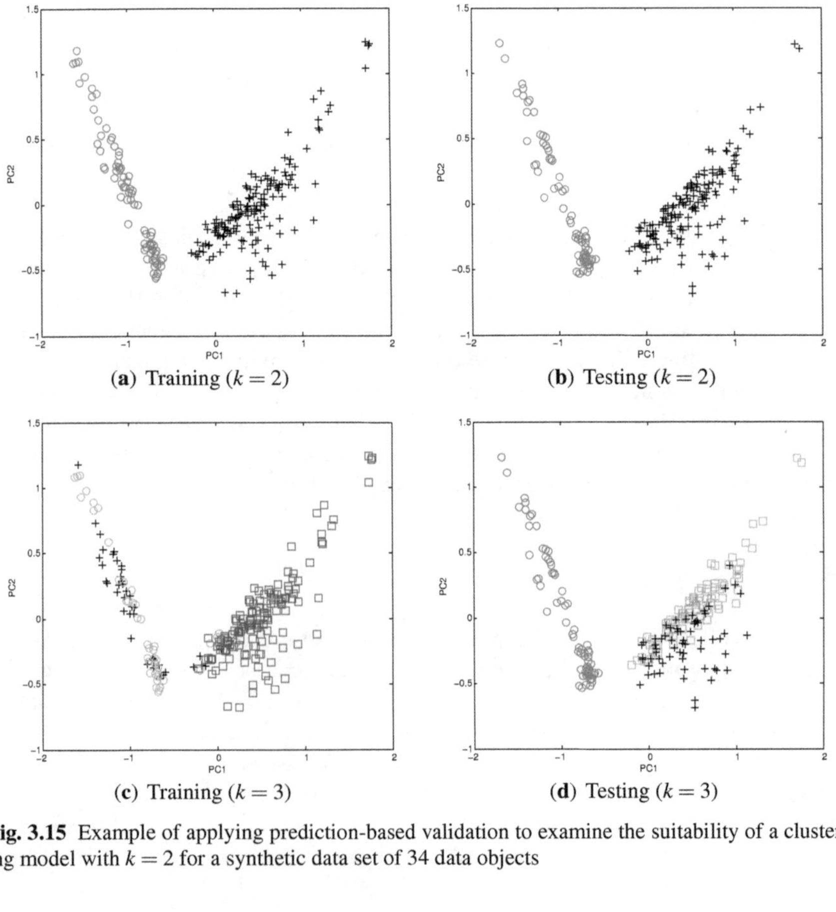

Fig. 3.15 Example of applying prediction-based validation to examine the suitability of a clustering model with $k = 2$ for a synthetic data set of 34 data objects

$$S(\mathscr{C}_b, \mathscr{P}_b) = \min_{1 \leq h \leq k} \left[\frac{1}{|C_h| (|C_h| - 1)} \sum_{x_i \neq x_j \in C_h} M_{ij} \right]. \tag{3.47}$$

The cross-validation process is repeated over τ runs for each candidate value k in the range $[k_{\min}, k_{\max}]$. The authors suggest a heuristic approach to select the final number of clusters, which is chosen to be the largest k for which the validation score is above a user-defined threshold. This can be viewed as the selection of the largest number clusters that can be reliably predicted for a given data set. They note that a threshold in the range $[0.8, 0.9]$ was appropriate for the data sets with which they evaluated the algorithm.

As a simple example, we consider a single cross-validation run for $k = 2$ applied to the set of 34 data objects shown in Fig. 3.15(a). This data set is randomly divided into two subsets containing 17 objects each. A training clustering $\mathscr{C}_a$ is generated on the first subset and a test clustering $\mathscr{C}_b$ is generated on the second, as shown in Fig. 3.15(b) and 3.15(c), respectively. The centroids μ_1 and μ_2 of $\mathscr{C}_a$ are subsequently used to build a nearest centroid classifier, which produces the predicted classification $\mathscr{P}_b$ for the set of test objects as illustrated in Fig. 3.15(d). By constructing a 17×17 co-assignment matrix $\mathbf{M}$ from $\mathscr{C}_b$ and $\mathscr{P}_b$ and applying (3.47), we can calculate that this run leads to a prediction strength of $S(\mathscr{C}_b, \mathscr{P}_b) = 0.43$, indicating that the clustering model is relatively unstable. In practice, multiple cross-validation runs would be applied to produce a result that is robust to the effects of unbiased random sampling.

3.6 Summary

This chapter provides a review of what we consider to be the main clustering techniques used for processing multimedia data. We have covered the classic techniques of k-means clustering and hierarchical clustering since these are still in widespread use. We have also reviewed the more modern techniques based on kernel techniques and spectral analysis of data. One of the most popular unsupervised techniques applied to multimedia is the SOM so the classic SOM architecture and variants have been reviewed in Sect. 3.4. The chapter concludes with a comprehensive review of cluster validation techniques given the challenge that quality assessment presents in unsupervised learning.

References

1. M. A. Aizerman, E. M. Braverman, and L. I. Rozonoer. Theoretical foundations of the potential function method in pattern recognition learning. *Automation and Remote Control*, 25(6):821–837, 1964.
2. A. Ben-Hur, A. Elisseeff, and I. Guyon. A stability based method for discovering structure in clustered data. In *Proceedings of the 7th Pacific Symposium on Biocomputing (PSB 2002)*, pp. 6–17, Lihue, HI, January 2002.

3. J. C. Bezdek and N. R. Pal. Cluster validation with generalized dunn's indices. In *ANNES '95: Proceedings of the 2nd New Zealand Two-Stream International Conference on Artificial Neural Networks and Expert Systems*, p. 190, Washington, DC, USA, 1995. IEEE Computer Society.
4. J. Blackmore and R. Miikkulainen. Incremental grid growing: encoding high-dimensional structure into a two-dimensional feature map. In *Proceedings of the ICNN'93, International Conference on Neural Networks*, Vol. I, pp. 450–455, Piscataway, NJ, 1993. IEEE Service Center.
5. N. Bolshakova and F. Azuaje. Cluster validation techniques for genome expression data. Technical Report TCD-CS-2002-33, Trinity College Dublin, September 2002.
6. M. Brand and K. Huang. A unifying theorem for spectral embedding and clustering. In *Proceedings of the 9th International Workshop on AI and Statistics*, January 2003.
7. T. Calinski and J. Harabasz. A dendrite method for cluster analysis. *Communications in Statistics*, 3:1–27, 1974.
8. N. Cristianini and J. Shawe-Taylor. *An Introduction to Support Vector Machines: and Other Kernel-Based Learning Methods*. Cambridge University Press, New York, NY, USA, 2000.
9. D. L. Davies and W. Bouldin. A cluster separation measure. *IEEE Transactions on Pattern Analysis and Machine Intelligence*, 1(2):224–227, 1979.
10. A. P. Dempster, N. M. Laird, and D. B. Rubin. Maximum likelihood from incomplete data via the em algorithm. *Journal of the Royal Statistical Society*, 39:1–38, 1977.
11. I. S. Dhillon. Co-clustering documents and words using bipartite spectral graph partitioning. In *Knowledge Discovery and Data Mining*, pp. 269–274, 2001.
12. I. S. Dhillon, Y. Guan, and B. Kulis. Kernel *k*-means: spectral clustering and normalized cuts. In *Proceedings of the 2004 ACM SIGKDD International conference on Knowledge Discovery and Data Mining*, pp. 551–556. New York, NY, 2004. ACM Press.
13. C. Ding and X. He. Cluster merging and splitting in hierarchical clustering algorithms. In *Proceedings of the 2002 IEEE International Conference on Data Mining (ICDM'02)*, p. 139. Washington, DC, 2002. IEEE Computer Society.
14. W. E. Donath and A. J. Hoffman. Lower bounds for the partitioning of graphs. *IBM Journal of Research and Development*, 17:420–425, 1973.
15. R. C. Dubes. How many clusters are best? – an experiment. *Pattern Recognition*, 20(6):645–663, 1987.
16. J. C. Dunn. A fuzzy relative of the ISODATA process and its use in detecting compact well-separated clusters. *Journal of Cybernetics*, 3:32–57, 1974.
17. J. C. Dunn. Well separated clusters and optimal fuzzy-partitions. *Journal of Cybernetics*, 4:95–104, 1974.
18. M. Fiedler. Algebraic connectivity of graphs. *Czechoslovak Mathematical Journal*, 23(98):298–305, 1973.
19. B. Fischer and J. M. Buhmann. Path-based clustering for grouping of smooth curves and texture segmentation. *Pattern Analysis and Machine Intelligence, IEEE Transactions*, 25(4):513–518, April 2003.
20. E. W. Forgy. Cluster analysis of multivariate data: efficiency vs interpretability of classifications. *Biometrics*, 21:768–769, 1965.
21. E. B. Fowlkes and C. L. Mallow. A method for comparing two hierarchical clusterings. *Journal of American Statistical Association*, 78:553–569, 1983.
22. B. Fritzke. Growing cell structures—a self-organizing network in *k* dimensions. In I. Aleksander and J. Taylor, editors, *Artificial Neural Networks, 2*, Vol. II, pp. 1051–1056, Amsterdam, Netherlands, 1992. North-Holland.
23. B. Fritzke. Growing grid – a self-organizing network with constant neighborhood range and adaptation strength. *Neural Processing Letters*, 2(5):9–13, 1995.
24. J. Ghosh. Scalable clustering methods for data mining. In N. Ye, editor, *Handbook of Data Mining*, chapter 10. Mahwah, NJ, 2003. Lawrence Erlbaum.
25. C. D. Giurcaneanu and I. Tabus. Cluster structure inference based on clustering stability with applications to microarray data analysis. *EURASIP Journal on Applied Signal Processing*, 1:64–80, 2004.

26. S. Guha, R. Rastogi, and K. Shim. CURE: an efficient clustering algorithm for large databases. In *Proceedings of the ACM SIGMOD International Conference on Management of Data*, pp. 73–84, 1998.
27. K. M. Hall. An r-dimensional quadratic placement algorithm. *Management Science*, 17(3):219–229, November 1970.
28. V. Hautamäki, S. Cherednichenko, I. Kärkkäinen, T. Kinnunen, and P. Fränti. Improving k-means by outlier removal. In *Image Analysis, 14th Scandinavian Conference, SCIA 2005*, pp. 978–987, 2005.
29. L. J. Hubert and P. Arabie. Comparing partitions. *Journal of Classification*, 2:193–218, 1985.
30. L. J. Hubert and J. R. Levin. A general statistical framework for accessing categorical. *Psychological Bulletin*, 83:1072–1082, 1976.
31. P. Jaccard. The distribution of flora in the alpine zone. *New Phytologist*, 11(2):37–50, 1912.
32. S. Kaski, J. Kangas, and T. Kohonen. Bibliography of self-organizing map (SOM) papers 1981–1997. *Neural Computing Surveys*, 1(3&4):1–176, 1998.
33. B. W. Kernighan and S. Lin. An efficient heuristic procedure for partitioning graphs. *The Bell System Technical Journal*, 49(2):291–308, 1970.
34. Y. Kluger, R. Basri, J. T. Chang, and M. Gerstein. Spectral biclustering of microarray data: coclustering genes and conditions. *Genome Research*, 13:703–716, April 2003.
35. T. Kohonen. *Self-Organizing Maps*. Springer-Verlag, New York, NY, 2001.
36. T. Kohonen, E. Oja, O. Simula, A. Visa, and J. Kangas. Engineering applications of the self-organizing map. *Proceedings of the IEEE*, 84(10):1358–1384, October 1996.
37. T. Lange, V. Roth, M. L. Braun, and J. M. Buhmann. Stability-based validation of clustering solutions. *Neural Computation*, 16(6):1299–1323, 2004.
38. B. Larsen and C. Aone. Fast and effective text mining using linear-time document clustering. In *KDD '99: Proceedings of the Fifth ACM SIGKDD International Conference on Knowledge Discovery and Data Mining*, pp. 16–22, New York, NY, USA, 1999. ACM Press.
39. M. Law and A. K. Jain. Cluster validity by bootstrapping partitions. Technical Report MSU-CSE-03-5, University of Washington, February 2003.
40. E. Levine and E. Domany. Resampling method for unsupervised estimation of cluster validity. *Neural Computation*, 13(11):2573–2593, 2001.
41. M. Meila. Comparing clusterings. Technical Report 418, University of Washington, 2002.
42. D. Merkl. Exploration of text collections with hierarchical feature maps. In *Research and Development in Information Retrieval*, pp. 186–195, 1997.
43. R. Miikkulainen. Script recognition with hierarchical feature maps. *Connection Science*, 2(1&2):83–101, 1990.
44. G. W. Milligan and M. C. Cooper. An examination of procedures for determining the number of clusters in a data set. *Psychometrika*, 50(2):159–179, 1985.
45. A. Ng, M. Jordan, and Y. Weiss. On spectral clustering: analysis and an algorithm. In *Proceedings of the Advances in Neural Information Processing*, 2001.
46. M. Oja, S. Kaski, and T. Kohonen. Bibliography of self-organizing map (SOM) papers: 1998–2001 addendum. *Neural Computing Surveys*, 3:1–156, 2003.
47. A. Pothen, H. D. Simon, and K.-P. Liou. Partitioning sparse matrices with eigenvectors of graphs. *SIAM Journal of Mathematical Analysis and Applications*, 11(3):430–452, 1990.
48. W. M. Rand. Objective criteria for the evaluation of clustering methods. *Journal of the American Statistical Association*, 66(66):846–850, 1971.
49. A. Rauber and D. Merkl. The SOMLib digital library system. In *Proceedings of the 3rd European Conference on Research and Advanced Technology for Digital Libraries (ECDL'99)*, Lecture Notes in Computer Science (LNCS 1696), pp. 323–342, Paris, France, September 22-24 1999. Springer.
50. A. Rauber, D. Merkl, and M. Dittenbach. The growing hierarchical self-organizing map: Exploratory analysis of high-dimensional data. *IEEE Transactions on Neural Networks*, 13(6):1331–1341, November 2002.
51. P. Rousseeuw. Silhouettes: a graphical aid to the interpretation and validation of cluster analysis. *Journal of Computational and Applied Mathematics*, 20(1):53–65, 1987.

52. J. W. Sammon Jr. A nonlinear mapping for data structure analysis. *IEEE Transactions on Computers*, C-18(5):401–409, May 1969.
53. B. Schölkopf, A. Smola, and K-R. Müller. Nonlinear component analysis as a kernel eigenvalue problem. *Neural Computation*, 10(5):1299–1319, 1998.
54. J. Shi and J. Malik. Normalized cuts and image segmentation. In *Proceedings of the 1997 Conference on Computer Vision and Pattern Recognition (CVPR '97)*, pp. 731–737. Huntsville, AL, 1997. IEEE Computer Society.
55. J. Shi and J. Malik. Normalized cuts and image segmentation. *IEEE Transactions on Pattern Analysis and Machine Intelligence (PAMI)*, 22(8):888–905, August 2000.
56. M. Steinbach, G. Karypis, and V. Kumar. A comparison of document clustering techniques. In *Proceedings of KDD Workshop on Text Mining 2000*, 2000.
57. A. Strehl and J. Ghosh. Cluster ensembles – a knowledge reuse framework for combining multiple partitions. *Journal of Machine Learning Research*, 3:583–617, December 2002.
58. R. Tibshirani, G. Walther, D. Botstein, and P. Brown. Cluster validation by prediction strength. Technical report, Statistics Department, Stanford University, 2001.
59. D. Verma and M. Meila. A comparison of spectral clustering algorithms. Technical report, University of Washington, 2003.
60. S. X. Yu and J. Shi. Multiclass spectral clustering. In *Proceedings of the 9th IEEE International Conference on Computer Vision*, p. 313, October 2003.
61. T. Zhang, R. Ramakrishnan, and M. Livny. BIRCH: an efficient data clustering method for very large databases. *Proceedings of the 1996 ACM SIGMOD International Conference on Management of Data*, pp. 103–114, 1996.
62. Y. Zhao and G. Karypis. Criterion functions for document clustering: experiments and analysis. Technical Report 01-040, University of Minnesota, November 2001.
63. Y. Zhao and G. Karypis. Evaluation of hierarchical clustering algorithms for document datasets. In *Proceedings of the Eleventh International Conference on Information and Knowledge Management*, pp. 515–524. New York, NY, 2002. ACM Press.

Chapter 4
Dimension Reduction

Pádraig Cunningham

Abstract When data objects that are the subject of analysis using machine learning techniques are described by a large number of features (i.e. the data are high dimension) it is often beneficial to reduce the dimension of the data. Dimension reduction can be beneficial not only for reasons of computational efficiency but also because it can improve the accuracy of the analysis. The set of techniques that can be employed for dimension reduction can be partitioned in two important ways; they can be separated into techniques that apply to *supervised* or *unsupervised* learning and into techniques that either entail *feature selection* or *feature extraction*. In this chapter an overview of dimension reduction techniques based on this organization is presented and the important techniques in each category are described.

4.1 Introduction

Data analysis problems where the data objects have a large number of features are becoming more prevalent in areas such as multimedia data analysis and bioinformatics. In these situations it is often beneficial to reduce the dimension of the data (describe it in less features) in order to improve the efficiency and accuracy of data analysis. Statisticians sometimes talk of problems that are "big p small n"; these are extreme examples of situations where dimension reduction (DR) is necessary because the number of explanatory variables p exceeds (sometimes greatly exceeds) the number of samples n [46]. From a statistical point of view it is desirable that the number of examples in the training set should significantly exceed the number of features used to describe those examples (see Fig. 4.1a). In theory the number of examples needs to increase exponentially with the number of features if inference is to be made about the data. In practice this is not the case as real high-dimension data will only occupy a manifold in the input space so the *implicit* dimension of the

Pádraig Cunningham
University College Dublin, Dublin, Ireland, e-mail: padraig.cunningham@ucd.ie

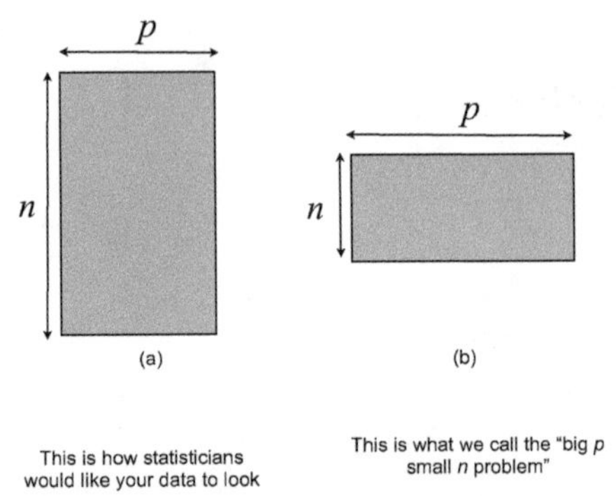

Fig. 4.1 Big p small n problems are problems where the number of features in a data set is large compared with the number of objects: (**a**) This is how statisticians would like your data to look. (**b**) This is what we call the "big p small n problem"

data will be less than the number of features p. For this reason data sets as depicted in Fig. 4.1b can still be analysed.

Nevertheless, traditional algorithms used in machine learning and pattern recognition applications are often susceptible to the well-known problem of the *curse of dimensionality* [4], which refers to the degradation in the performance of a given learning algorithm as the number of features increases. To deal with this issue, dimension reduction techniques are often applied as a data pre-processing step or as part of the data analysis to simplify the data model. This typically involves the identification of a suitable low-dimensional representation for the original high-dimensional data set. By working with this reduced representation, tasks such as classification or clustering can often yield more accurate and readily interpretable results, while computational costs may also be significantly reduced. The motivation for dimension reduction can be summarized as follows:

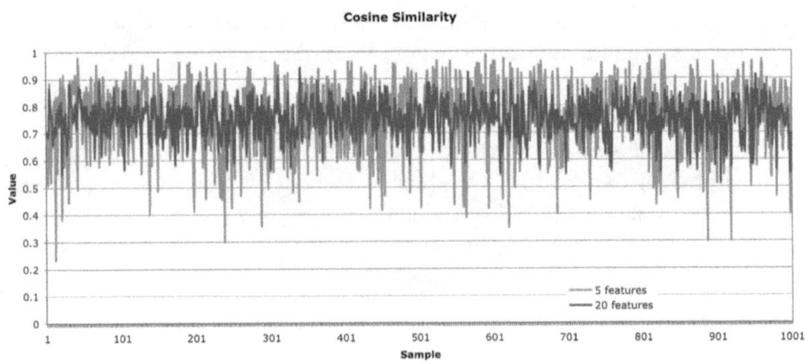

Fig. 4.2 The more dimensions used to describe objects the more similar on average things appear. This figure shows the cosine similarity between randomly generated data objects described by 5 and by 20 features. It is clear that in 20 dimensions similarity has a lower variance than in 5

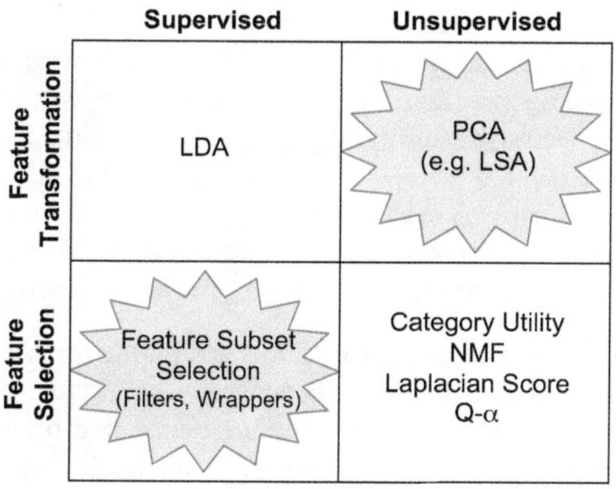

Fig. 4.3 The two key distinctions in dimension reduction research are the distinction between supervised and unsupervised techniques and the distinction between feature transformation and feature extraction techniques. The dominant techniques are feature subset selection and principal component analysis

- The identification of a reduced set of features that are predictive of outcomes can be very useful from a knowledge discovery perspective.
- For many learning algorithms, the training and/or classification time increases directly with the number of features.
- Noisy or irrelevant features can have the same influence on classification as predictive features so they will impact negatively on accuracy.
- Things look more similar on average the more features used to describe them (see Fig. 4.2). The example in the figure shows that the resolution of a similarity measure can be worse in 20D than in a 5D space.

Research on dimension reduction has itself two dimensions as shown in Fig. 4.3. The first design decision is whether to *select* a subset of the existing features or to *transform* to a new reduced set of features. The other dimension in which DR strategies differ is the question of whether the learning process is *supervised* or *unsupervised*. The dominant strategies used in practice are principal components analysis (PCA) which is an *unsupervised* feature *transformation* technique and *supervised* feature *selection* strategies such as the use of information gain for feature ranking/selection. This chapter proceeds with subsections dedicated to each of the four (2×2) possible strategies.

4.2 Feature Transformation

Feature transformation refers to a family of data pre-processing techniques that transform the original features of a data set to an alternative, more compact set of dimensions, while retaining as much information as possible. These techniques can be sub-divided into two categories:

1. Feature extraction involves the production of a new set of features from the original features in the data, through the application of some mapping. Well-known unsupervised feature extraction methods include *principal component analysis* (PCA) [22] and spectral clustering (e.g. [36]). The important corresponding supervised approach is *linear discriminant analysis* (LDA) [24].
2. Feature generation involves the discovery of missing information between features in the original data set, and the augmentation of that space through the construction of additional features that emphasize the newly discovered information.

Recent work in the literature has primarily focused on the former approach, where the number of extracted dimensions will generally be significantly less than the original number of features. In contrast, feature generation often expands the dimensionality of the data, though feature selection techniques can subsequently be applied to select a smaller subset of useful features.

For feature transformation let us assume that we have a dataset D made up of $(\mathbf{x}_i)_{i\in[1,n]}$ training samples. The examples are described by a set of features $F(p = |F|)$ so there are n objects described by p features. This can be represented by a feature–object matrix $\mathbf{X}_{p\times n}$ where each column represents an object (this is the transpose of what is shown in Fig. 4.1). The objective with feature transformation is to transform the data into another set of features F' where $k = |F'|$ and $k < p$, i.e. $\mathbf{X}_{p\times n}$ is transformed to $\mathbf{X}'_{k\times n}$. Typically this is a linear transformation $\mathbf{W}_{k\times p}$ that will transform each object $\mathbf{x}_i$ to $\mathbf{x}'_i$ in k dimensions:

$$\mathbf{x}'_i = \mathbf{W}\mathbf{x}_i. \tag{4.1}$$

The dominant feature transformation technique is principal components analysis (PCA) that transforms the data into a reduced space that captures most of the variance in the data (see Sect. 4.2.1). PCA is an unsupervised technique in that it does not take class labels into account. By contrast linear discriminant analysis (LDA) seeks a transformation that maximizes between-class separation (Sect. 4.2.2).

4.2.1 Principal Component Analysis

In PCA the transformation described in (4.1) is achieved so that feature f'_1 is in the dimension in which the variance on the data is maximum, f'_2 is in an orthogonal dimension where the remaining variance is maximum and so on (see Fig. 4.4).

Central to the whole PCA idea is the covariance matrix of the data $\mathbf{C} = 1/n-1\ \mathbf{X}\mathbf{X}^\mathsf{T}$ [22]. The diagonal terms in $\mathbf{C}$ capture the variance in the individual features and the off-diagonal terms quantify the covariance between the corresponding pairs of features. The objective with PCA is to transform the data so that the covariance terms are zero, i.e. $\mathbf{C}$ is diagonalized to produce $\mathbf{C}_{\mathrm{PCA}}$. The data is transformed by $\mathbf{Y} = \mathbf{P}\mathbf{X}$ where the rows of $\mathbf{P}$ are the eigenvectors of $\mathbf{X}\mathbf{X}^\mathsf{T}$, then

$$\mathbf{C}_{\mathrm{PCA}} = \frac{1}{n-1}\mathbf{Y}\mathbf{Y}^\mathsf{T} \tag{4.2}$$

$$= \frac{1}{n-1}(\mathbf{P}\mathbf{X})(\mathbf{P}\mathbf{X})^\mathsf{T}. \tag{4.3}$$

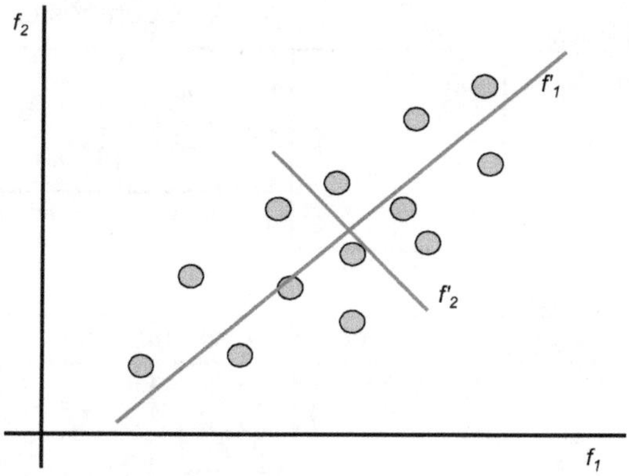

Fig. 4.4 In this example of PCA in 2D the feature space is transformed to f_1' and f_2' so that the variance in the f_1' direction is maximum

The *i*th diagonal entry in $\mathbf{C}_{\mathrm{PCA}}$ quantifies the variance of the data in the direction of the corresponding principal component. Dimension reduction is achieved by discarding the lesser principal components, i.e. **P** has dimension $(k \times p)$ where k is the number of principal components retained.

In multimedia data analysis a variant on the PCA idea called singular value decomposition or latent semantic analysis (LSA) has become popular – this will be described in the next section.

4.2.1.1 Latent Semantic Analysis

LSA is a variant on the PCA idea presented by Deerwester et al. in [8]. LSA was originally introduced as a text analysis technique so the objects are documents and the features are terms occurring in these text documents – hence the feature–object matrix $\mathbf{X}_{p\times n}$ is a term–document matrix. LSA is a method for identifying an informative transformation of documents represented as a bag-of-words in a vector space. It was developed for information retrieval to reveal semantic information from document co-occurrences. Terms that did not appear in a document may still associate with a document. LSA derives uncorrelated index factors that might be considered artificial concepts, i.e. the *latent semantics*. LSA is based on a singular value decomposition of the term–document matrix as follows:

$$\mathbf{X} = \mathbf{TSV}^{\mathsf{T}}, \tag{4.4}$$

where:

- $\mathbf{T}_{p\times m}$ is the matrix of eigenvectors of $\mathbf{XX}^{\mathsf{T}}$; m is the rank of $\mathbf{XX}^{\mathsf{T}}$
- $\mathbf{S}_{m\times m}$ is a diagonal matrix containing the squareroot of the eigenvalues of $\mathbf{XX}^{\mathsf{T}}$
- $\mathbf{V}_{n\times m}$ is the matrix of eigenvectors of $\mathbf{X}^{\mathsf{T}}\mathbf{X}$

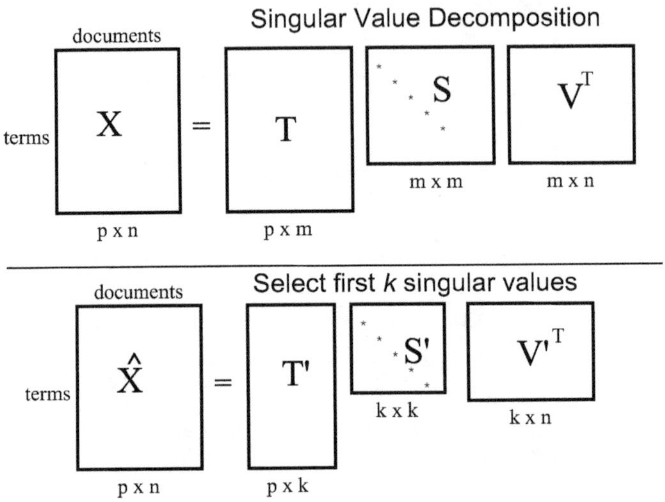

Fig. 4.5 Latent semantic analysis is achieved by performing a singular value decomposition on the term–document matrix and dropping the least significant singular values, in this scenario k singular values are kept

In this representation the diagonal entries in S are the singular values and they are normally ordered with the largest singular value (largest eigenvalue) first. Dimension reduction is achieved by dropping all but k of these singular values as shown in Fig. 4.5. This gives us a new decomposition:

$$\hat{\mathbf{X}} = \mathbf{T}'\mathbf{S}'\mathbf{V}'^{\mathsf{T}}, \tag{4.5}$$

where $\mathbf{S}'$ is now $(k \times k)$ and corresponding columns have been dropped in $\mathbf{T}'$ and $\mathbf{D}'$. In this situation $\mathbf{V}'\mathbf{S}'$ is a $(n \times k)$ matrix that gives us the coordinates of the n documents in the new k-dimension space. Reducing the dimension of the data in this way may remove noise and make some of the relationships in the data more apparent. Furthermore the transformation

$$\mathbf{q}' = \mathbf{S}'^{-1}\mathbf{T}'^{\mathsf{T}}\mathbf{q} \tag{4.6}$$

will transform any new query $\mathbf{q}$ to this new feature space. This transformation is a linear transformation of the form outlined in (4.1).

It is easy to understand the potential benefits of LSA in the context of text documents. The LSA process exploits co-occurrences of terms in documents to produce a mapping into a *latent semantic* space where documents can be associated even if they have very few terms in common in the original term space. LSA is particularly appropriate for the analysis of text documents because the term–document matrix provides an excellent basis on which to perform the singular value decomposition. It has also been employed on other types of media despite the difficulty in identifying a base representation to take the place of the term–document matrix. LSA has been employed on image data [21], video [43] and music and audio [44]. It has also been applied outside of multimedia on gene expression data [38]. More generally PCA is often a key data pre-processing step across a range of disciplines, even if it is not couched in the terms of latent semantic analysis.

The fact that PCA is constrained to be a linear transformation would be considered a shortcoming in many applications. Kernel PCA [33] has emerged as the dominant technique to overcome this. With kernel PCA the dimension reduction occurs in the kernel-induced feature space with the algorithm operating on the kernel matrix representation of the data. The introduction of the kernel function opens up a range of possible non-linear transformations that may be appropriate for the data.

4.2.2 Linear Discriminant Analysis

PCA is *unsupervised* in that it does not take class labels into account. In the supervised context the training examples have class labels attached, i.e. data objects have the form $(\mathbf{x}_i, y_i)$ where $y_i \in C$, a set of class labels or simply $y_i \in \{-1, +1\}$, the binary classification situation. In situations where class labels are available we are often interested in discovering a transformation that emphasizes the separation in the data rather than one that discovers dimensions that maximize the variance in the data as happens with PCA. This distinction is illustrated in Fig. 4.6. In this 2D scenario PCA projects the data onto a single dimension that maximizes variance; however, the two classes are not well separated in this dimension. By contrast Fisher's linear discriminant analysis (LDA) discovers a projection on which the two classes are better separated [14, 15]. This is achieved by uncovering a transformation that maximizes between-class separation.

While the mathematics underpinning LDA are more complex than those on which PCA is based the principles involved are fairly straightforward. The objective is to uncover a transformation that will maximize between-class separation and minimize within-class separation. To do this we define two scatter matrices, $\mathbf{S_B}$ for between-class separation and $\mathbf{S_W}$ for within-class separation:

$$\mathbf{S_B} = \sum_{c \in C} n_c (\mu_c - \mu)(\mu_c - \mu)^\mathsf{T}, \tag{4.7}$$

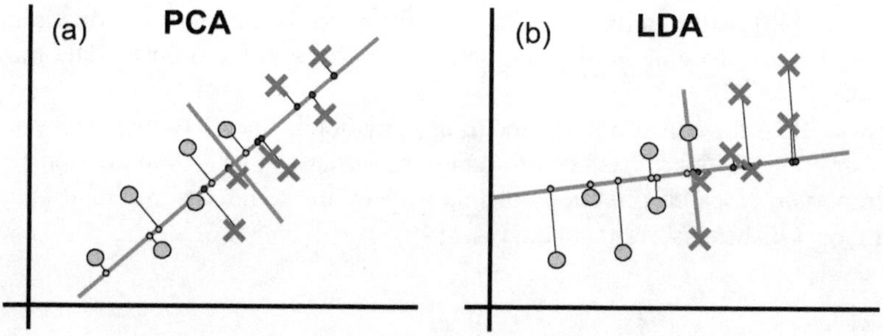

Fig. 4.6 In (**a**) it is clear that PCA will not necessarily provide a good separation when there are two classes in the data. In (**b**) LDA seeks a projection that maximizes the separation in the data

$$\mathbf{S_W} = \sum_{c \in C} \sum_{j:y_j=c} (x_j - \mu_c)(x_j - \mu_c)^\mathsf{T}, \tag{4.8}$$

where n_c is the number of objects in class c, μ is the mean of all examples and μ_c is the mean of all examples in class c:

$$\mu_c = \frac{1}{n}\sum_{i=1}^{n} x_i \qquad \mu_c = \frac{1}{n_c}\sum_{j:y_j=c} x_j. \tag{4.9}$$

The components within these summations (μ, μ_c, x_j) are vectors of dimension p so $\mathbf{S_B}$ and $\mathbf{S_W}$ are matrices of dimension $p \times p$.

The objectives of maximizing between-class separation and minimizing within-class separation can be combined into a single maximization called the Fisher criterion [14, 15]:

$$\mathbf{W}_{\mathrm{LDA}} = \arg\max_{\mathbf{W}} \frac{|\mathbf{W}^\mathsf{T}\mathbf{S_B}\mathrm{W}|}{|\mathbf{W}^\mathsf{T}\mathbf{S_W}\mathbf{W}|}, \tag{4.10}$$

i.e. find $\mathbf{W} \in \mathrm{I\!R}_{p \times k}$ so that this fraction is maximized ($|\mathbf{A}|$ denotes the determinant of matrix $\mathbf{A}$). This matrix $\mathbf{W}_{\mathrm{LDA}}$ provides the transformation described in (4.1). While the choice of k is again open to question it is sometimes selected to be $k = |C| - 1$, i.e. one less than the number of classes in the data.

It transpires that $\mathbf{W}_{\mathrm{LDA}}$ is formed by the eigenvectors $(\mathbf{v}_1|\mathbf{v}_2|...\mathbf{v}_k|)$ of $\mathbf{S_W}^{-1}\mathbf{S_B}$. The fact that this requires the inversion of $\mathbf{S_W}$ which can be of high dimension can be problematic so the alternative approach is to use simultaneous diagonalization [29], i.e. solve:

$$\mathbf{W}^\mathsf{T}\mathbf{S_W}\mathbf{W} = \mathbf{I} \qquad \mathbf{W}^\mathsf{T}\mathbf{S_B}\mathbf{W} = \mathbf{\Lambda}. \tag{4.11}$$

Here $\mathbf{\Lambda}$ is a diagonal matrix of eigenvalues $\{\lambda\}_{i=1}^{k}$ that solve the generalized eigenvalue problem:

$$\mathbf{S_B}\mathbf{v}_i = \lambda_i \mathbf{S_W}\mathbf{v}_i. \tag{4.12}$$

Most algorithms that are available to solve this simultaneous diagonalization problem require that $\mathbf{S_W}$ be non-singular [23, 29]. This can be a particular issue if the data is of high dimension because more samples than features are required if $\mathbf{S_W}$ is to be non-singular. Addressing this topic is a research issue in its own right [23]. Even if $\mathbf{S_W}$ is non-singular there may still be issues as the "small p large n" problem [46] may manifest itself by overfitting in the dimension reduction process, i.e. dimensions that are discriminating *by chance* in the training data may be selected.

As with PCA the constraint that the transformation is linear is sometimes considered restricting and there has been research on variants of LDA that are non-linear. Two important research directions in this respect are kernel discriminant analysis [3] and local Fisher discriminant analysis [45].

4.3 Feature Selection

Feature selection (FS) algorithms take an alternative approach to dimension reduction by locating the "best" minimum subset of the original features, rather than transforming the data to an entirely new set of dimensions. For the purpose of knowledge discovery, interpreting the output of algorithms based on feature extraction can often prove to be problematic, as the transformed features may have no physical meaning to the domain expert. In contrast, the dimensions retained by a feature selection procedure can generally be directly interpreted.

Feature selection in the context of supervised learning is a reasonably well-posed problem. The objective can be to identify features that are correlated with or predictive of the class label. Or more comprehensively, the objective may be to select features that will construct the most accurate classifier. In unsupervised feature selection the object is less well posed and consequently it is a much less explored area.

4.3.1 Feature Selection in Supervised Learning

In supervised learning, selection techniques typically incorporate a search strategy for exploring the space of feature subsets, including methods for determining a suitable starting point and generating successive candidate subsets, and an evaluation criterion to rate and compare the candidates, which serves to guide the search process. The evaluation schemes used in both supervised and unsupervised feature selection techniques can generally be divided into three broad categories [5, 25]:

1. Filter approaches attempt to remove irrelevant features from the feature set prior to the application of the learning algorithm. Initially, the data is analysed to identify those dimensions that are most relevant for describing its structure. The chosen feature subset is subsequently used to train the learning algorithm. Feedback regarding an algorithm's performance is not required during the selection process, though it may be useful when attempting to gauge the effectiveness of the filter.
2. Wrapper methods for feature selection make use of the learning algorithm itself to choose a set of relevant features. The wrapper conducts a search through the feature space, evaluating candidate feature subsets by estimating the predictive accuracy of the classifier built on that subset. The goal of the search is to find the subset that maximizes this criterion.
3. Embedded approaches apply the feature selection process as an integral part of the learning algorithm. The most prominent example of this are the decision tree building algorithms such as Quinlan's C4.5 [40]. There are a number of neural network algorithms that also have this characteristic, e.g. optimal brain damage from Le Cun et al. [27]. Breiman [6] has shown recently that random forests, an ensemble technique based on decision trees, can be used for scoring the importance of features. He shows that the increase in error due to perturbing feature

values in a data set and then processing the data through the random forest is an effective measure of the relevance of a feature.

4.3.1.1 Filter Techniques

Central to the filter strategy for feature selection is the criterion used to score the predictiveness of the features. In this section we will outline three of the most popular techniques for scoring the predictiveness of features – these are the chi-square measure, information gain and odds ratio. The overall scenario is described in Table 4.1. In this scenario the feature being assessed has r possible values and the table shows the distribution of those values across the classes. Intuitively, the closer these values are to an even distribution the less predictive that feature is of the class. It happens that all three of these techniques as described here require that the features under consideration are discrete valued. These techniques can be applied to numeric features by *discretizing* the data. Summary descriptions of the three techniques are as follows:

1. Chi-square measure: The chi-square measure is based on a statistical test for comparing proportions [48]. It produces a score that follows a χ^2 distribution, however, this aspect is not that relevant from a feature selection perspective as the objective is simply to rank the set of input features. The chi-square measure for scoring the *relatedness* of feature f to class c based on data D is as follows:

$$\chi^2(D,c,f) = \sum_{i=1}^{r} \left(\frac{(n_{i+} - \mu_{i+})^2}{\mu_{i+}} + \frac{(n_{i-} - \mu_{i-})^2}{\mu_{i-}} \right). \qquad (4.13)$$

In essence this scores the deviation of counts in each feature-value category against expected values if the feature were not correlated with the class (e.g. n_{i+} is the number of objects that have positive class and feature value v_i, μ_{i+} is the expected value if there were no relationship between f and c).

Table 4.1 The objective in supervised feature selection is to identify how well the distribution of feature values predicts a class variable. In this example the class variable is binary $\{c_+, c_-\}$ and the feature under consideration has r possible values. n_{i+} is the number of positive examples with feature value i and μ_{i+} is the *expected* value for that figure if the data were uniformly distributed, i.e. $\mu_{i+} = \frac{n_i n_+}{n}$

Feature value	c_+	c_-	
v_1	$n_{1+}(\mu_{1+})$	$n_{1-}(\mu_{1-})$	n_1
...	...	...	
v_i	$n_{i+}(\mu_{i+})$	$n_{i-}(\mu_{i-})$	n_i
...	...	...	
v_r	$n_{r+}(\mu_{r+})$	$n_{r-}(\mu_{r-})$	n_r
	n_+	n_-	n

2. Information gain: In recent years information gain (IG) has become perhaps the most popular criterion for feature selection. The IG of a feature is a measure of the amount of information that a feature brings to the training set [40]. It is defined as the expected reduction in entropy caused by partitioning the training set D using the feature f as shown in (4.14) where D_v is that subset of the training set D where feature f has value v:

$$\mathrm{IG}(D,c,f) = \mathrm{Entropy}(D,c) - \sum_{v \in \mathrm{values}(f)} \frac{|D_v|}{|D|} \mathrm{Entropy}(D_v,c). \tag{4.14}$$

   Entropy is a measure of how much randomness or impurity there is in the data set. It is defined in terms of the notation presented in Table 4.1 for binary classification as follows:

$$\mathrm{Entropy}(D,c) = -\sum_{i=1}^{r} \left(\frac{n_{i+}}{n_i} \log_2 \frac{n_{i+}}{n_i} + \frac{n_{i-}}{n_i} \log_2 \frac{n_{i-}}{n_i} \right). \tag{4.15}$$

   Given that for each feature the entropy of the complete data set $\mathrm{Entropy}(D,c)$ is constant, the set of features can be ranked by IG by simply calculating the remainder term – the second term in (4.14). Predictive features will have small remainders.
3. Odds ratio: The odds ratio (OR) [34] is an alternative filtering criterion that is popular in medical informatics. It is really only meaningful to calculate the odds ratio when the input features are binary; we can express this in the notation presented in Table 4.1 by assigning v_1 to the positive feature value and v_2 to the negative feature value:

$$\mathrm{OR}(D,c_+,f) = \frac{n_{1+}/n_{1-}}{n_{2+}/n_{2-}} = \frac{n_{1+}n_{2-}}{n_{2+}n_{1-}}. \tag{4.16}$$

   For feature selection, the features can be ranked according to their OR with high values indicating features that are very predictive of the class. The same can be done for the negative class to highlight features that are predictive of the negative class. Where a specific feature does not occur in a class, it can be assigned a small fixed value so that the OR can still be calculated.

Filtering Policy

The three filtering measures (chi-square, information gain and odds ratio) provide us with a *principle* on which a feature set might be filtered; we still require a filtering *policy*. There is a variety of policies that can be employed:

1. Select the top m of n features according to their score on the filtering criterion (e.g. select the top 50%).

2. Select all features that score above some threshold T on the scoring criterion (e.g. select all features with a score within 50% of the maximum score).
3. Starting with the highest scoring feature, evaluate using cross-validation the performance of a classifier built with that feature. Then add the next highest ranking feature and evaluate again; repeat until no further improvements are achieved.

This third strategy is simple but quite straightforward. An example of this strategy in operation is presented in Fig. 4.7. The graph shows the IG scores of the features in the UCI segment data set [35] and the accuracies of classifiers built with the top feature, the top two features and so on. It can be seen that after the ninth feature (saturation−mean) is added the accuracy drops slightly so the process would stop after selecting the first eight features. While this strategy is straightforward and effective it does have some potential shortcomings. The features are scored in isolation so two highly correlated features can be selected even if one is redundant in the presence of the other. The full space of possible feature subsets is not explored so there may be some very effective feature subsets that act in concert that are not discovered.

While these strategies are effective for feature selection they have the drawback that features are considered in isolation so redundancies or dependencies are ignored as already mentioned. Two strongly correlated features may both have high IG scores but one may be redundant once the other is selected. More sophisticated filter techniques that address these issues using mutual information to score *groups* of features have been researched by Novovičová et al. [39] and have been shown to be more effective than these simple filter techniques.

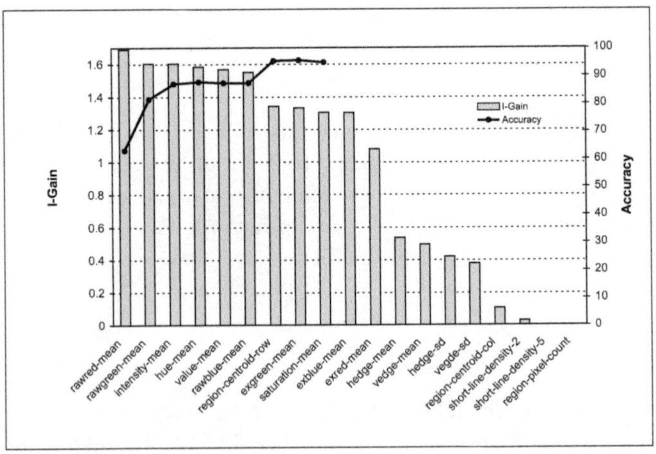

Fig. 4.7 This graph shows features from the UCI segment data set scored by IG and also the accuracies of classifiers built with the top ranking sets of features

4.3.1.2 Wrapper Techniques

The obvious criticism of the filter approach to feature selection is that the filter criterion is separate from the induction algorithm used in the classifier. This is overcome in the wrapper approach by using the performance of the classifier to guide search in feature selection – the classifier is *wrapped* in the feature selection process [26]. In this way the merit of a feature subset is the generalization accuracy it offers as estimated using cross-validation on the training data. If 10-fold cross-validation is used then 10 classifiers will be built and tested for each feature subset evaluated – so the wrapper strategy is very computationally expensive. If there are p features under consideration then the search space is of size 2^p so it is an exponential search problem.

A simple example of the search space for feature selection where $p = 4$ is shown in Fig. 4.8. Each node is defined by a feature mask; the node at the top of the figure has no features selected while the node at the bottom has all features selected. For large values of p an exhaustive search is not practical because of the exponential nature of the search. Four popular strategies are:

- **Forward selection (FS)** which starts with no features selected, evaluates all the options with just one feature, selects the best of these and considers the options with that feature plus one other, etc.
- **Backward elimination (BE)** starts with all features selected, considers the options with one feature deleted, selects the best of these and continues to eliminate features.
- **Genetic search** uses a genetic algorithm (GA) to search through the space of possible feature sets. Each state is defined by a feature mask on which crossover and mutation can be performed [30]. Given this convenient representation, the use of a GA for feature selection is quite straightforward although the evaluation of the fitness function (classifier accuracy as measured by cross-validation) is expensive.
- **Simulated annealing** is an alternative stochastic search strategy to GAs [31]. Unlike GAs, where a population of solutions is maintained, only one solution (i.e. feature mask) is maintained in simulated annealing (SA). SA implements a stochastic search since there is a chance that some deteriorations in solution are accepted – this allows a more effective exploration of the search space.

The first two strategies will terminate when adding (or deleting) a feature will not produce an improvement in classification accuracy as assessed by cross-validation. Both of these are greedy search strategies and so are not guaranteed to discover the best feature subset. More sophisticated search strategies such as GA or SA can be employed to better explore the search space; however, Reunanen [41] cautions that more intensive search strategies are more likely to overfit the training data.

A simple example of BE is shown in Fig. 4.9. In this example there are just four features (A,B,C and D) to consider. Cross-validation gives the full feature set a score of 71%, the best feature set of size 3 is (A,B,D) and the best feature set of size 2 is (A,B) and the feature sets of size 1 are no improvement on this.

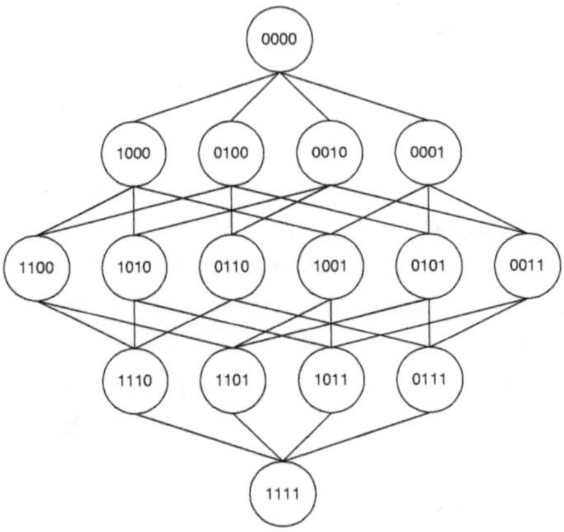

Fig. 4.8 The search space of feature subsets when $n = 4$. Each node is represented by a feature mask; in the topmost node no features are selected and in the bottom node all features are selected

Of the two simple wrapper strategies (BE and FS) BE is considered to be more effective as it more effectively considers features in the context of other features [2].

4.3.2 Unsupervised Feature Selection

Feature selection in a supervised learning context is a well-posed problem in that the objective can be clearly expressed. The objective can be to identify features that are correlated with the outcome or to identify a set of features that will build an accurate classifier – in either case the objective is to discover a reduced set of the original features in which the classes are well separated. By contrast feature selection in an unsupervised context is ill posed in that the overall objective is less

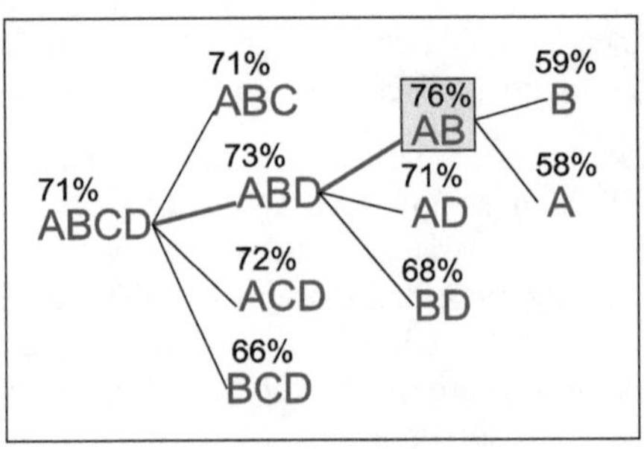

Fig. 4.9 This graph shows a wrapper-based search employing backward elimination. The search starts with all features (A,B,C,D) selected – in this example this is judged to have a score of 71%. The best feature subset uncovered in this example would be (A,B) which has a score of 76%

clear. The difficulty is further exacerbated by the fact that the number of clusters in the data is generally not known in advance; this further complicates the problem of finding a reduced set of features that will help *organize* the data.

If we think of unsupervised learning as clustering then the objective with feature selection for clustering might be to select features that produce clusters that are well separated. This objective can be problematic as different feature subsets can produce different well-separated clusterings. This can produce a "chicken and egg" problem where the question is "which comes first, the feature selection or the clustering?". A simple example of this is shown in Fig. 4.10; in this 2D example selecting feature f_1 produces the clustering $\{C_a, C_b\}$ while selecting f_2 produces the clustering $\{C_x, C_y\}$. So there are two alternative and very different valid solutions. If this data is initially clustered in 2D with $k = 2$ then it is likely that partition $\{C_x, C_y\}$ will be selected and then feature selection would select f_2.

This raises a further interesting question, does this clustering produced on the original (full) data description have special status? The answer to this is surely problem dependent; in problems such as text clustering, there will be many irrelevant features and the clustering on the full vocabulary might be quite noisy. On the other hand, in carefully designed experiments such as gene expression analysis, it might be expected that the clustering on the full data description has special merit. This co-dependence between feature selection and clustering is a big issue in feature selection for unsupervised learning; indeed Dy and Brodley [12] suggest that research in this area can be categorized by where the feature selection occurs in the clustering process:

Before clustering: To perform feature selection prior to clustering is analogous to the filter approach to supervised feature selection. A simple strategy would be to employ variance as a ranking criterion and select the features in which the data have the highest variance [10]. A more sophisticated strategy in this category is the Laplacian score [19] described in Sect. 4.3.2.1.

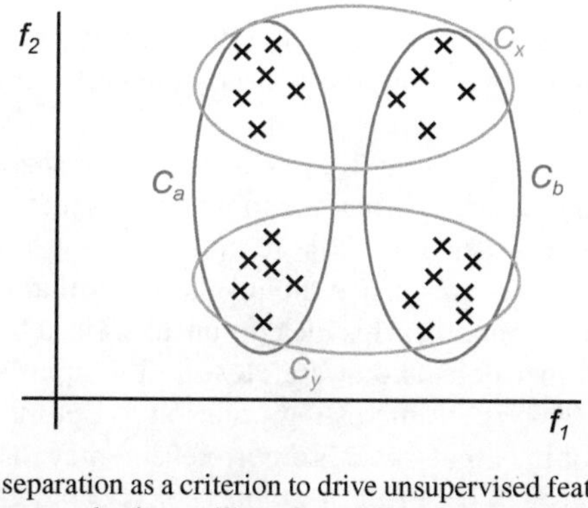

Fig. 4.10 Using cluster separation as a criterion to drive unsupervised feature selection is problematic because different feature selections will produce different clusterings with good separation. In this example if f_1 is selected then the obvious clustering is $\{C_a, C_b\}$, if f_2 is selected then $\{C_x, C_y\}$ is obvious

During clustering: Given the co-dependence between the clustering and the feature selection process, it makes sense to integrate the two processes if that is possible. Three strategies that do this are; the strategy based on category utility described in Sect. 4.3.2.2, the $Q-\alpha$ algorithm [47] described in Sect. 4.3.2.3 and biclustering [7].

After clustering: If feature selection is postponed until after clustering then the range of supervised feature selection strategies can be employed as the clusters can be used as class labels to provide the *supervision*. However, the strategy of using a set of features for clustering and then deselecting some of those features because they are deemed to be not relevant will not make sense in some circumstances.

One of the reasons why unsupervised feature selection is a challenging problem is because the success criterion is ill posed as stated earlier. This is particularly an issue if the feature selection stage is to be integrated into the clustering process. Two criteria that can be used to quantify a good partition are the criterion based on the scatter matrices presented in Sect. 4.2.2 and category utility which is explained in Sect. 4.3.2.2. The objective with the criterion based on scatter is to maximize $trace(\mathbf{S}_W^{-1}\mathbf{S}_B)$ [12] – this is particularly appropriate when the data is numeric. For categorical data the category utility measure described in the next section is applicable.

In the remainder of this section on unsupervised feature selection we will describe a variety of unsupervised feature selection techniques that have emerged in recent research. These techniques will be organized into the categories of filter, wrapper and embedded in the same manner as in the section on supervised feature selection (Sect. 4.3.1). However, the distinction between these categories is less clear-cut in the unsupervised case.

4.3.2.1 Unsupervised Filters

The defining characteristic of a filter-based feature selection technique is that features are scored or ranked by a criterion that is separate from the classification or clustering process.

A prominent example of such a strategy is the Laplacian score that can be used as a criterion in dimension reduction when the motivation is that *locality* is preserved. Such locality-preserving projections [20] are appropriate in image analysis where images that are similar in the input space should also be similar in the reduced space. The Laplacian score (LS) embodies this idea for unsupervised feature selection [19]. LS selects features so that objects that are close in the input space are still close in the reduced space. This is an interesting criterion to optimize as it contains the implication that none of the input features are irrelevant; they may be just redundant.

The calculation of LS is based on a graph G that captures nearest neighbour relationships between the n data points. G is represented by a square matrix $\mathbf{S}$ where $\mathbf{S}_{ij} = 0$ unless x_i and x_j are neighbours, in which case

$$\mathbf{S}_{ij} = \exp\left(\frac{||x_i - x_j||^2}{t}\right), \tag{4.17}$$

where t is a bandwidth parameter. The neighbourhood idea introduces another parameter k which is the number of neighbours used to construct $\mathbf{S}$. $\mathbf{L} = \mathbf{D} - \mathbf{S}$ is the Laplacian of this graph where $\mathbf{D}$ is a degree diagonal matrix $\mathbf{D}_{ii} = \sum_j \mathbf{S}_{ij}, D_{ij,i \neq j} = 0$ [42]. If $\mathbf{m}_i$ is the vector of values in the data set for the ith feature then the LS is defined using the following calculations [19]:

$$\widetilde{\mathbf{m}}_i = \mathbf{m}_i - \frac{\mathbf{m}_i^\mathsf{T}\mathbf{D}\mathbf{1}}{\mathbf{1}^\mathsf{T}\mathbf{D}\mathbf{1}}\mathbf{1}, \tag{4.18}$$

where $\mathbf{1}$ is a vector or 1s of length n. Then the Laplacian score for the ith feature is

$$\mathrm{LS}_i = \frac{\widetilde{\mathbf{m}}_i^\mathsf{T}\mathbf{L}\widetilde{\mathbf{m}}_i}{\widetilde{\mathbf{m}}_i^\mathsf{T}\mathbf{D}\widetilde{\mathbf{m}}_i}. \tag{4.19}$$

This can be used to score all the features in the data set on how effective they are in preserving locality. This has been shown to be an appropriate criterion for dimension reduction in applications such as image analysis where locality preservation is an effective motivation [19]. However, if the data contains irrelevant features, as can occur in text classification or the analysis of gene expression data, then locality preservation is not a sensible motivation. In such circumstances selecting the features in which the data has the highest variance [10] might be a more appropriate filter.

4.3.2.2 Unsupervised Wrappers

The defining characteristic of a wrapper-based feature selection technique is that the classification or clustering process is used to evaluate feature subsets. This is more problematic in clustering than in classification as there is no single criterion that can be used to score cluster quality and many cluster validity indices have biases – e.g. toward small numbers of clusters or balanced cluster sizes [11, 17]. Nevertheless there has been work on unsupervised wrapper-like feature selection techniques and two such techniques – one based on category utility and the other based on the EM clustering algorithm – are described here.

Category Utility

Devaney and Ram [9] proposed a wrapper-like unsupervised feature subset selection algorithm based on the notion of category utility (CU) [16]. This was implemented in the area of conceptual clustering, using Fisher's [13] COBWEB system as the underlying concept learner. Devaney and Ram demonstrate that if feature selection is performed as part of the process of building the concept hierarchy (i.e. concepts are defined by a subset of features) then a better concept hierarchy is developed. As with the original COBWEB system, they use CU as their evaluation function to guide the process of creating concepts – the CU of a clustering C based on a feature set F is defined as follows:

$$\mathrm{CU}(C,F) = \frac{1}{k} \sum_{c_l \in C} \left[\sum_{f_i \in F} \sum_{j=1}^{r_i} p(f_{ij}|C_l)^2 - \sum_{f_i \in F} \sum_{j=1}^{r_i} p(f_{ij})^2 \right], \tag{4.20}$$

where $C = \{C_1, ..., C_l, ..., C_k\}$ is the set of clusters and $F = \{F_1, ..., F_i, ..., F_p\}$ is the set of features. CU measures the difference between the conditional probability of a feature i having value j in cluster l and the prior probability of that feature value. The inner most sum is over r feature values, the middle sum is over p features and the outer sum is over k clusters. This function measures the increase in the number of feature values that can be predicted correctly given a set of concepts, over those which can be predicted without using any concepts.

Their approach was to generate a set of feature subsets (using either FS or BE as described in Sect. 4.3.1.2), run COBWEB on each subset and then evaluate each resulting concept hierarchy using the category utility metric on the first partition. BSS starts with the full feature set and removes the least useful feature at each stage until utility stops improving. FSS starts with an empty feature set and adds the feature providing the greatest improvement in utility at each stage. At each stage the algorithm checks how many feature values can be predicted correctly by the partition – i.e. if the value of each feature f can be predicted for most of the clusters C_l in the partition, then the features used to produce this partition were informative or relevant. The highest scoring feature subset is retained and the next larger (or smaller) subset is generated using this subset as a starting point. The process continues until no higher CU score can be achieved.

The key idea here is that CU is used to score the quality of clusterings in a wrapper-like search. It has been shown by Gluck and Corter [16] that CU corresponds to mutual information so this is quite a principled way to perform unsupervised feature selection.

Devaney and Ram improved upon the time it takes to reconstruct a concept structure by using their own concept learner, attribute-incremental concept creator (AICC), instead of COBWEB. AICC can add features without having to rebuild the concept hierarchy from scratch and shows large speedups.

Expectation Maximization (EM)

Dy and Brodley present a comprehensive analysis of unsupervised wrapper-based feature selection in [12]. They present their analysis in the context of the EM clustering algorithm [32]. Specifically, they consider wrapping the EM clustering algorithm where feature subsets are evaluated with criteria based on cluster separability and maximum likelihood. However, they emphasize that the approach is general and can be used with any clustering algorithm by selecting an appropriate criterion for scoring the clusterings produced by different feature subsets. They discuss the biases associated with cluster validation techniques (e.g. biases on cluster size, data dimension, a balanced cluster sizes) and propose ways in which some of these issues can be ameliorated.

4.3.2.3 The Embedded Approach

The final category of feature selection technique mentioned in Sect. 4.3.1 is the embedded approach, i.e. feature selection is an integral part of the classification algorithm as happens for instance in the construction of decision trees [40] or in some types of neural network [27]. In the unsupervised context this general approach is a good deal more prominent. There are a number of clustering techniques that have dimension reduction as a by-product of the clustering process, for example, non-negative matrix factorization (NMF) [28], biclustering [18] and projected clustering [1]. These approaches have in common that they discover clusters in the data that are defined by a subset of the features and different clusters can be defined by different feature subsets. Thus these are implicitly *local* feature selection techniques.

The alternative to this is a *global* approach where the same feature subset is used to describe all clusters. A representative example of this is Q-α algorithm presented by Wolf and Shashua [47].

The Q-α Approach

A well-motivated criterion of cluster quality is cluster coherence, in graph theoretic terms this is expressed by the notion of objects within clusters being well connected and individual clusters being weakly linked. The whole area of spectral clustering captures these ideas in a well-founded family of clustering algorithms based on the idea of minimizing the *graph-cut* between clusters [37].

The principles of spectral clustering have been extended by Wolf and Shashua [47] to produce the Q-α algorithm that simultaneously performs feature subset selection and discovers a good partition of the data. As with spectral clustering, the fundamental data structure is the affinity matrix $\mathbf{A}$ where each entry $\mathbf{A}_{ij}$ captures the similarity (in this case as a dot-product) between data points i and j. In order to facilitate feature selection the affinity matrix for Q-α is expressed as $\mathbf{A}_\alpha = \sum_{i=1}^{p} \alpha_i \mathbf{m}_i \mathbf{m}_i^\mathsf{T}$ where $\mathbf{m}_i$ is the ith row in the data matrix that has been normalized so as to be centred on 0 and be of unit L_2 norm (this is the set of values in the data set for feature i). $\mathbf{m}_i \mathbf{m}_i^\mathsf{T}$ is the *outer*-product of $\mathbf{m}_i$ with itself. α is the weight vector for the p features – ultimately the objective is for most of these weight terms to be set to 0.

In spectral clustering $\mathbf{Q}$ is an $n \times k$ matrix composed of the k eigenvectors of $\mathbf{A}$ corresponding to the largest k eigenvalues. Wolf and Shashua show that the relevance of a feature subset as defined by the weight vector α can be quantified by

$$\mathrm{Rel}(\alpha) = \mathrm{trace}(\mathbf{Q}^\mathsf{T} \mathbf{A}_\alpha^\mathsf{T} \mathbf{A}_\alpha \mathbf{Q}). \tag{4.21}$$

They show that feature selection and clustering can be performed as a single process by optimizing

$$\max_{\mathbf{Q}\alpha} \operatorname{trace}(\mathbf{Q}^\mathsf{T}\mathbf{A}_\alpha^\mathsf{T}\mathbf{A}_\alpha\mathbf{Q}) \tag{4.22}$$

subject to $\alpha^\mathsf{T}\alpha = 1$ and $\mathbf{Q}^\mathsf{T}\mathbf{Q} = \mathbf{I}$.

Wolf and Shashua show that this can be solved by solving two inter-linked eigenvalue problems that produce solutions for α and $\mathbf{Q}$. They show that a process of iteratively solving for α then fixing α and solving for $\mathbf{Q}$ will converge. They also show that the process has the convenient property that the α_i weights are biased to be positive and sparse, i.e. many of them will be zero.

So the Q-α algorithm performs feature selection in the spirit of spectral clustering, i.e. the motivation is to increase cluster coherence. It discovers a feature subset that will support a partitioning of the data where clusters are well separated according to a graph-cut criterion.

4.4 Conclusions

The objective of this chapter was to provide an overview of the variety of strategies that can be employed for dimension reduction when processing high dimension data. When feature transformation is appropriate then PCA is the dominant technique if the data are not labelled. If the data are labelled then LDA can be applied to discover a projection of the data that separates the classes. When feature selection is required and the data are labelled then the problem is well posed. A variety of filter and wrapper-based techniques for feature selection are described in Sect. 4.3.1. The chapter concludes with a review of unsupervised feature selection in Sect. 4.3.2. This is a more difficult problem than the supervised situation in that the success criterion is less clear. Nevertheless this is an active research area at the moment, and a variety of unsupervised feature selection strategies have emerged.

References

1. C.C. Aggarwal, J.L. Wolf, P.S. Yu, C. Procopiuc, and J.S. Park. Fast algorithms for projected clustering. In *Proceedings of the 1999 ACM SIGMOD International Conference on Management of Data*, pages 61–72, 1999. http://www.sigmod.org/sigma.
2. D.W. Aha and R.L. Bankert. A comparative evaluation of sequential feature selection algorithms. In *Proceedings of the Fifth International Workshop on Artificial Intelligence and Statistics*, pages 1–7, 1995.
3. G. Baudat and F. Anouar. Generalized discriminant analysis using a kernel approach. *Neural Computation*, 12(10):2385–2404, 2000.
4. R. Bellman. *Adaptive Control Processes: A Guided Tour*. Princeton University Press, 1961.
5. A. Blum and P. Langley. Selection of relevant features and examples in machine learning. *Artificial Intelligence*, 97(1–2):245–271, 1997.
6. L. Breiman. Random forests. *Machine Learning*, 45(1):5–32, 2001.
7. K. Bryan, P. Cunningham, and N. Bolshakova. Biclustering of expression data using simulated annealing. In *CBMS*, pages 383–388. IEEE Computer Society, 2005.

8. S.C. Deerwester, S.T. Dumais, T.K. Landauer, G.W. Furnas, and R.A. Harshman. Indexing by latent semantic analysis. *Journal of the American Society of Information Science*, 41(6):391–407, 1990.
9. M. Devaney and A. Ram. Efficient feature selection in conceptual clustering. In D.H. Fisher, editor, *ICML*, pages 92–97. Morgan Kaufmann, 1997.
10. M. Doyle and P. Cunningham. A dynamic approach to reducing dialog in on-line decision guides. In E. Blanzieri and L. Portinale, editors, *EWCBR*, volume 1898 of *Lecture Notes in Computer Science*, pages 49–60. Springer, 2000.
11. R.C. Dubes. How many clusters are best?—an experiment. *Pattern Recognition*, 20(6):645–663, 1987.
12. J.G. Dy and C.E. Brodley. Feature selection for unsupervised learning. *The Journal of Machine Learning Research*, 5:845–889, 2004.
13. D.H. Fisher. Knowledge acquisition via incremental conceptual clustering. *Machine Learning*, 2(2):139–172, 1987.
14. R.A. Fisher. The use of multiple measurements in taxonomic problems. *Annals of Eugenics*, 7:179–188, 1936.
15. K. Fukunaga. *Introduction to Statistical Pattern Recognition*. Academic Press, Inc, 2nd edition, 1990.
16. M.A. Gluck and J.E. Corter. Information, uncertainty, and the utility of categories. In *Proceedings of the Seventh Annual Conference of the Cognitive Science Society*, pages 283–287, Hillsdale, NJ, 1985. Lawrence Earlbaum.
17. J. Handl, J. Knowles, and D.B. Kell. Computational cluster validation in post-genomic data analysis. *Bioinformatics*, 21(15):3201–3212, 2005.
18. J.A. Hartigan. Direct clustering of a data matrix. *Journal of the American Statistical Association*, 67(337):123–129, 1972.
19. X. He, D. Cai, and P. Niyogi. Laplacian score for feature selection. In *NIPS*, 2005.
20. X. He and P. Niyogi. Locality preserving projections. In S. Thrun, L.K. Saul, and B. Schölkopf, editors, *NIPS*. MIT Press, 2003.
21. D.R. Heisterkamp. Building a latent semantic index of an image database from patterns of relevance feedback. In *ICPR (4)*, pages 134–137, 2002.
22. H. Hotelling. Analysis of a complex of statistical variables into principal components. *Journal of Educational Psychology*, 24:417–441, 1933.
23. R. Huang, Q. Liu, H. Lu, and S. Ma. Solving the small sample size problem of LDA. In *ICPR (3)*, pages 29–32, 2002.
24. A. Hyvärinen, J. Karhunen, and E. Oja. *Independent Component Analyis*. John Wiley & Sons, Inc, 2001.
25. G.H. John, R. Kohavi, and K. Pfleger. Irrelevant features and the subset selection problem. In *Proceedings of the 11th International Conference on Machine Learning*, pages 121–129, New Brunswick, NJ, 1994. Morgan Kaufmann.
26. R. Kohavi and G. H. John. Wrappers for feature subset selection. *Artificial Intelligence*, 97(1–2):273–324, 1997.
27. Y. LeCun, J. Denker, S. Solla, R.E. Howard, and L.D. Jackel. Optimal brain damage. In D.S. Touretzky, editor, *Advances in Neural Information Processing Systems II*, San Mateo, CA, 1990. Morgan Kauffman.
28. D.D. Lee and H.S. Seung. Learning the parts of objects by non-negative matrix factorization. *Nature*, 401(6755):788–791, 1999.
29. W. Liu, Y. Wang, S.Z. Li, and T. Tan. Null space approach of fisher discriminant analysis for face recognition. In D. Maltoni and A.K. Jain, editors, *ECCV Workshop BioAW*, volume 3087 of *Lecture Notes in Computer Science*, pages 32–44. Springer, 2004.
30. J. Loughrey and P. Cunningham. Overfitting in wrapper-based feature subset selection: The harder you try the worse it gets. In *24th SGAI International Conference on Innovative Techniques and Applications of Artificial Intelligence (AI-2004)*, pages 33–43, 2004.
31. J. Loughrey and P. Cunningham. Using early-stopping to avoid overfitting in wrapper-based feature subset selection employing stochastic search. In M. Petridis, editor, In *10th UK Workshop on Case-Based Reasoning*, pages 3–10. CMS Press, 2005.

32. G.J. McLachlan and T. Krishnan. *The EM Algorithm and Extensions*. Wiley, New York 1997.
33. S. Mika, B. Schölkopf, A.J. Smola, K.R. Müller, M. Scholz, and G. Rätsch. Kernel PCA and de-noising in feature spaces. In M.J. Kearns, S.A. Solla, and D.A. Cohn, editors, *NIPS*, pages 536–542. The MIT Press, 1998.
34. D. Mladenic. Feature subset selection in text-learning. In C. Nedellec and C. Rouveirol, editors, *ECML*, volume 1398 of *Lecture Notes in Computer Science*, pages 95–100. Springer, Berlin, 1998.
35. D.J. Newman, S. Hettich, C.L. Blake, and C.J. Merz. UCI repository of machine learning databases, 1998. http://www.ics.uci.edu/~mlearn/MLRepository.html.
36. A. Ng, M. Jordan, and Y. Weiss. On spectral clustering: Analysis and an algorithm. In *Proceedings of Advances in Neural Information Processing*, 2001. http://books.nips.cc.
37. A.Y. Ng, M. Jordan, and Y. Weiss. On spectral clustering: Analysis and an algorithm. *Advances in Neural Information Processing Systems*, 14(2):849–856, 2001.
38. S.K. Ng, Z. Zhu, and Y.S. Ong. Whole-genome functional classification of genes by latent semantic analysis on microarray data. In Y.-P. Phoebe Chen, editor, *APBC*, volume 29 of *CRPIT*, pages 123–129. Australian Computer Society, 2004.
39. J. Novovičová, A. Malík, and P. Pudil. Feature selection using improved mutual information for text classification. In A.L. N. Fred, T. Caelli, R.P.W. Duin, A.C. Campilho, and D. de Ridder, editors, *SSPR/SPR*, volume 3138 of *Lecture Notes in Computer Science*, pages 1010–1017. Springer, Berlin, 2004.
40. J.R. Quinlan. *C4.5: Programs for Machine Learning*. Morgan Kaufmann, San Francisco, CA, 1993.
41. J. Reunanen. Overfitting in making comparisons between variable selection methods. *Journal of Machine Learning Research*, 3:1371–1382, 2003.
42. M. Saerens, F. Fouss, L. Yen, and P. Dupont. The principal components analysis of a graph, and its relationships to spectral clustering. In *Proceedings of the 15th European Conference on Machine Learning (ECML 2004). Lecture Notes in Artificial Intelligence*, 3201:371–383, 2004.
43. E. Sahouria and A. Zakhor. Content analysis of video using principal componets. In *ICIP (3)*, pages 541–545, IEEE Computer Society, 1998.
44. P. Smaragdis, B. Raj, and M. Shashanka. A probabilistic latent variable model for acoustic modeling. In *Workshop on Advances in Models for Acoustic Processing at NIPS 2006*, 2006. http://www.idiap.ch/amac.
45. M. Sugiyama. Local Fisher discriminant analysis for supervised dimensionality reduction. In W.W. Cohen and A. Moore, editors, *ICML*, pages 905–912. ACM, 2006.
46. M. West. Bayesian factor regression models in the "large p, small n" paradigm. *Bayesian Statistics*, 7:723–732, 2003.
47. L. Wolf and A. Shashua. Feature selection for unsupervised and supervised inference: The emergence of sparsity in a weight-based approach. *Journal of Machine Learning Research*, 6:1855–1887, 2005.
48. S. Wu and P.A. Flach. Feature selection with labelled and unlabelled data. In *Proceedings of ECML/PKDD'02 Workshop on Integration and Collaboration Aspects of Data Mining, Decision Support and Meta-Learning*, pages 156–167, 2002.

Part II
Multimedia Applications

Chapter 5
Online Content-Based Image Retrieval Using Active Learning

Matthieu Cord and Philippe-Henri Gosselin

Abstract Content-based image retrieval (CBIR) has attracted a lot of interest in recent years. When considering visual information retrieval in image databases, many difficulties arise. Learning is definitively considered as a very interesting issue to boost the efficiency of information retrieval systems. Different strategies, such as offline supervised learning or semi-supervised learning, have been proposed. Active learning methods have been considered with an increased interest in the statistical learning community. Initially developed in a classification framework, a lot of extensions are now proposed to handle multimedia applications. The purpose of this chapter is to present an overview of the online image retrieval systems based on supervised classification techniques. This chapter also provides algorithms in a statistical framework to extend active learning strategies for online content-based image retrieval.

5.1 Introduction

Large collections of digital images are being created in different fields and many applicative contexts. Some of these collections are the product of digitizing existing collections of analog photographs, paintings, etc., and others result from digital acquisitions. Potential applications include web searching, cultural heritage, geographic information systems, biomedicine, surveillance systems.

The traditional way of searching these collections is by keyword indexing, or simply by browsing. Digital image databases, however, open the way to content-based searching. Content-based image retrieval (CBIR) has attracted a lot of research interest in recent years. A common scheme to search the database is to

Matthieu Cord
LIP6, UPMC Paris 6, Paris, France, e-mail: Matthieu.Cord@lip6.fr

Philippe-Henri Gosselin
ETIS, UCP Cergy, France, e-mail: gosselin@ensea.fr

automatically extract different types of features (for instance color or texture) and image descriptors (indexes). These indexes are then used in a search engine strategy to compare, classify, and rank the images.

Major sources of difficulties in CBIR are variable imaging conditions, complex and hard-to-describe image content, and the gap between arrays of numbers representing images and conceptual information perceived by humans. In CBIR field, the semantic gap usually refers to this separation between the low-level information extracted from images and the semantics [37, 38]: the user is looking for one image or an image set representing a concept, for instance a type of landscape, whereas current processing strategies deal with color or texture features!

Learning is definitively considered as the most interesting issue to reduce the semantic gap. Different learning strategies, such as offline supervised learning, online active learning, semi-supervised learning, may be considered to improve the efficiency of retrieval systems:

- Some offline learning methods focus on the feature extraction or on the similarity function improvement. Using experiments, a similarity function may be trained in order to better represent the distance between semantic categories [30]. Thanks to local primitives and descriptors, such as salient points or regions, supervised learning may be introduced to learn object or region categories [5]. The classification function is next used to retrieve images from the learnt category in large databases.
- Human interactive systems have attracted a lot of research interest in recent years, especially for content-based image retrieval systems. In CBIR, the retrieval may be initiated using a query as example. The top rank similar images are then presented to the user. Next, the interactive process allows the user to refine his query as much as necessary. Many kinds of interaction between the user and the system have been proposed, but most of the time, user information consists of binary labels (labels) indicating whether or not the image belongs to the desired concept. The positive labels indicate *relevant* images for the current concept and the negative labels *irrelevant* images.

The main scope of this chapter is to present modern online retrieval approaches of *concepts* within a large image collection. We assume that a user is looking for a set of documents, the *query concept*, within a database. Performing an estimation of the query concept can be seen as a statistical learning problem, and more precisely as a binary classification task between the relevant and irrelevant classes [4]. In image retrieval, techniques based on statistical learning have been proposed, as for instance Bayes classification [47], k-nearest neighbors, Gaussian mixtures [31], Gaussian random fields, or support vector machines [4].

In this chapter, we focus on these statistical learning techniques for interactive image retrieval. Set of solutions to deal with the CBIR specificities are proposed. The RETIN search engine [10] will be used to illustrate this chapter. For doing this, we present in the following the three essential components of interactive CBIR statistical systems: feature and similarity computation, classification, and active learning framework.

5.2 Database Representation: Features and Similarity

5.2.1 Visual Features

To index images, basic visual components must be selected: pixels, patches, regions, points of interest, edges, etc. Next, to describe these visual components, color and texture features are commonly used. Color is definitively the most used feature for image indexing. The building of color features is based on the choice of a color space, which can include properties such as invariance with illumination. Many systems simply use the RGB color space [32]. However, transformed color spaces such as HSV or CIELab, differentiating hue from intensity, are also widely used [6, 41]. They have been introduced to better represent human perceptual similarity. When using visual components defined on local area, color distribution may be computed as presented in Chap. 8. Many techniques of texture extraction and description are proposed in literature, for instance based on the Wold decomposition [26], co-occurrence matrixes [1], and Gabor or wavelet filter banks [28, 36].

Using color and texture descriptors is not always discriminant enough for some applications. To embed invariance properties, several strategies have been proposed as local derivative JET analysis, scale-invariant feature transform (SIFT) descriptors [27], moment analysis as for instance using Zernike moments [20].

5.2.2 Signature Based on Visual Pattern Dictionary

Building a visual codebook is an effective mean of extracting the relevant visual content of an image database, which is used by most of the retrieval systems.

A first approach is to perform a *static* clustering, like [41] where 166 regular colors are a priori defined. These techniques directly provide an index and a similarity for comparing images, but the visual codebook is far from being optimal, except in very specific applications.

A second approach is to perform a *dynamic* or *adaptive* clustering, using a standard clustering algorithm such as k-means. In this case, the visual codebook is adapted to the image database. Using a k-means algorithm leads to a sub-optimal codebook, where codewords are under- or over-representing visual content. Actually, the standard k-means algorithm uses a random initial codebook and thus converges to a local minimum in most of the cases. A usual way to find a good visual codebook is to train several times the clustering algorithm and keep the best codebook. Other improvements about the initialization have been proposed, like the k-means splitting or LBG [24]. The algorithm starts with only one codeword, and step after step, split the clusters into two sub-clusters. Patanè proposes ELBG, an enhanced LBG algorithm, that introduces an heuristic in order to jump from a local minimum to a better one [33]. This heuristic swaps codewords so that their respective distortions are as much equal as possible. However, because of the large number of vectors to be clustered, this strategy has a very high cost in computational time.

All image indexing systems have proposed strategies to deal with this quantification, as the text database systems done before. For instance, in the RETIN system, two techniques have been experimented. The first method, called retinl in experiments, is based on an adaptive k-means by sub-sampling images [15]. The second one, called retin2, exploits a two-stage k-means, both performing ELBG:

- *First stage*: quantization of each image independently;
- *Second stage*: quantization of the whole database from the dictionary obtained at the first stage.

Once color, texture, shape-based codebooks are carried out, image signatures are computed. For each pixel, the closest codeword is detected and a pixel distribution is generated for each image over the visual codebooks. For instance, the RETIN image signature is composed of a 50-length vector $\mathbf{x}_i$, gluing two distributions over two 25-length color and texture codebooks. These final dimensions have been obtained with an empirical maximization of the mean average precision (MAP) over test databases (see Appendix for details about MAP statistic measurement and databases). Results are reported in Tables 5.1 and 5.2. As we are looking for the maximum of the MAP values, the best sizes of the codebooks have been selected to 25 for both color and texture analyses.

In the visual codebooks computed from local feature extraction and characterization as sift descriptors, the resulting codebook dimension is usually much larger, around 1000 [13]. There is no consensus about the best way to handle this problem of dimension, and there are a lot of attempts to deal with very high dimension giving promising results.

$\mathbf{X} = \{\mathbf{x}_1, \dots, \mathbf{x}_n\}$ will be the notation for all the signatures of any database having n images of dimension p.

5.2.3 Similarity

Once signatures are computed, a metric or a similarity function has to be defined to compare images. Basically, the Euclidean distance is used to compute the similarity (or dissimilarity) between histograms, or more generally a Minkowski distance. However, these metrics are not necessarily relevant for histograms. Alternatives such as histogram intersections [43], Kullback–Leibler divergence, have been used and are discussed in Chap. 2.6. These metrics independently compare each value of the histograms, and does not address the problem of correlation between

Table 5.1 Mean average precision for different values of the codebook dimension, using two methods for adaptive database color feature quantization

Codebook size	12	25	50	100	200
retin2	13.47%	16.92%	17.12%	15.37%	14.89%
retin1	13.96%	16.24%	15.87%	14.72%	13.68%

Table 5.2 Mean average precision for different values of the codebook dimension, using two methods for adaptive database texture feature quantization

Codebook size	12	25	50	100	200
retin2	11.97%	12.22%	13.43%	11.23%	10.52%
retin1	2.86%	3.47%	3.92%	3.65%	3.42%

axes. More robust metrics have been proposed, like in [42], earth mover's distance [36], or generalized quadratic distances, to solve this problem. They are fully detailed in Chap. 2.6.

Whenever these metrics are efficient for histograms, they usually lead to non-linear problems, and handcraft learning techniques are developed in order to solve them. Kernel framework offers flexible set of solutions to represent similarities between images and complex data.

5.2.4 Kernel Framework

As explained in Chap. 2.6, the approach consists in finding a mapping Φ from an input space X (here our histogram space) to a Hilbert space $\mathscr{H}$. Such a mapping should verify: $k(\mathbf{x}_i, \mathbf{x}_j) = \langle \Phi(\mathbf{x}_i), \Phi(\mathbf{x}_j) \rangle$, so that efficient learning schemes may be considered in the induced space $\mathscr{H}$. Working in a vectorial space, it is easy to express the binary classification as a linear discrimination optimization, and that leads to fast and efficient algorithms such as the popular support vector machine classifiers (SVMs). Another interest of the kernel function framework is that we do not need to directly express the mapping function Φ. The idea is to work not on the mapped vectors $\Phi(\mathbf{x}_i)$ and $\Phi(\mathbf{x}_j)$, but on their dot products, that is evaluated by $k(\mathbf{x}_i, \mathbf{x}_j)$ in the original space.

The kernel function framework is interesting only if one can find a relevant mapping for the application. In our case, since we are working on histograms, an interesting kernel function is the Gaussian one, with the χ^2 distance: $\chi^2(\mathbf{x}_i, \mathbf{x}_j) = \sum_{r=1}^{p} \frac{(x_{ri} - x_{rj})^2}{x_{ri} + x_{rj}}$. Recent works present some kernel design trying to build a kernel function from dedicated distances already validated in some specific fields. For instance, in [11] Cuturi proposes to elaborate kernels on the well-known dynamic time warping (DTW) family of distances by considering the same set of elementary operations, namely substitutions and repetitions of tokens, to map a sequence onto another.

The input space X is of course not limited to Euclidean space, and another interesting extension is to consider functions to deal with bags of vectors, when local feature vectors are computed from images for instance. In this context, kernels may be designed, and let us reproduce here a basic way to do it for bags B_k of unordered vectors $\mathbf{b}_{rk}$ [40]: $K(B_i, B_j) \triangleq \sum_r \sum_s k(\mathbf{b}_{ri}, \mathbf{b}_{sj})$. Based on the kernel function k on vectors (called *minor* kernel), associated to the mapping function ϕ, it is easy to prove that K is also a kernel:

$$\begin{aligned} K(B_i,B_j) &= \sum_r \sum_s k(\mathbf{b}_{ri},\mathbf{b}_{sj}) \\ &= \sum_r \sum_s \langle \phi(\mathbf{b}_{ri}),\phi(\mathbf{b}_{sj})\rangle = \langle \sum_r \phi(\mathbf{b}_{ri}),\sum_s \phi(\mathbf{b}_{sj})\rangle. \end{aligned}$$

It follows that the mapping function $\Phi(B_i)$ defined as $\Phi(B_i) = \sum_r \phi(\mathbf{b}_{ri})$ allows to write K as the corresponding dot product in the induced space: $K(B_i,B_j) = \langle \Phi(B_i),\Phi(B_j)\rangle$.

When considering this class of kernels, one problem is the computational complexity with large bags. Several propositions have been made to reduce this complexity. A common strategy is to build a model of the bags, and next propose a kernel function for these models. For instance, in [21], Gaussian functions are used to represent bags, and a Bhattacharyya kernel to compare them. Another way is to represent a bag as a distribution of prototypes of points of interest [17, 19]. The main drawback is that the mapping is explicit in that case.

From the kernel framework, similarity functions s are simply expressed in the induced space, as for example as the dot product in the induced space: $s(\Phi(\mathbf{x}_i),\Phi(\mathbf{x}_j)) = \langle \Phi(\mathbf{x}_i),\Phi(\mathbf{x}_i)\rangle = k(\mathbf{x}_i,\mathbf{x}_j)$ meaning $s = k$, but other measures, as for instance the angle between two vectors, may be used:

$$s(\Phi(\mathbf{x}_i),\Phi(\mathbf{x}_j)) = \frac{|k(\mathbf{x}_i,\mathbf{x}_j)|}{\sqrt{k(\mathbf{x}_i,\mathbf{x}_i)k(\mathbf{x}_j,\mathbf{x}_j)}}.$$

When working in a Hilbert space, several standard operators may be expressed using k, as for instance the Euclidean distance: $d(\Phi(\mathbf{x}_i),\Phi(\mathbf{x}_j))^2 = ||\Phi(\mathbf{x}_i) - \Phi(\mathbf{x}_j)||^2 = k(\mathbf{x}_i,\mathbf{x}_i) + k(\mathbf{x}_j,\mathbf{x}_j) - 2k(\mathbf{x}_i,\mathbf{x}_j)$.

Kernel functions k will be used as similarity functions in this chapter. As k is a kernel function, the matrix $\mathbf{K}$ with $\mathbf{K}_{ij} = k(\mathbf{x}_i,\mathbf{x}_j)$ is symmetric and semi-definite positive (sdp), that is to say a Gram matrix. This matrix embeds the index information $\mathbf{X} = \{\mathbf{x}_1,\dots,\mathbf{x}_n\}$ and the similarity function k related to the whole database. Any further step of data mining, classification, ranking, will be only based on this Gram matrix data. The advantage of this framework is to well separate the similarity definition from the learning problem.

5.2.5 Experiments

In order to evaluate the different kernels, it is possible to compute classification and retrieval using the same classifier, and then to compare performances. We report here experiments for vectorial features on two databases described in the Appendix. A SVM classifier has been used as refereed classifier for all kernels. Results are shown in Fig. 5.1.

The linear kernel, which can be seen as the "no kernel" strategy, gives the worst performances. It is followed by the polynomial kernel (of degree 3), which was

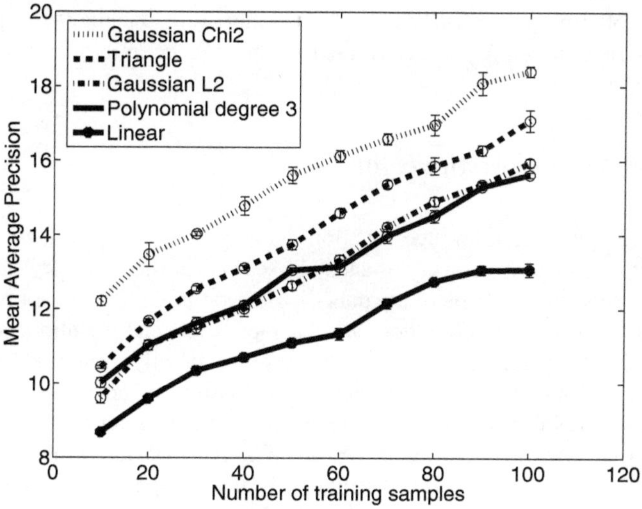

Fig. 5.1 Mean average precision for a classification by SVM according to the number of training data, for several kernel functions on the Corel photo database

originally tuned for the tracking of high-level correlations of data. Close to this one is the Gaussian kernel, with an Euclidean distance, and next is the triangle kernel, which is invariant to scale variation. Finally, the Gaussian kernel with a χ^2 distance gives the best performances, results which are consistent with the use of histograms as index. Thus, in the following experiments, we will use a Gaussian kernel with a χ^2 distance.

Remark 5.1. Whenever the Gaussian distance χ^2 is the most interesting for our indexes, it will be no longer true on non-histogram ones. However, assuming that one can find a kernel function relevant for one's indexes, all the results about the learning techniques we present in the next sections are still valid, since they are made to work in an Hilbert space induced by a kernel function.

5.3 Classification Framework for Image Collection

In the classification framework for CBIR, retrieving image concepts is modeled as a two classes problem: the relevant class, the set of images in the searched concept, and the irrelevant class, composed by the rest of the database.

Remember that $\{\mathbf{x}_i\}_{1,n}$ be the n image indexes of the database, a training set is expressed from any user label retrieval session as

$$\mathscr{A}_{\mathbf{y}} = \{(\mathbf{x}_i, y_i)_{i=1,n} \mid y_i \neq 0\},$$

where $y_i = 1$ if the image $\mathbf{x}_i$ is labeled as relevant, $y_i = -1$ if the image $\mathbf{x}_i$ is labeled as irrelevant (otherwise $y_i = 0$). The classifier is then trained using these labels, and a relevance function $f_{\mathscr{A}_\mathbf{y}}(\mathbf{x}_i)$ is determined in order to be able to rank the whole image database.

5.3.1 Classification Methods for CBIR

In order to compute the relevance function $f_{\mathscr{A}_\mathbf{y}}(\mathbf{x}_i)$ of any image, a classification method has to estimate the density of each classes and/or the boundary. This task is mainly dependent on the shape of the data distribution in the feature space.

Bayes and probabilistic classifiers are the most used classification method for CBIR [12, 47]. They are interesting since they are directly "relevance oriented", i.e., no modifications are required to get the probability of an image to be in the concept. However, since they focus on the estimation the center of each class, they are less accurate near the boundary than discriminative methods, which is an important aspect for active learning. Furthermore, because of the imbalance of training data, the tuning of the irrelevant class is not trivial.

In CBIR context, relevant images may be distributed in a single mode for one concept, as well as in a large number of modes for another concept, producing non-linear classification problems. Gaussian mixtures are highly used in the classification, since they can model complex distributions. However, in order to get an optimal estimation of the density of a concept, data have to be Gaussian distributed. Furthermore, the large number of parameters required for Gaussian mixtures leads to high computational complexity.

Another approach consists in using the kernel function framework previously introduced. One of the main aims of that framework is to linearize the learning problem, by mapping image indexes using the kernel function. It also distinguishes itself from other linearization techniques by their ability to work on this linear space, without explicitly computing the features in the induced space. This also allows to work with very large or infinite spaces. Furthermore, learning tasks and distribution of data can be separated. Under the assumption that we have a kernel function which maps our data into a linear space, all linear learning techniques can be used on any image database.

Support vector machines are very successful classifiers that have been introduced with kernel functions in the statistical community [46]. They aim at discriminating the data with the largest margin between the classes. Their high ability to be used with a kernel function allows to move some of the problems in the tuning of the feature space instead of the classification itself.

Fisher discriminant analysis are also hyperplan classifiers, but aims at minimizing the size of the classes and maximizing the distance between the classes, with simple extension to kernelized version [29]. Let us note that in experiments, they give results close to SVMs [39].

k-Nearest neighbors technique set the class of a document considering its nearest neighbors. Despite its simplicity, this approach is usually among the best ones, especially for the processing of huge databases.

5.3.2 Query Updating Scheme

Information retrieval framework has been applied to CBIR. Strategies consider the initial query and refine or improve it using additional relevant and irrelevant images, that is the set $\mathscr{A}_{\mathrm{y}}$ in the current context. A simple approach, called *query modification*, computes a new query by averaging the feature vectors of relevant images [37]. Another approach, the *query reweighting*, consists in computing a new similarity function between the query and any picture in the database. An usual heuristic is to weight the axes of the feature space. In order to perform a better refinement of the similarity function, optimization-based techniques can be used. They are based on a mathematical criterion for computing the reweighting, for instance Bayes error [34], or average quadratic error [14] computed on $\mathscr{A}_{\mathrm{y}}$. Although these techniques are efficient for target search and monomodal concept retrieval, they fell to track complex image concepts.

5.3.3 Experiments

We have experimented several classification methods for CBIR: Bayes classification [47] with a Parzen density estimation, k-nearest neighbors, support vector machines [4], kernel Fisher discriminant [29], and also a query-reweighting strategy [14]. ANN and Corel databases, scenario, evaluation protocol, and quality measurement as mean average precision are detailed in the Appendix. The results in terms of mean average precision are shown in Fig. 5.2 according to the training set size (we omit the KFD which gives results very close to inductive SVMs) for both ANN and Corel Databases.

One can see that the classification-based methods give the best results, showing the interest of statistical methods against geometrical approaches, like the one

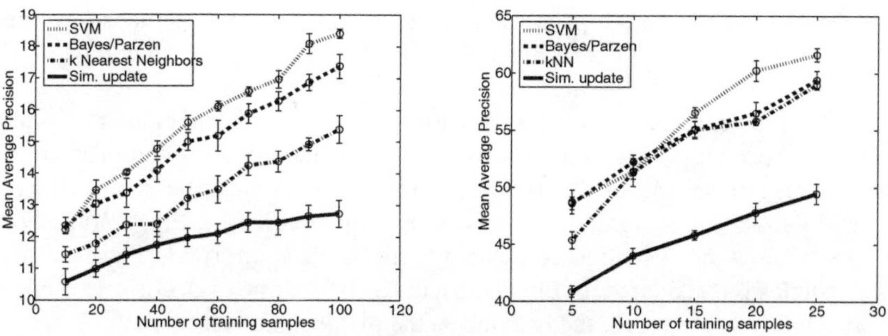

Corel photo database ANN database

Fig. 5.2 Mean average precision (%) for a classification by SVM for various learning set sizes on the Corel photo database (left) and the ANN photo database (right)

reported here (similarity refinement method). The SVM technique performs slightly better than others in this context. As they have a strong mathematical framework and efficient algorithmic implementation, we will use SVM as the default method in RETIN classification.

We also made some comparisons with different kernels (linear, polynomial, Gaussian $L2$, Gaussian χ^2), to select the Gaussian χ^2 which gives the best performances.[1] In the following experiments, we will use a Gaussian kernel with a χ^2 distance.

Remark 5.2. As soon as the whole data set is available during the training, using unlabeled data for training seems to be interesting in a semi-supervised or transductive framework. For instance, there are extended versions of Gaussian mixtures [12, 31] and transductive SVM [18]. We have experimented these methods. The computational time is very high and no improvement has been observed. Transduction does not seem to be an interesting approach for CBIR as already observed in [3].

Anyway, the global performances stay very low for any proposed methods for the Corel experiments. The MAP is under 20% in any case. even if the number of training data is going up to 100 for Corel. The Corel database is much bigger than ANN and the simulated concepts more difficult. The training set remains too small to allow classifiers to efficiently learn the query concept. Active learning is now considered to overcome this problem.

5.4 Active Learning for CBIR

Active classification is an extension of semi-supervised learning. More than exploiting the unlabeled data, the active method challenge[2] is to find the image(s) in the pool of unlabeled data that, once labeled, will provide the best classification result.

The process is close to supervised classification, except that after labeling, the selected subset is added to the training set. That means the training data are no more iid, but depend on the process, user labeling, etc. These methods have been introduced to perform good classifications with few training data in comparison to the standard supervised scheme.

In Fig 5.3, an example of interactive learning is proposed. Images are represented by 2D feature vectors, the white circles are images the user is looking for, and the black circles are the images the user is not interested in. At the beginning, the user provided two labels, represented in figures by larger circles (*cf.* Fig. 5.3(a)). These two labels allow the system to compute a first classification. Next, the selection of new examples is considered. In Fig. 5.3(b), the active learning selection maximizing the uncertainty is proposed: the user labels the pictures the closest to the boundary, resulting in an enhanced classification in that case (Fig. 5.3(b)).

[1] It is not a surprising result because, in our experiments, histograms are used as image signatures, and the χ^2 distance is dedicated for comparing distributions.

[2] In this chapter, we only consider selective sampling approaches in the active classification framework.

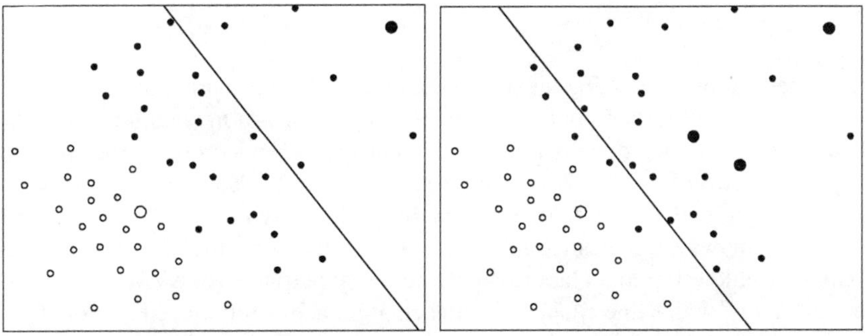

(**a**) Initial training set (**b**) Label the most uncertain

Fig. 5.3 Active learning illustration. A linear classifier is computed for white (relevant) and black (irrelevant) data classification. Only the large circles are used for training, the small ones represent unlabeled data. The line is the boundary between classes after training. (**a**) represents the initial boundary; in (**b**), the most uncertain data (closest to the boundary) are added to the training; the boundary significantly moved (further the best separation between B&W data)

5.4.1 Notations for Selective Sampling Optimization

We consider here the *pool-based* active learning framework in which the learner has available a large pool of unlabeled images, and the new image candidates may be chosen only from this pool.

As already defined in Sect. 5.3, let X be the input space, $\{\mathbf{x}_i\}_{1,n}$ image indexes, $\mathscr{A}_\mathbf{y} = \{(\mathbf{x}_i, y_i)_{i=1,n} \mid y_i \neq 0\}$ the training set defined on the whole database but represented by the $\mathbf{y} \neq 0$ data subset, a relevance function $f_{\mathscr{A}_\mathbf{y}} : X \to [-1, 1]$ trained with $\mathscr{A}_\mathbf{y}$, and a teacher $s : X \to \{-1, 1\}$ that labels documents as -1 or 1. We also denote by I the indices of the labeled documents, $\bar{I}$ the unlabeled ones.

Hence, starting from a training set $\mathscr{A}_\mathbf{y}$, the active learning within this context aims at selecting the unlabeled data $\mathbf{x}_{i^\star}$ that will enhance the most the relevance function f trained with the label $s(\mathbf{x}_{i^\star})$ added to $\mathscr{A}_\mathbf{y}$. We propose to formalize this choice as the minimization of a cost function $g_{\mathscr{A}_\mathbf{y}}$. According to a particular active learning method, the chosen image to label is the argument of the minimum of $g_{\mathscr{A}_\mathbf{y}}(\mathbf{x})$ evaluated over the set of unlabeled documents.

5.4.2 Active Learning Methods

We present here two of the most popular active learning strategies: the uncertainty-based sampling, that selects the documents for which the relevance function is most uncertain about (see initial example), and the error reduction strategy, that aims at minimizing the generalization error of the classifier.

5.4.2.1 Uncertainty-Based Sampling

This strategy aims at selecting unlabeled documents that the learner of the relevance function is most uncertain about. The first solution consists in computing a probabilistic output for each document and selecting the unlabeled documents with the probabilities closest to 0.5 [23]. Similar strategies have also been proposed with SVM classifier, with a theoretical justification [45], and with nearest neighbor classifier [25]. Anyway, a relevance function is computed. This function can be a distribution, a fellowship to a class (distance to the hyperplane for SVM), or a utility function. Thus, with some adaptation of each approach, a relevance function $f_{\mathscr{A}_{\mathbf{y}}}$ is trained, where the most uncertain documents have an output close to 0. The cost function to minimize is then

$$g_{\mathscr{A}_{\mathbf{y}}}(\mathbf{x}) = |f_{\mathscr{A}_{\mathbf{y}}}(\mathbf{x})|. \tag{5.1}$$

The efficiency of these methods depends on the accuracy of the relevance function estimation close to 0. Unfortunately, it is the area where it is the most difficult to perform a good evaluation.[3]

Another approach consist of using several models to represent classes and selecting elements where the models contradict the most. For instance, Cohn [8] propose to train several classifiers with the same training set and to select the elements whose classifications contradict the most. Similar approaches are proposed using neural networks [7, 22].

5.4.2.2 Error Reduction-Based Strategy

Active learning strategies based on error reduction select documents that, once added to the training set, minimize the error of generalization [35].

Let $P(y|\mathbf{x})$ be the (unknown) probability of an image $\mathbf{x}$ to be in class y (relevant or irrelevant) and $P(\mathbf{x})$ the (also unknown) distribution of the images. With $\mathscr{A}_{\mathbf{y}}$, the training provides the estimation $\hat{P}_{\mathscr{A}_{\mathbf{y}}}(y|\mathbf{x})$ of $P(y|\mathbf{x})$. Hence, the expected error of generalization is

$$E_{\hat{P}_{\mathscr{A}_{\mathbf{y}}}} = \int_{\mathbf{x}} L(P(y|\mathbf{x}), \hat{P}_{\mathscr{A}_{\mathbf{y}}}(y|\mathbf{x}))\mathrm{d}P(x) \tag{5.2}$$

with L a loss function which evaluates the loss between the estimation $\hat{P}_{\mathscr{A}_{\mathbf{y}}}(y|\mathbf{x})$ and the true distribution $P(y|\mathbf{x})$.

The optimal pair $(\mathbf{x}_i^\star, y_i^\star)$ is the one which minimizes this expectation over the pool of available images, i.e. unlabeled data $\bar{I}$:

$$\forall (\mathbf{x}_i, y_i), i \in \bar{I} \quad E_{\hat{P}_{\mathscr{A}_{\mathbf{y}}^\star}} < E_{\hat{P}_{\mathscr{A}_{\mathbf{y}+(\mathbf{x}_i, y_i)}}} \tag{5.3}$$

with $\mathscr{A}_{\mathbf{y}}^\star = \mathscr{A}_{\mathbf{y}+(\mathbf{x}_i^\star, y_i^\star)}$.

[3] In the context of human interactive system, where only few training data is available, this is a major problem.

Of course, the expected error is not accessible, and we approximate the integral over $P(\mathbf{x})$ using again the unlabeled set, in the manner of Roy and McCallum [35]. They also propose to estimate the probability $P(y|\mathbf{x})$ with the relevance function provided by the current classifier. With a $0/1$ loss function L, the estimation of the expectation is expressed for any $\mathscr{A}$:

$$\hat{E}_{\hat{P}_{\mathscr{A}}} = \frac{1}{|\bar{I}|} \sum_{\mathbf{x}_i, i \in \bar{I}} \left(1 - \max_{y \in \{-1,1\}} \hat{P}_{\mathscr{A}}(y|\mathbf{x}_i)\right).$$

Furthermore, as the labels $s(\mathbf{x}_i)$ on $\bar{I}$ are unknown, they are estimated by computing the expectation for each possible label. Hence, the cost function g is given:

$$g_{\mathscr{A}_{\mathbf{y}}}(\mathbf{x}) = \sum_{y \in \{-1,1\}} \hat{E}_{\hat{P}_{\mathscr{A}+(\mathbf{x},y)}} \hat{P}_{\mathscr{A}}(y|\mathbf{x}). \tag{5.4}$$

The following relation is used between $\hat{P}_{\mathscr{A}_{\mathbf{y}}}(y|\mathbf{x})$ and $f_{\mathscr{A}_{\mathbf{y}}}(\mathbf{x})$:

$$\hat{P}_{\mathscr{A}_{\mathbf{y}}}(y|\mathbf{x}) = \frac{y}{2}(f_{\mathscr{A}_{\mathbf{y}}}(\mathbf{x}) + y).$$

In order to propose efficient methods of selection, we have to deal with some characteristics of CBIR that we are going to detail in the next section. Several aspects of the active learning strategies are discussed as the estimation of the boundary between classes or the batch selection, based on the optimization of the g function, as proposed in (5.1) or (5.4).

5.5 Further Insights on Active Learning for CBIR

Active learning methods have been generally proposed in classification problems, where the aim is to divide the database in two classes of same size. However, in our context the class of relevant image (the searched concept) is generally 20–100 times smaller than the class of irrelevant images. As a result, the boundary used to be very inaccurate, especially in the first iterations of relevance feedback, where the size of the training set is dramatically small. Many methods become inefficient in that context, and selection is then close to random. An active correction technique is present to prevent this problem.

The computation time is also an important criterion for CBIR in generalist applications. The user will not wait more than one minute between two feedback steps. This point prohibits methods that compute all possibilities. Furthermore, a fast selection allows the user to provide more labels within the same retrieval session time. Thus, it is more interesting to use less efficient but fast method than a more efficient but highly computational one. To boost the CBIR systems, the first step aims at reducing the computational time, by pre-selecting pictures $\mathbf{x}_i$, $i \in J$ which may be in the optimal selection set. For example, the pre-selection of a large set of pictures (but small in comparison to the database size) close to the boundary is often proposed.

This process is computed very fast, and the uncertainty-based selection method have proved their interest in CBIR context.

In active classification, the images are selected to enhance the current classification. Whenever this choice is interesting for CBIR, this criterion does not completely reflect the user satisfaction. Other utility criteria closer to the final aim, such as average precision, should provide more efficient selections. This aspect will be illustrated before comparing several active strategies on real image search problems.

5.5.1 Active Boundary Correction

The boundary plays an important role for selection. Large variations can dramatically change the selection. During the firsts steps of relevance feedback, classifiers are trained with very few data, about 0.1% of the database size. At these steps, classifiers are not even able to perform a good estimation of the size of the concept. Their natural behavior in these cases is to divide the database in two parts of close sizes. Each new sample changes the class of hundreds, sometimes thousands of images. Selection is then close to a random selection.

A simple solution is to assume that the retrieval session is initialized with a minimum of example. For instance, Tong proposes to initialize with 20 examples [44]. More than the question of the minimum number of examples required for a good starting boundary, this assumption is not always feasible with no third-party knowledge.

A method to correct the boundary has been proposed in [16], that we extend here:

$$\hat{f}_{\mathscr{A}_{\mathbf{y}_t}}(\mathbf{x}_i) = f_{\mathscr{A}_{\mathbf{y}_t}}(\mathbf{x}_i) - b_t$$

with $f_{\mathscr{A}_{\mathbf{y}_t}}$ the relevance function and b_t the correction at feedback step t.

There is several ways to set the boundary. We want $f_{\mathscr{A}_{\mathbf{y}_t}}()$ be positive for any image in the concept, and negative otherwise. Considering the ranking of the database,

$$O_{\mathscr{A}_{\mathbf{y}_t}} = \text{argsort} f_{\mathscr{A}_{\mathbf{y}_t}}(X)$$

with $argsort(\mathbf{v})$ a function returning the indexes of the sorted values of $\mathbf{v}$ and r_t the rank of the image whose relevance is the most uncertain:

$$\underbrace{\mathbf{x}_{O_1}, \mathbf{x}_{O_2},}_{\text{Concept center}} \ \underbrace{\ldots, \mathbf{x}_{O_{r_t-1}}, \mathbf{x}_{O_{r_t}}, \mathbf{x}_{O_{r_t+1}}, \ldots,}_{\text{Zone of uncertainty}} \ \underbrace{\mathbf{x}_{O_{n-1}}, \mathbf{x}_{O_n}}_{\text{Less relevant images}} \ .$$

The correction is then expressed by

$$b_t = f(\mathbf{x}_{O_{r_t}}).$$

To compute this correction, the tuning of r_t is necessary.

Adaptive tuning during the feedback steps has been proposed [10] analyzing the set of labels provided by the user at the tth iteration in order to determine the next value r_{t+1}.

Actually, the technique supposes that the best threshold corresponds to the searched boundary. Such a threshold allows to present as many relevant images as irrelevant ones. Thus, if and only if the set of the selected images is well balanced (between relevant and irrelevant images), then the threshold r_t is good. This property is used to tune r_t:

- At the tth feedback step, the system is able to classify images using the current training set.
- The user gives new labels for images $\mathbf{x}_{O_{r_t-q/2}}, \mathbf{x}_{O_{r_t}}, \mathbf{x}_{O_{r_t+q/2}}$.
- They are compared to the current classification.
- If user mostly gives. relevant labels, the system should propose new images for labeling around a higher rank to get more irrelevant labels.
- If user mostly gives irrelevant labels, thus classification does not seem to be good to the rank r_t, and new images for labeling should be selected around a lower rank (to get more relevant labels).

In order to get this behavior, the following update rule is used

$$r_{t+1} = r_t + h(\hat{f}_{\mathscr{A}_{\mathbf{y}_t}}, I_t^\star, \mathbf{y}_{t+1})$$

with $I_t^\star$ the set of indexes of selected images at step t and

$$h(\hat{f}_{\mathscr{A}_{\mathbf{y}_t}}, I_t^\star, \mathbf{y}_{t+1}) = \sum_{i \in I_t^\star} \big(y_{i,t+1} - \hat{f}_{\mathscr{A}_{\mathbf{y}_t}}(\mathbf{x}_i)\big).$$

The function h computes the sum of the differences between new labels and their relevance according to the current classification. Such a function h allows to work as well with uncertainty-based approaches as error-based strategies with batch selection. Indeed, in the case where images are close to the boundary, i.e., $\forall i \in I^\star$, $\hat{f}_{\mathscr{A}_{\mathbf{y}_t}}(\mathbf{x}_i)$ is close to 0, the function h is the sum of the new labels. Thus, if all new labels are positives, h has a positive value, and the new rank r_{t+1} will be higher that the previous rank r_t. On the contrary, if all new labels are negatives, the new rank will be lesser that the current one.

In the case where images are far from the boundary, the news labels $y_{i,t+1}$ are compared to their current relevance value $\hat{f}_{\mathscr{A}_{\mathbf{y}_t}}(\mathbf{x}_i)$. If $y_{i,t+1}$ matches with $\hat{f}_{\mathscr{A}_{\mathbf{y}_t}}(\mathbf{x}_i)$, no change will appear. However, if an image is labeled as negative whereas its current relevance is positive, the new rank will be highly reduced, since the uncertain area seems to be closer to the center of the relevant class. A similar reasoning can be made for an image labeled as positive but irrelevant for the current classifier.

Remark 5.3. This correction process is close to the computation of the b parameter in the SVM decision function:

$$f_{\mathscr{A}_{\mathbf{y}_t}}^{SVM}(\mathbf{x}_i) = \sum_j \alpha_j y_j k(\mathbf{x}_j, \mathbf{x}_i) + b. \tag{5.5}$$

The parameter b can be computed using the KKT conditions [39], for instance if $\mathbf{x}_i$ is a support vector,

$$b = y_i - \sum_j \alpha_j y_j k(\mathbf{x}_j, \mathbf{x}_i). \tag{5.6}$$

5.5.2 MAP vs Classification Error

Active learning methods have been built to select samples which decrease the error of classification. However, in interactive CBIR context, users are more interested in similarity ranking than in classification. A usual metric to evaluate this ranking is the average precision.

We made some experiments to evaluate the difference between the direct optimization of the average precision instead of the classification error.

In the case of the minimization of the classification error, the optimal pair $(\mathbf{x}_i^\star, y_i^\star)$ is the one which satisfies the following equation over the unlabeled data $\bar{I}$ (cf. 5.3):

$$E_{\hat{P}_{\mathscr{A}_{\mathbf{y}+(\mathbf{x}_i^\star, y_i^\star)}}} < E_{\hat{P}_{\mathscr{A}_{\mathbf{y}+(\mathbf{x}_i, y_i)}}}, \tag{5.7}$$

where $E_{\hat{P}_{\mathscr{A}_{\mathbf{y}}}}$ is the expected error of generalization of the classifier trained with the set $\mathscr{A}_{\mathbf{y}}$ (cf. 5.2).

In the case of the maximization of the average precision, the expression is close to the previous one, except that we consider the average precision $M_{\hat{O}_{\mathscr{A}_{\mathbf{y}}}}$:

$$M_{\hat{O}_{\mathscr{A}_{\mathbf{y}+(\mathbf{x}_i^\star, y_i^\star)}}} > M_{\hat{O}_{\mathscr{A}_{\mathbf{y}+(\mathbf{x}_i, y_i)}}}, \tag{5.8}$$

where $\hat{O}_{\mathscr{A}_{\mathbf{y}}}$ is the ranking of the database computed from the classifier output trained with the set $\mathscr{A}_{\mathbf{y}}$.

We compared these two criteria on a reference database where all the labels are known.[4] Results are shown in Fig. 5.4. Whenever the classification error method increases the MAP, the technique maximizing the average precision performs significantly better with a gain around 20%. These results are a motivation to develop average precision-based strategies even if the true MAP is not available and has to be approximated. The aim here is to select the images that will maximize the average precision, and hence estimate $M_{\hat{O}_{\mathscr{A}_{\mathbf{y}}}}$.

5.5.3 Batch Selection

Few active learning methods directly address the problem of batch selection. Usually, they simply extend the selection for more than one image, as for instance, by

[4] These two criteria cannot be used for real applications, where most of the labels are unknown, and are only used here for evaluation purpose.

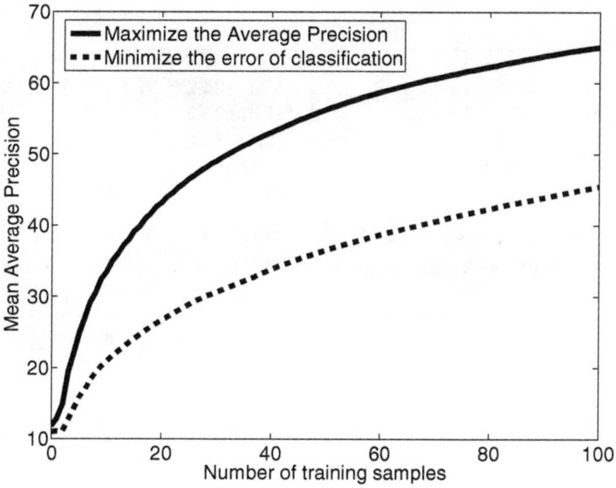

Fig. 5.4 Comparison using MAP between two active learning optimizations. The dot curves represent the classification error minimization performances and the solid curves the average precision optimization ones

selecting the q best images that minimize the g function. This is a batch processing. Ideally, the general formulation should be to select the set that minimizes the corresponding g function over all the sets of q unlabeled images.

As for the selection of one image, this minimization have to be estimated. Because of the constraint in terms of complexity of the CBIR context, a direct estimation cannot be performed. Iterative selection strategies will be developed in CBIR. Whereas this strategy is under-optimal, it requires few computational time, and gives a real boost.

With the power of current classifiers, the labeling of close pictures in the feature space gives the same classification than with the label of only one of them. Hence, the main idea of batch selection is to prevent from selecting two close images. A weight factor to the function g that penalizes close images is usually introduced [2]. The resulting algorithm selects q images one by one, and prevents from the selection of too close images:

$$I^\star = \{\}$$
for $l \in [1..q]$
$$i^\star = \text{Argmin } i \in \bar{I} - I^\star \left(g_{\mathscr{A}_\mathbf{y}}(\mathbf{x}_i) + \max_{j \in I \cup I^\star} s(\mathbf{x}_i, \mathbf{x}_j) \right)$$
$$I^\star = I^\star \cup \{i^\star\}$$
endfor

with $s(\mathbf{x}_i, \mathbf{x}_j)$ the similarity induced by the kernel function between image $\mathbf{x}_i$ and image $\mathbf{x}_j$ presented in Sect. 5.2.4.

Table 5.3 Mean average precision(%) for different active learners on the Corel photo database

Training samples	10	20	30	40	50	60	70	80	90	100
RETIN AL	14.2	17.5	20.5	23.0	25.3	27.3	29.0	30.7	32.3	33.8
SVMactive	11.5	15.4	18.1	20.4	22.5	24.2	26.0	27.5	28.9	30.3
Roy & McCallum	11.5	14.4	17.1	19.4	21.5	23.2	25.0	26.5	27.9	29.3
No active	12.2	13.1	14.0	14.5	15.4	16.0	16.5	16.8	17.7	18.2

Batch selection does not depend on a particular technique of selection of one image, hence may be used with all the previously described active learning processes.

5.5.4 Experiments

Experiments have been carried out on the databases described in the Appendix. In the following, we use a SVM classifier with a Gaussian kernel with a χ^2 distance.

We compare several active strategies: the uncertainty-based method $\text{SVM}_{\text{active}}$ of Tong [44], one implementation of the generalization error minimization method of Roy and McCallum [35], the RETIN strategy combining active correction and batch selection [10]. We also add a non-active method, which randomly selects the images.

Global results are shown in Table 5.3. First, one can see the benefit of active learning in our context. In these experiments, the gain goes from 11 to 15% for the Corel database. The method which aims at minimizing the error of generalization is the less efficient active learning method. The most efficient method is the precision-oriented method RETIN, especially in the first iterations, where the number of samples is small. About computational time per feedback, the $\text{SVM}_{\text{active}}$ method needs about 20 ms, the method of [35] several minutes, and the one proposed in this Chap. 45 ms.

The most interesting result is the gap in performance between active and non-active learning methods. The main improvements for CBIR active learning approaches concern the boundary correction to make the retrieval process more robust, the batch processing and speed-up processes.

The framework introduced in this chapter may be extended. Developing kernel functions for object classes retrieval based on bags of features is currently one of the research directions the most investigated by the computer science community. The implementation of such kernel functions is fully compatible with the active learning scheme described in this chapter.

5.6 CBIR Interface: Result Display and Interaction

An example of a CBIR graphical user interface is presented in Fig. 5.5. The main window displays the top ranking of the database. Usually, images are sorted from the most similar to less one (the top left of Fig. 5.5), but other methods of display have

Fig. 5.5 Typical graphical user interface for CBIR: the images are presented to the user using the similarity ranking (from top left to bottom right). Labeling is obtained by clicking on the data. Active learning window is also proposed on the bottom part in this interface

been proposed. Two-dimensional distribution of the best selected images presents the data in a meaningful way and may offer more complex user interaction than just binary labeling of the images.

The user gives new labels by clicking the left or right mouse button. Once new labels are given, the retrieval is updated and a new ranking is displayed in the main part.

We show in Figs. 5.6 and 5.7 one example of a retrieval session step by step using the RETIN visual search engine: initialization by choosing one query example, first results, labeling, feedback, etc., and after few iterations, for the concept "airplane", the final ranking. One can see that the system is able to retrieve many images of the concept, while discriminating irrelevant pictures with close visual characteristics.

(**a**) Query initialization: airplane (**b**) First screen similarity ranking

Fig. 5.6 Example of CBIR search (Caltech database) for a concept of airplane: initialization

(**a**) First screen results

(**b**) Eighth screen results (the 140 top pictures of the 3 first screens are relevant)

Fig. 5.7 Example of CBIR search for a concept of airplane (continuation). After few feedback iterations with active learning-based strategy (using RETIN engine)

This concept is very complex and overlaps with several other concepts and will necessitate more iterations to be completely learnt.

Appendix: Databases and Features for Experiments

Databases
The choice of the database for experiments is not easy and has a great impact on results. A comparative quality assessment is usually preferred to highlight the performance difference between techniques more than the absolute scores. Many databases have been recently proposed. They differ in content and size, but also in the type of objectives, object recognition, classification, ranking, etc. Let us give some of the most used, for which information is very easy to find: COIL-100 Columbia object image library, Corel dataset, ANN database, Caltech image collection, Graz dataset, Pascal Voc challenge, ImagEVAL campaign, TrecVid conference series, ImageClef campaign.

In the present experiments, two databases have been considered. The first one is the Corel photo database. This dataset contains more than 50,000 pictures organized in categories. Each category has about 100 images. To get tractable computation for the statistical evaluation, we randomly selected 10% of the Corel categories. We obtained 50 categories and the corresponding database is composed of 6000 images. These categories have been directly extracted from the initial categories or obtained by merging some of them (to get sets with different sizes and complexities). The resulting sets range from small and monomodal categories to large and multimodal categories. The second database is the ANN collection from the Washington university, with 500 images.

Features
We use an histogram of 25 colors and 25 textures for each image, computed from a vector quantization [9].

Concepts
For the ANN database, we used the already existing 11 concepts. For the Corel database, we used the 50 concepts of various complexity previously extracted. The concept sizes are from 50 to 300. The set of all the concepts covers the whole database, and many of them share common images.

Quality Assessment
The CBIR system performances are measured using precision (P), recall (R) and statistics computed on P and R for each category. We use the mean average precision (MAP) which represents the value of the P/R integral function.[5]

The performances are evaluated by simulating retrieval scenario. For each simulation, an image category is randomly chosen and 100 images of the category, drawn at random or with active learning, constitute the learning set. After each classification of the database, the mean average precision (MAP) is computed. These

[5] http://www-nlpir.nist.gov/projects/trecvid/.

simulations are repeated 1000 times, and all values of MAP are averaged. Next, we repeat ten times these simulations to get the mean and the standard deviation of the MAP.

References

1. S. Aksoy and R.M. Haralick. Feature normalization and likelihood-based similarity measures for image retrieval. *Pattern Recognition Letters*, 22:563–582, 2001.
2. K Brinker. Incorporating diversity in active learning with support vector machines. In *International Conference on Machine Learning*, pages 59–66, February 2003.
3. E. Chang, B.T. Li, G. Wu, and K.S. Goh. Statistical learning for effective visual information retrieval. In *IEEE International Conference on Image Processing*, pages 609–612, Barcelona, Spain, September 2003.
4. O. Chapelle, P. Haffner, and V. Vapnik. Svms for histogram based image classification. *IEEE Transactions on Neural Networks*, 10:1055–1064, 1999.
5. Y. Chen and J.Z. Wang. Image categorization by learning and reasoning with regions. *International Journal on Machine Learning Research*, 5:913–939, 2004.
6. L. Cinque, S. Levialdi, and A. Pellican? Color-based image retrieval using spatial-chromatic histograms. In *IEEE International Conference on Multimedia Computing and Systems*, 1999.
7. D. Cohn, L. Atlas, and R. Ladner. Improving generalization with active learning. *Machine Learning*, pages 15(2):201–221, May 1994.
8. D. Cohn, Z. Ghahramani, and M. Jordan. Active learning with statistical models. *Journal of Artificial Intelligence Research*, 4:129–145, 1996.
9. M. Cord, J. Fournier, P.H. Gosselin, and S. Philipp-Foliguet. Interactive exploration for image retrieval. *EURASIP Journal on Applied Signal Processing*, 14:2173–2186, 2006.
10. M. Cord, P.H. Gosselin, and S. Philipp-Foliguet. Stochastic exploration and active learning for image retrieval. *Image and Vision Computing*, 25:14–23, 2006.
11. M. Cuturi, J.P. Vert, O. Birkenes, and T. Matsui. A kernel for time series based on global alignments. In *IEEE ICASSP*, volume 2, pages 413–416, 2007.
12. A. Dong and B. Bhanu. Active concept learning in image databases. *IEEE Transactions on Systems, Man, and Cybernetics – Part B: Cybernetics*, 35:450–466, 2005.
13. M. Everingham, A. Zisserman, C. Williams, L. Van Gool, M. Allan, C. Bishop, O. Chapelle, N. Dalal, T. Deselaers, G. Dorko, S. Duffner, J. Eichhorn, J. Farquhar, M. Fritz, C. Garcia, T. Griffiths, F. Jurie, D. Keysers, M. Koskela, J. Laaksonen, D. Larlus, B. Leibe, H. Meng, H. Ney, B. Schiele, C. Schmid, E. Seemann, J. Shawe-Taylor, A. Storkey, S. Szedmak, B. Triggs, I. Ulusoy, V. Viitaniemi, and J. Zhang. The 2005 pascal visual object classes challenge. In *Proceedings of the First PASCAL Challenges Workshop*, 2006.
14. J. Fournier, M. Cord, and S. Philipp-Foliguet. Back-propagation algorithm for relevance feedback in image retrieval. In *International Conference in Image Processing (ICIP'01)*, volume 1, pages 686–689, Thessaloniki, Greece, October 2001.
15. J. Fournier, M. Cord, and S. Philipp-Foliguet. Retin: A content-based image indexing and retrieval system. *Pattern Analysis and Applications Journal, Special issue on image indexation*, 4(2/3):153–173, 2001.
16. P.H. Gosselin and M. Cord. RETIN AL: An active learning strategy for image category retrieval. In *IEEE International Conference on Image Processing*, volume 4, pages 2219–2222, Singapore, October 2004.
17. K. Grauman and T. Darell. Efficient image matching with distribution of local invariant features. In *IEEE International Conference on Computer Vision and Pattern Recognition (CVPR)*, San Diego, CA, June 2005.
18. T. Joachims. Transductive inference for text classification using support vector machines. In *Proceedings of 16th International Conference on Machine Learning*, pages 200–209. Morgan Kaufmann, San Francisco, CA, 1999.

19. F. Jurie and B. Triggs. Creating efficient codebooks for visual recognition. In *International Conference on Computer Vision and Pattern Recognition*, 2005.
20. A. Khotanzad and Y.H. Hong. Invariant image recognition by zernike moments. *IEEE Trans on. Pattern Analysis and Machine Intelligence*, 12(5):489–497, 1990.
21. R. Kondor and T. Jebara. A kernel between sets of vectors. In *International Conference on Machine Learning (ICML)*, 2003.
22. A. Krogh and J. Vedelsby. Neural network ensembles, cross validation, and active learning. In *Advances in Neural Information Processing Systems*, pages 7:231–238, 1995.
23. D. D. Lewis and J. Catlett. Heterogeneous uncertainly sampling for supervised learning. In W. W. Cohen and H. Hirsh, editors, *International Conference on Machine Learning*, pages 148–56. Morgan Kaufmann Publishers, San Francisco July 1994.
24. Y. Linde, A. Buzo, and R.M. Gray. An algorithm for vector quantizer design. *IEEE Transaction on Communication*, 28:84–94, 1980.
25. M. Lindenbaum, S. Markovitch, and D. Rusakov. Selective sampling for nearest neighbor classifiers. *Machine Learning*, pages 54(2):125–152, February 2004.
26. F. Liu and R. W. Picard. Periodicity, directionality, and randomness: Wold features for image modeling nad retrieval. *IEEE Transactions on Pattern Analysis and Machine Intelligence*, 18(7):722–733, July 1996.
27. D. Lowe. Distinctive image features from scale-invariant keypoints. *International Journal on Computer Vision (IJCV)*, 2(60):91–110, 2004.
28. B.S. Manjunath and W.Y. Ma. Texture features for browsing and retrieval of image data. *IEEE Transactions on Pattern Analysis and Machine Intelligence (Special Issue on Digital Libraries)*, 18(8):837–842, August 1996.
29. S. Mika, G. Rätsch, J. Weston, B. Schölkopf, and K.-R. Müller. Fisher discriminant analysis with kernels. In Y.-H. Hu, J. Larsen, E. Wilson, and S. Douglas, editors, *Neural Networks for Signal Processing IX*, pages 41–48. IEEE, 1999.
30. A. Mojsilovic and B. Rogowitz. Capturing image semantics with low-level descriptors. In *International Conference in Image Processing (ICIP'01)*, volume 1, pages 18–21, Thessaloniki, Greece, October 2001.
31. N. Najjar, J.P. Cocquerez, and C. Ambroise. Feature selection for semi supervised learning applied to image retrieval. In *IEEE ICIP*, volume 2, pages 559–562, Barcelena, Spain, Sept. 2003.
32. G. Pass, R. Zabih, and J. Miller. Comparing images using color coherence vectors. In *ACM International Multimedia Conference*, 1996.
33. G. Patanè and M. Russo. The enhanced LBG algorithm. *Neural Networks*, 14(9):1219–1237, November 2001.
34. J. Peng, B. Bhanu, and S. Qing. Probabilistic feature relevance learning for content-based image retrieval. *Computer Vision and Image Understanding*, 75(1-2):150–164, July-August 1999.
35. N. Roy and A. McCallum. Toward optimal active learning through sampling estimation of error reduction. In *International Conference on Machine Learning*, 2001.
36. Y. Rubner. *Perceptual Metrics for Image Database Navigation*. PhD thesis, Stanford University, May 1999.
37. Y. Rui, T. Huang, S. Mehrotra, and M. Ortega. A relevance feedback architecture for content-based multimedia information retrieval systems. In *IEEE Workshop on Content-Based Access of Image and Video Libraries*, pages 92–89, 1997.
38. S. Santini, A. Gupta, and R. Jain. Emergent semantics through interaction in image databases. *IEEE Transactions on Knowledge and Data Engineering*, 13(3):337–351, 2001.
39. B. Schölkopf and A. Smola. *Learning with Kernels*. MIT Press, Cambridge, MA, 2002.
40. J. Shawe-Taylor and N. Cristianini. *Kernel methods for Pattern Analysis*. Cambridge University Press, ISBN 0-521-81397-2, 2004.
41. J.R. Smith and S.F. Chang. VisualSEEK: a fully automated content-based image query system. In *ACM Multimedia Conference*, pages 87–98, Boston, USA, November 1996.
42. M.A. Stricker and M. Orengo. Similarity of color images. In *SPIE, Storage and Retrieval for Image Video Databases III*, volume 2420, pages 381–392, 1995.

43. M.J. Swain and D.H. Ballard. Color indexing. *International Journal of Computer Vision*, 7(1):11–32, 1991.
44. S. Tong and E. Chang. Support vector machine active learning for image retrieval. In *ACM Multimedia*, pages 107–118, 2001.
45. S Tong and D Koller. Support vector machine active learning with application to text classification. *Journal of Machine Learning Research*, 2:45–66, November 2001.
46. V.N. Vapnik. *Statistical learning theory*. Wiley, New York, 1998.
47. N. Vasconcelos. *Bayesian models for visual information retrieval*. PhD thesis, Massachusetts Institute of Technology, 2000.

Chapter 6
Conservative Learning for Object Detectors*

Peter M. Roth and Horst Bischof

Abstract In this chapter we will introduce a new effective framework for learning an object detector. The main idea is to minimize the manual effort when learning a classifier and to combine the power of a discriminative classifier with the robustness of a generative model. Starting with motion detection an initial set of positive examples is obtained by analyzing the geometry (aspect ratio) of the motion blobs. Using these samples a discriminative classifier is trained using an online version of AdaBoost. In fact, applying this classifier nearly all objects are detected but there is a great number of false positives. Thus, we apply a generative classifier to verify the obtained detections and to decide if a detected patch represents the object of interest or not. As we have a huge amount of data (video stream) we can be very conservative and use only patches for (positive or negative) updates if we are very confident about our decision. Applying this update rules, an incrementally better classifier is obtained without any user interaction. Moreover, an already trained classifier can be retrained online and can therefore easily be adapted to a completely different scene. We demonstrate the framework on different scenarios including pedestrian and car detection.

Peter M. Roth
Institute for Computer Graphics and Vision, Graz University of Technology, Graz, Austria, e-mail: `roth@icg.tugraz.at`

Horst Bischof
Institute for Computer Graphics and Vision, Graz UConservative Learning for Object Detectorsniversity of Technology, Austria, Graz, e-mail: `bischof@icg.tugraz.at`

* This work has been supported by the Austrian Joint Research Project Cognitive Vision under projects S9103-N04, the S9104-N04 EU FP6-507752 NoE MUSCLE IST project, and the FIT-IT project AUTOVISTA funded by the Austrian Research Promotion Agency (www.ffg.at).

6.1 Introduction

In the recent years, the demand for automatic methods analyzing large amounts of visual data has been increasing. Starting with face detection [32, 38] there has been a considerable interest in visual object detection, e.g., pedestrians [39], cars [1], bikes [23], etc. All these trends have been encouraging research in the area of automatic detection, recognition, categorization, and interpretation of objects, scenes, and events.

These visual systems have to encompass a certain level of knowledge about the world they are observing and analyzing. The most convenient way of knowledge acquisition is its accumulation through learning. Thus, the core of most systems is usually a classifier, e.g., AdaBoost [8], Winnow [18], neural network [32], or support vector machine [37] that is obtained by learning (see also Chap. 2).

These approaches have achieved considerable success in the above-mentioned applications but a requirement of all these methods is a training set which in some cases needs to be quite large. The problem of obtaining enough training data increases even further because the methods are view based, i.e., if the viewpoint of the camera changes the classifier needs to encompass this variability (e.g., car from the side and car from the back). Training data are usually obtained by hand labeling a large number of images which is a time-consuming and tedious task. For illustration consider the pictures shown in Fig. 6.1. To get a considerable training set the task would be to mark all persons that occur within this scene. Clearly, this is not practicable for applications requiring a large number of different viewpoints (e.g., video surveillance by large camera networks). Thus, acquiring training samples and as a consequence learning should be as automatic as possible (especially if the amount of data required for learning is huge).

Negative examples (i.e., examples of images not containing the object) are usually obtained by a bootstrap approach [35]. One starts with a few negative examples and trains a classifier. The obtained classifier is applied to images not containing the object. Those sub-images where a (wrong) detection occurs are added to the set

(a) (b)

Fig. 6.1 Hand labeling: labeling all occurring persons as positive examples is a time-consuming task if a great amount of training data is required of negative examples and the classifier is retrained. This process can be repeated several times.

Obtaining reliable positive examples is, however, a more difficult problem, since discriminant classifiers are very sensitive to false training data. In addition, if the visual system's environment is constantly changing, the system has to keep adapting to these changes. It has, therefore, to keep continuously learning and updating its knowledge. When implementing continuous learning mechanisms, two main issues have to be addressed. First, the representation, which is used for modeling the observed world, has to allow updating with newly acquired information. This update step should be efficient and should not require access to the previously observed data while still preserving the previously acquired knowledge. And secondly, a crucial issue is the quality of updating, which highly depends on the correctness of the interpretation of the current visual input. In this context, several learning strategies can be used, ranging from a completely supervised learning approach (when the correct interpretation of the current visual input is given by a tutor) to a completely unsupervised approach (when the visual system has to interpret the current input without any additional assistance). Obviously, the latter approach is preferable, especially when the amount of data to be processed is large.

Having these two issues in mind, one has to carefully select the type of representations of the objects (or subjects or scenes) that can be used. Discriminative representations are compact, task dependent, efficient, and effective, but usually not very robust. On the other hand, reconstructive representations are usually less efficient and less effective, but more general and robust. Thus, it would be desirable to combine these two representations to achieve the best of both worlds which would lead to efficient and effective, while still general and robust continuous learning techniques.

In fact, all of these requirements are met by the *online conservative learning framework* [29, 31]. A preliminary version of this approach [30] was based on batch methods. Thus, it was not suitable for online learning which we show is a beneficial extension for unsupervised learning. To avoid hand labeling of input data, we want the visual system to label data automatically. During the learning process a sufficient knowledge is required for reliably evaluating the visual input. Thus, the process of labeling should strongly be intertwined with the process of continuous learning, which could provide enough redundant information to determine statistically consistent data. Only the sufficiently consistent data would then be used to build the representations, enabling robust learning (and updating of the representations) under non-ideal real-world conditions. We refer to this approach as conservative learning. Also importantly, such a conservative approach assures that non-relevant (corrupted, inexact, false) data are not included into the model and that the model is not degraded.

The proposed framework is depicted in Fig. 6.2. The basic idea is to use a huge amount of unlabeled data that is readily available for most detection task (i.e., just mount a video camera and observe the scene) to avoid hand labeling of training data for object detection tasks.

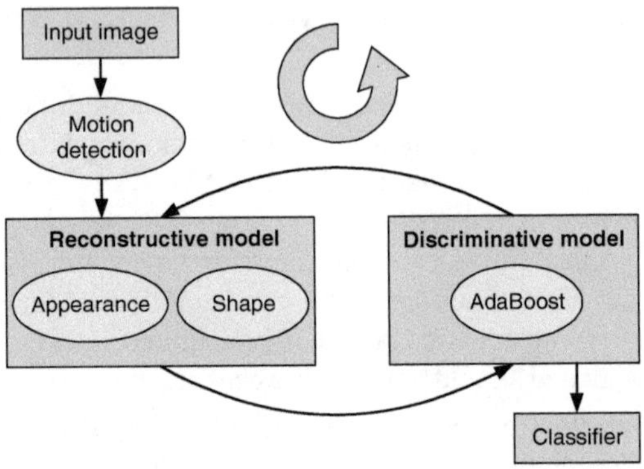

Fig. 6.2 The proposed online learning framework

We use two types of models, a reconstructive one which assures robustness and serves for verification and a discriminative one which actually performs the detection. To get the whole process started we use a simple motion detector to detect potential objects of interest. In fact, we miss a considerable amount of objects (which can be compensated by just using longer sequences) and we will also get miss-detections (which will be reduced in the subsequent steps). The output from the motion detector can be used to robustly build a first initial reconstructive representation. To further increase the robustness we are combining multiple cues – we use one representation on shape and the other on appearance. In particular, we use the robust incremental PCA [34] at this stage such that most of the miss-detections (background, false detections, over-segmentations, etc.) are not incorporated in the reconstructive model. This is very crucial as the discriminative classifier needs to be trained with "clean" images to produce good classification results. The discriminative model, the incremental AdaBoost [9], is then used to detect new objects in new images. The output of the discriminative classifier is then verified by the reconstructive model, and detected false positives are fed back into the discriminative classifier as negative examples and true positives as positive examples to further improve the discriminative model. In fact, it has been shown in the active learning community [26] that it is more effective to sample the current estimate of the decision boundary than the unknown true boundary.[1] This is exactly achieved by our combination of reconstructive and discriminative classifiers. Exploiting the huge amount of video data this process can be iterated to produce a stable and robust classifier. Since all methods (PCA, AdaBoost, Background) operate in an incremental manner, every image can be discarded immediately after it is captured and used for updating the model.

The outlined approach is similar to the recent work of Nair and Clark [21] and Levin et al. [15] who also proposed methods for automatic labeling of new training data. Nair and Clark propose to use motion detection for obtaining the initial

[1] For a detailed discussion of active learning see Chap. 5.

training set and then Winnow as a final classifier. Their approach does not include reconstructive classifiers, nor does it iterate the process to obtain more accurate results. In that sense our framework is more general. Levin et al. use the so-called co-training framework to start with a small training set and to increase it by using a co-training of two classifiers operating on different features. We show that using a combination of reconstructive and discriminative classifiers helps to increase the performance of the discriminative one.

6.2 Online Conservative Learning

6.2.1 Motion Detection

To get the process started we use motion detection (or more generally change detection). Having a stationary camera a common approach to detect foreground objects is to pixel-wise threshold the difference image between the currently processed image and a background image. Let $\mathbf{B}_t$ be the current estimated background image, $\mathbf{I}_t$ the current input image, and θ a threshold, then a pixel is classified as foreground if

$$|\mathbf{B}_t(x,y) - \mathbf{I}_t(x,y)| > \theta. \tag{6.1}$$

For realistic applications different environmental conditions (e.g., changing lightening conditions or foreground models moving to background and vice versa) have to be handled. Therefore, several adaptive methods for estimating a background model $\mathbf{B}_t$ have been proposed that update the existing model with respect to the current input frame $\mathbf{I}_t$ (e.g., running average [12], temporal median filter [19], approximated median filter [20]). A more detailed discussion of these different background models can be found in [4, 10, 27].

To minimize the computational costs and to have an adaptive online method we apply the approximated median method [20] to estimate the background model. The approximated median filter computes an approximation of the temporal median by incrementing the current estimate by one if the input pixel value is larger than the estimate and decreasing it by one if smaller:

$$\mathbf{B}_{t+1}(x,y) = \begin{cases} \mathbf{B}_t(x,y) + 1 & \mathbf{B}_t(x,y) < \mathbf{I}_t(x,y) \\ \mathbf{B}_t(x,y) - 1 & \mathbf{B}_t(x,y) > \mathbf{I}_t(x,y) \,. \end{cases} \tag{6.2}$$

This estimate (only one reference image has to be stored!) eventually converges to the real median. To avoid that non-moving foreground objects are accumulated into the background model and to preserve a robust median estimation spatial and temporal weights may be assigned for the updates.

The obtained motion blobs can be labeled as persons if the aspect ratio of their bounding box is within the pre-specified limits. Examples of accepted patches are shown in Fig. 6.3 where the corresponding blob regions are marked by a bounding

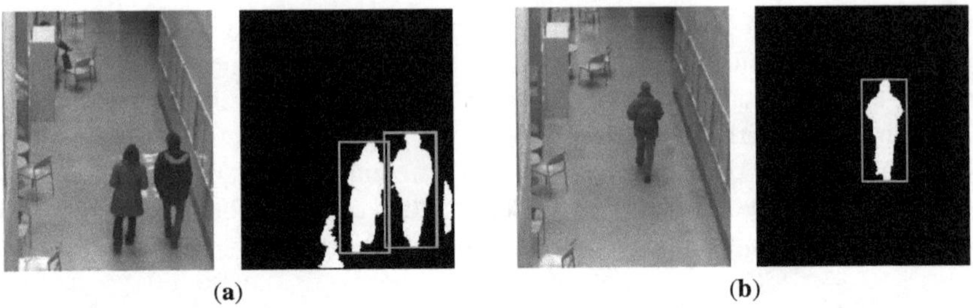

Fig. 6.3 Motion blobs obtained from background subtraction. Blobs that fulfill the constraints are accepted as positive samples (bounding boxes)

box. As can be seen from Fig. 6.3(a) motion blobs that do not fit to the given constraints are not included into the set of positive training samples.

6.2.2 Reconstructive Model

As reconstructive model we use a PCA-based subspace representation (see also Chap. 4 for different subspace representations). This low-dimensional representation captures the essential reconstructive characteristics by exploiting the redundancy in the visual data. As such, it enables "hallucinations" and comparison of the visual input with the stored model [13]. In this way the inconsistent data can be rejected and the discriminative model can be trained from clear data only.

During the learning process, however, a sufficient knowledge is required for a reliable evaluation of the visual input that is still to be acquired. Nevertheless, by considering the reconstruction error, the robust learning procedure can discard inconsistencies in the input data and train the model from only the consistent data [6, 33]. Furthermore, this can also be done in an incremental way. A bunch of methods for incremental building of eigenspaces have been proposed [3, 11, 16]. Some of them also address the problem of robust incremental learning [17, 34]. We use a simplified version of [34] and by checking the consistency of the input images (patches) we keep continuously accepting or rejecting potential patches as positive or negative training examples for the discriminative learner (and updating the reconstructive model). In the following this algorithm is summarized.

For batch PCA all training images are processed simultaneously. A fixed set of input images $\mathbf{X} = [\mathbf{x}_1, \ldots, \mathbf{x}_n] \in \mathbb{R}^{m \times n}$ is given, where $\mathbf{x}_i \in \mathbb{R}^m$ is an individual image represented as a vector. It is assumed that $\mathbf{X}$ is mean normalized. Let $\mathbf{Q} \in \mathbb{R}^{m \times m}$ be the covariance matrix of $\mathbf{X}$, then the subspace $\mathbf{U} = [\mathbf{u}_1, \ldots, \mathbf{u}_n] \in \mathbb{R}^{m \times n}$ can be computed by solving the eigenproblem for $\mathbf{Q}$ or more efficiently by solving SVD of $\mathbf{X}$.

In contrast, for incremental learning, the training images are given sequentially. Assuming that an eigenspace was already built from n images, at step $n+1$ the current eigenspace can be updated in the following way [34]: First, the new image $\mathbf{x}$ is projected in the current eigenspace $\mathbf{U}^{(n)}$ and the image is reconstructed: $\tilde{\mathbf{x}}$. The residual vector $\mathbf{r} = \mathbf{x} - \tilde{\mathbf{x}}$ is orthogonal to the current basis $\mathbf{U}^{(n)}$. Thus, a new basis $\mathbf{U}'$ is obtained by enlarging $\mathbf{U}^{(n)}$ with $\mathbf{r}$. $\mathbf{U}'$ represents the current images as well as the new sample. Next, batch PCA is performed on the corresponding low-dimensional space $\mathbf{A}'$ and the eigenvectors $\mathbf{U}''$, the eigenvalues λ'', and the mean μ'' are obtained. To update the subspace the coefficients are projected in the new basis $\mathbf{A}^{(n+1)} = \mathbf{U}''^{\mathrm{T}}\left(\mathbf{A}' - \mu''\mathbf{1}\right)$ and the subspace is rotated: $\mathbf{U}^{(n+1)} = \mathbf{U}'\mathbf{U}''$. Finally, the mean $\mu^{(n+1)} = \mu^{(n)} + \mathbf{U}'\mu''$ and the eigenvalues $\lambda^{(n+1)} = \lambda''$ are updated. In each step the dimension of the subspace is increased by one. To preserve the dimension of the subspace the least significant principal vector may be discarded [11]. The method is summarized in more detail in Algorithm 1.

To obtain an initial model, the batch method may be applied to a smaller set of training images. Alternatively, to have a fully incremental algorithm, the eigenspace may be initialized using the first training image $\mathbf{x}$: $\mu^{(1)} = \mathbf{x}$, $\mathbf{U}^{(1)} = \mathbf{0}$, and $\mathbf{A}^{(1)} = \mathbf{0}$.

This method can easily be extended in a robust manner, i.e., corrupted input images may be used for incrementally updating the current model. To achieve this, outliers in the current image are detected and replaced by more confident values: First, an image is projected to the current eigenspace using the robust approach [13] and the image is reconstructed. Second, outliers are detected by pixel-wise thresholding (based on the expected reconstruction error) the original image and its

---

Algorithm 1 Incremental PCA

Input: mean vector $\mu^{(n)}$, eigenvectors $\mathbf{U}^{(n)}$, coefficients $\mathbf{A}^{(n)}$, input image $\mathbf{x}$
Output: mean vector $\mu^{(n+1)}$, eigenvectors $\mathbf{U}^{(n+1)}$, coefficients $\mathbf{A}^{(n+1)}$, eigenvalues $\lambda^{(n+1)}$

1: Project image $\mathbf{x}$ to current eigenspace:
$\mathbf{a} = \mathbf{U}^{(n)\mathrm{T}}\left(\mathbf{x} - \mu^{(n)}\right)$
2: Reconstruct image:
$\tilde{\mathbf{x}} = \mathbf{U}^{(n)}\mathbf{a} + \mu^{(n)}$
3: Compute residual vector:
$\mathbf{r} = \mathbf{x} - \tilde{\mathbf{x}}$
4: Append $\mathbf{r}$ as new basis vector to $\mathbf{U}$:
$\mathbf{U}' = \left(\mathbf{U}^{(n)} \; \frac{\mathbf{r}}{\|\mathbf{r}\|}\right)$
5: Determine the coefficients in the new basis:
$\mathbf{A}' = \begin{pmatrix} \mathbf{A}^{(n)} & \mathbf{a} \\ \mathbf{0} & \|\mathbf{r}\| \end{pmatrix}$
6: Perform PCA on $\mathbf{A}'$ and obtain the mean μ'', the eigenvectors $\mathbf{U}''$, and the eigenvalues λ''.
7: Project coefficients to new basis:
$\mathbf{A}^{(n+1)} = \mathbf{U}''^{\mathrm{T}}\left(\mathbf{A}' - \mu''\mathbf{1}\right)$
8: Rotate subspace:
$\mathbf{U}^{(n+1)} = \mathbf{U}'\mathbf{U}''$
9: Update mean:
$\mu^{(n+1)} = \mu^{(n)} + \mathbf{U}'\mu''$
10: Update eigenvalues:
$\lambda^{(n+1)} = \lambda''$

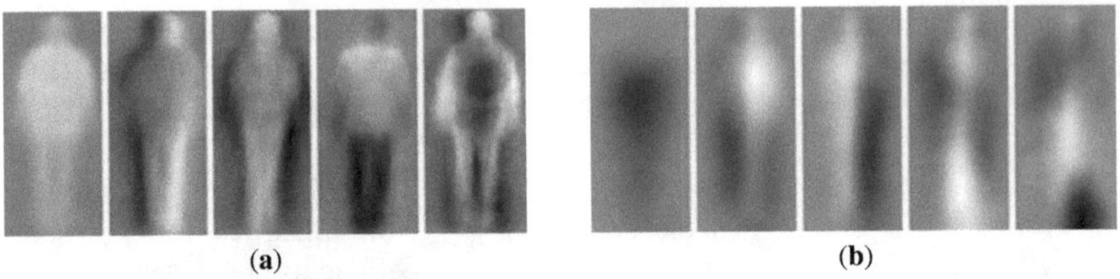

Fig. 6.4 Reconstructive model – first five principal vectors: (**a**) appearance-based model and (**b**) shape model

robust reconstruction. Finally, the outlying pixel values are replaced by the robustly reconstructed values.

To further increase the robustness of the reconstructive model, we build two subspace representations in parallel: appearance-based and shape-based representations. The former is created from the cropped and resized appearance patches, which are detected by the motion detector. Since the output of this detector is also a binary segmentation mask, this mask is used to calculate the shape images based on the Euclidean distance transform [2]. The first five eigenvectors of the thus obtained appearance-based and shape-based models are shown in Fig. 6.4(a,b). In addition, an example of such binary masks and the corresponding shapes obtained by distance transform is shown in Fig. 6.5.

6.2.3 Online AdaBoost for Feature Selection

In general, boosting converts (boosts) a weak learning algorithm into a strong one. Various variants of boosting have been developed (e.g., Real-Boost [8], LP-Boost [7]) but here we focus on the discrete AdaBoost (adaptive boosting) algorithm introduced by Freund and Shapire [8] (for a discussion on boosting and other ensemble methods see also Chap. 2). It adaptively re-weights the training samples instead of re-sampling them and trains a weak classifier with respect to this weight distribution. Finally, a strong classifier is built from a linear combination of all the trained weak classifiers.

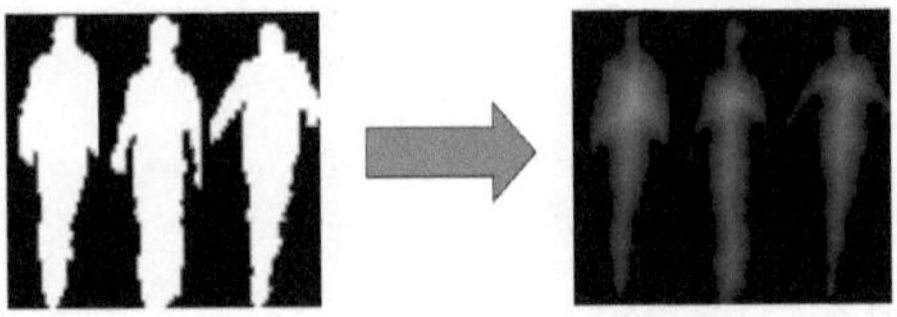

Fig. 6.5 Shape model obtained by Euclidean distance transform of binary shape images

Boosting for feature selection was first introduced by Tieu and Viola [36] and later used by Viola and Jones [38] for face detection. The main idea is that each feature corresponds to a single weak classifier and that boosting selects an informative subset of N features from all possible features $\mathscr{F} = \{f_1, ..., f_M\}$. In each iteration step $n = 1, ..., N$ all features f_j are evaluated on all positive and negative training samples $\mathscr{X} = \{(\mathbf{x}_1, y_1), ..., (\mathbf{x}_L, y_L)\}$, where $\mathbf{x}_i \in \mathbb{R}^m$ is a sample and $y_i \in \{-1, +1\}$ is the corresponding label. A hypothesis is generated by applying the supervised learning algorithm $\mathscr{L}$ (e.g., perceptron, SVM, etc., see also Chap. 2) with respect to the weight distribution $p(\mathbf{x}_i)$ over the training samples which is initialized uniformly. The best hypothesis is selected and forms the weak classifier h_n^{weak}. The weight distribution $p(\mathbf{x}_i)$ is updated according to the error of the current weak classifier. The process is repeated until N features are selected (i.e., N weak classifiers are trained). Finally, a strong classifier h^{strong} is computed as a weighted linear combination of all weak classifiers where the weights α_n are calculated according to the errors of h_n^{weak}.

In our work we use three different types of features, which are Haar-like features [38], orientation histograms [5, 14], and a simple version of local binary patterns (LBP) [22]. Note that the computation of all feature types can be done very efficiently using integral images [38] and integral histograms [28] as data structures. This allows to do exhaustive template matching when scanning the whole image.

The method as described above works off-line; all training samples must be given in advance which is not the case for many applications. In order to get an online algorithm, all steps have to be online. Therefore, we use the online AdaBoost algorithm proposed by Oza and Russell [24, 25]. In contrast to the off-line version, the weak classifiers are updated whenever a new training sample arrives. Since we cannot calculate the weight distribution the basic idea of online boosting is that the importance λ of a sample can be estimated by propagating it through the set of weak classifiers (for more details see [9]). This allows to adaptively train a detector and to efficiently generate the training set. In addition, a classifier is available at any time. Moreover, Oza and Russell [25] have proved that if off-line and online boosting are given to the same training set the weak classifiers returned by online boosting converge statistically to the one obtained by off-line boosting as the number of iterations $N \to \infty$.

In the following we will shortly discuss online boosting for feature selection that was proposed by Grabner and Bischof [9]. They introduce "selectors" and perform online boosting on these selectors and not directly on the weak classifiers. Each selector $h^{\text{sel}}(\mathbf{x})$ holds a set of M weak classifiers $\{h_1^{\text{weak}}(\mathbf{x}), \ldots, h_M^{\text{weak}}(\mathbf{x})\}$ and selects one of them

$$h^{\text{sel}}(\mathbf{x}) = h_m^{\text{weak}}(\mathbf{x}), \tag{6.3}$$

according to an optimization criterion (using the estimated error e_i of each weak classifier h_i^{weak} such that $m = \arg\min_i e_i$).

Thus, online boosting for feature selection can be summarized as follows: first, a fixed set of N selectors $h_1^{\text{sel}}, \ldots, h_N^{\text{sel}}$ is initialized randomly with weak classifier (i.e., features). When a new training sample $\langle \mathbf{x}, y \rangle$ arrives the selectors are updated. This update is done with respect to the importance weight λ of the current sample. For updating the weak classifiers any online learning algorithm can be used. To obtain the weak classifiers h_j^{weak} the feature f_j is evaluated on the sample image $\mathbf{x}$. The weak classifier with the smallest estimated error $e_n^\star$ is selected by the selector and the corresponding voting weight $\alpha_n = \frac{1}{2}\ln\left(\frac{1-e_n^\star}{e_n^\star}\right)$ and the importance weight λ of the sample are updated and passed to the next selector h_{n+1}^{sel}. The weight is increased if the example is misclassified by the current selector or decreased otherwise. Finally, a strong classifier is obtained by linear combination of N selectors:

$$h^{\mathrm{strong}}(\mathbf{x}) = \mathrm{sign}\left(\sum_{n=1}^{N} \alpha_n \cdot h_n^{\mathrm{sel}}(\mathbf{x})\right) . \tag{6.4}$$

6.2.4 Conservative Update Rules

Having the reconstructive models described in Sect. 6.2.2 each image patch can be checked for whether it is consistent with them or not. Figure 6.6(a,b) depicts an image and its appearance and shape reconstructions in the case of a correct and a false detection. In the latter case, the reconstruction error for both the appearance-based and the shape models are significantly larger (i.e., the original image and its reconstruction differ significantly) whereas we get small reconstruction errors for the correct example.

Thus, the reconstruction error can be considered a meaningful criterion for our update decisions. In the conservative learning framework we perform updates only if we are very confident; in particular we use the following update rule: the current discriminative classifier is applied to a training image and all patches that were labeled as object are verified by the reconstructive model. If the reconstruction error for each of the appearance and the shape is very low there is a positive update of the classifier; if the reconstruction error for at least one is large a negative update is applied. Otherwise there is no update. As these decisions are based on very

(a) (b)

Fig. 6.6 Appearance image, its reconstruction, shape image, its reconstruction: (**a**) in case of correct detection, (**b**) in case of false detection conservative thresholds most patches are not considered at all. This update rule is illustrated in Fig. 6.7.

Finally, in Fig. 6.8 we illustrate the incremental learning process by visualizing the updates. Therefore, the 1st, the 3rd, the 34th, and the 64th frames of a sequence of total 300 frames are shown. A dark bounding box indicates a negative update, a light gray bounding box a positive update, and a white bounding box indicates a detection that is not used for updating the existing classifier. It can be seen that there is a great number of negative updates within the very first frames (see Fig. 6.8(a,b)). If the incremental learning process is running for a longer time we finally get a stable classifier and only a small number of updates are necessary (see Fig. 6.8(c,d)).

6.3 Experimental Results

6.3.1 Description of Experiments

For testing our framework we used two different surveillance video sequences. For the first one (*CoffeeCam*), showing a corridor in a public building near a coffee dispenser, we have recorded images over several days. A simple motion detector triggers the camera and then for each second one image is recorded. In total we have recorded over 35,000 images. To train the classifiers a sequence containing 1200 frames has been used. For evaluation purposes we have generated a challenging independent test set of 300 frames (containing groups of persons, persons partially occluding each other, and persons walking in different directions) and a corresponding ground truth.

The second sequence (*Caviar*), showing a corridor in a shopping center, was taken by the CAVIAR project and is publicly available.[2] There is a great number of short sequences that have been joined to a single one. To avoid redundancy the frame rate was reduced to approximately 1 fps (only every 25th frame was stored). For evaluation purposes an independent test set of 144 frames was created. Note that CAVIAR provides a ground truth in XML format which we used for our evaluation. This ground truth also annotates persons that are only partially visible (e.g., only a hand or head) that are not detected by our system because we have trained a whole person detector.

To monitor the progress during online training after several training images the classifier obtained up to now is evaluated on the same independent test sequence. Therefore, the experiments are split into two main parts: First, we trained and evaluated classifiers on the *CoffeeCam* sequence. Second, to demonstrate the online adaptation capability, a classifier trained on the *CoffeeCam* sequence was applied to the *Caviar* test set. In addition, we show detection results obtained from classifiers that were trained using conservative learning. Therefore, we evaluated a person detector on a very complex sequence (a crowd in a lobby of a public building) and a car detector on a sequence showing a tunnel.

[2] http://homepages.inf.ed.ac.uk/rbf/CAVIARDATA1; January 29, 2007.

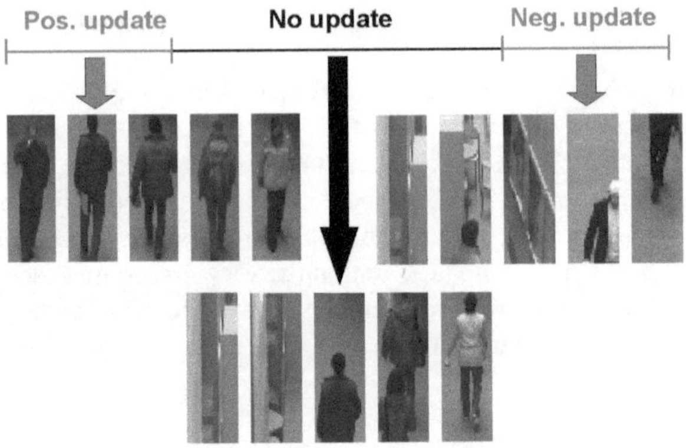

Fig. 6.7 Conservative updating of the discriminant classifier

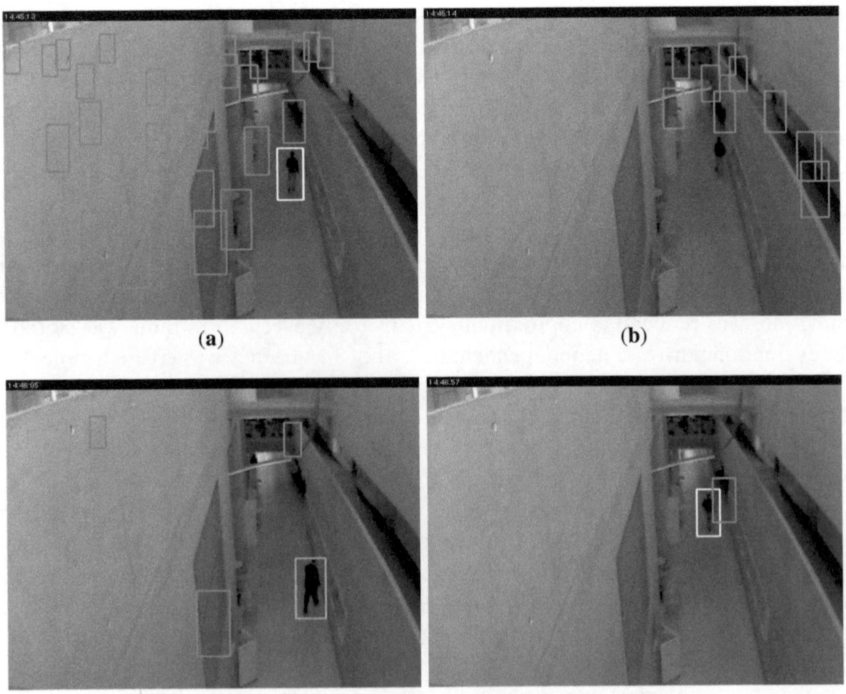

Fig. 6.8 Online conservative learning: (**a**) updates for 1st frame, (**b**) updates for 3rd frame, (**c**) updates for 34th frame, and (**d**) update for 64th frame

6.3.2 CoffeeCam

First, we need an initialization stage to collect positive and negative samples applying a motion-based classifier. All patches where motion detection has detected an object are selected as positive examples. The negative examples are obtained by randomly sampling regions where no motion was detected (*AdaBoost1*). Since the motion detector returns approximately 10% false positives a robust reconstructive representation (PCA on appearance and shape) is computed from the output of the motion detector. Thus, the false positives can be filtered out and may be used as negative examples (*AdaBoost2*). Next, we can use the thus obtained data sets to train an initial AdaBoost classifier and PCA models for appearance and shape and start the online training. The PCA models were updated using the incremental PCA later on.

Let us have a look at the results obtained using online learning. As an evaluation criterion we used similar to [1], precision, recall, and the F-measure that can be considered as trade-off between recall and precision. Figure 6.9 depicts the performance curves if we start online training from *AdaBoost2*. One can see a clear improvement (especially in the first 100 steps) where a lot of false positives can be eliminated. The sudden decrease in performance around frame no. 900 is caused by a single background patch that is detected as a false positive over the whole test sequence; after the next update the curves get back to the previous level.

Figure 6.9(b) depicts the performance curves if we start online training from *AdaBoost1*. Since this initial classifier is worse there is a greater number of false positives in the beginning. Therefore more than 300 frames are necessary to obtain comparable results to the previous classifier but finally we get the same performance as for the first test case! This example demonstrates that it is beneficial (1) first to perform a few steps of off-line learning at the beginning and then switch to an online version and (2) to use clean data for training.

Figure 6.10 shows the detections by three different online classifiers (initial, after 300, and after 1200 training frames) on the test sequence. There are many false

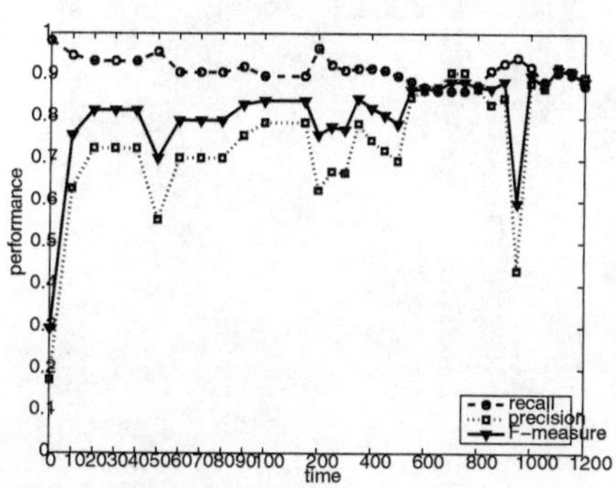

(a)

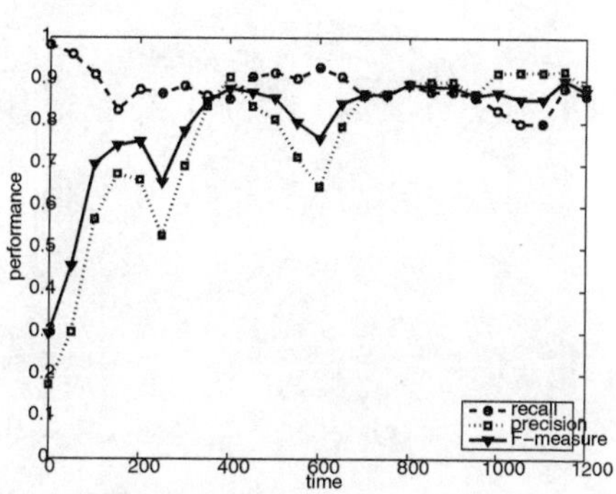

(b)

Fig. 6.9 (**a**) Online learning started from *AdaBoost2*, (**b**) online learning started from *AdaBoost1*

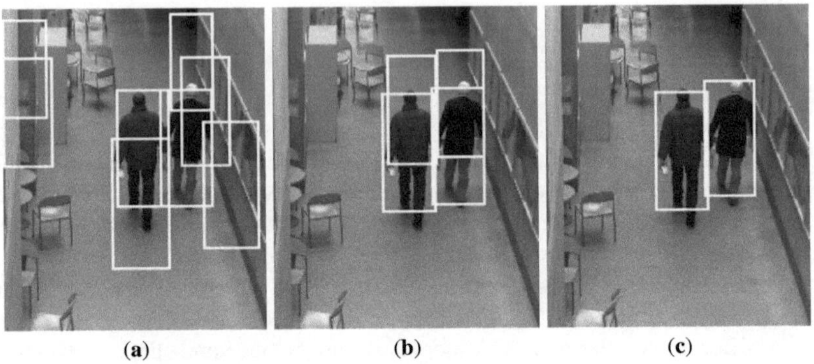

Fig. 6.10 Improvement of online classifier: (**a**) initial classifier, (**b**) after 300 frames, and (**c**) after 1200 frames

positives in (a) that can be completely removed by online training, as can be seen in (b) and (c). Figure 6.11 depicts some more examples of persons correctly detected by the final classifier that was trained with 1200 frames.

Finally, we want to show that the online algorithms are comparable to the off-line versions of the methods. Therefore we have trained classifiers of different stages using the off-line framework we have proposed in [30] and evaluated them on the test sequence. The initialization stage (collect patches and build an initial classifier) is the same as described above (*Off-line AdaBoost1* and *Off-line AdaBoost2*). To train a new classifier the current classifier is evaluated on another sequence. Thus, new positive and negative examples are added to the current training set; a new classifier is trained (*Off-line AdaBoost3*). Table 6.1 depicts these increasingly better results compared to the final classifiers obtained by the online framework.

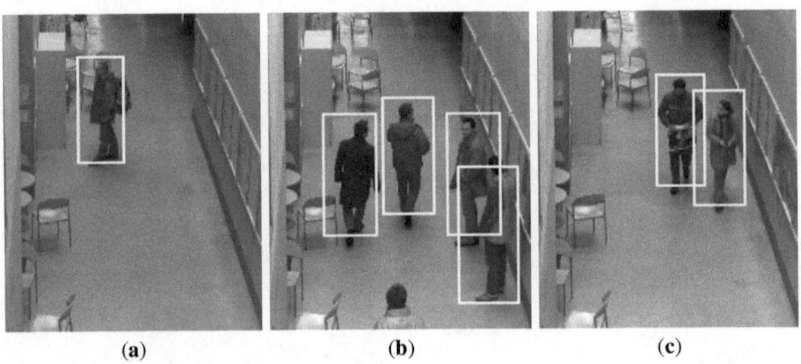

Fig. 6.11 Detections by the final classifier trained with 1200 frames

Table 6.1 Experimental results of the off-line and the online frameworks

Method	Recall (%)	Prec. (%)	F-m. (%)
Off-line AdaBoost1	97.4	27.5	42.9
Off-line AdaBoost2	91.9	57.4	70.7
Off-line AdaBoost3	93.6	94.8	94.2
AdaBoost1	86.4	88.3	87.4
AdaBoost2	87.7	89.7	88.7

6.3.3 Switch to Caviar

After we have shown that the online learning framework is working on the *CoffeeCam* data we want to demonstrate two interesting aspects: First, we show that the classifiers trained on the *CoffeeCam* data describe a generalized person model. Therefore Fig. 6.12 depicts the performance of the classifiers while training them on *CoffeeCam* training data and evaluating them on the *Caviar* test sequence. One can clearly see that the precision (and therefore the F-measure) is improved by online training while the recall is roughly constant. This shows that the learnt person model generalizes over specific setups.

Second, we demonstrate that the online learning framework is able to adapt to a completely different scene. Therefore we perform online training on the *Caviar* data set starting with a classifier obtained by the *CoffeeCam* training. Due to the compression noise a new model for appearance is required. Furthermore motion detection cannot be applied to the *Caviar* data set, because the quality of the motion blobs is too bad for classifying based on size and aspect ratio restrictions only. Since the shape model is more robust (holes, etc.), the shape model estimated from the *CoffeeCam* data can be used to collect patches for generating an appearance-based PCA model. The PCA models (appearance and shape) were updated using the incremental PCA later on.

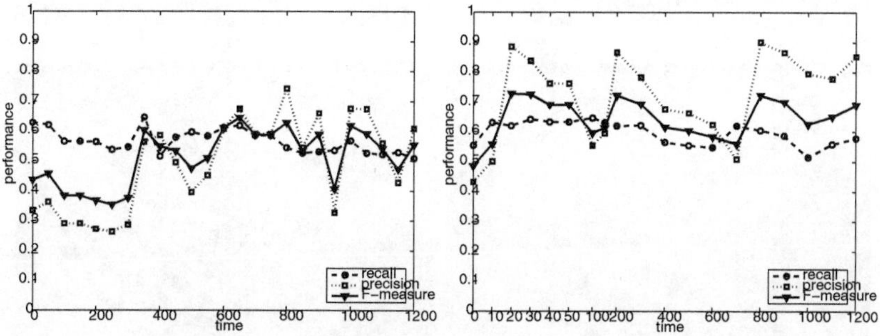

(a) (b)

Fig. 6.12 (**a**) Performance of *CoffeeCam* classifier evaluated on *Caviar* test set, (**b**) performance of Caviar *online* learning

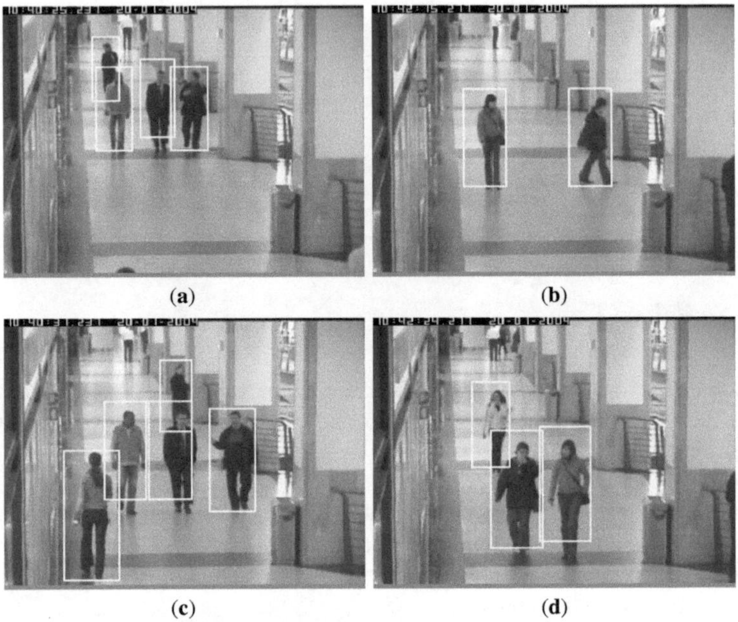

Fig. 6.13 Detections by the final classifier trained with 1200 frames

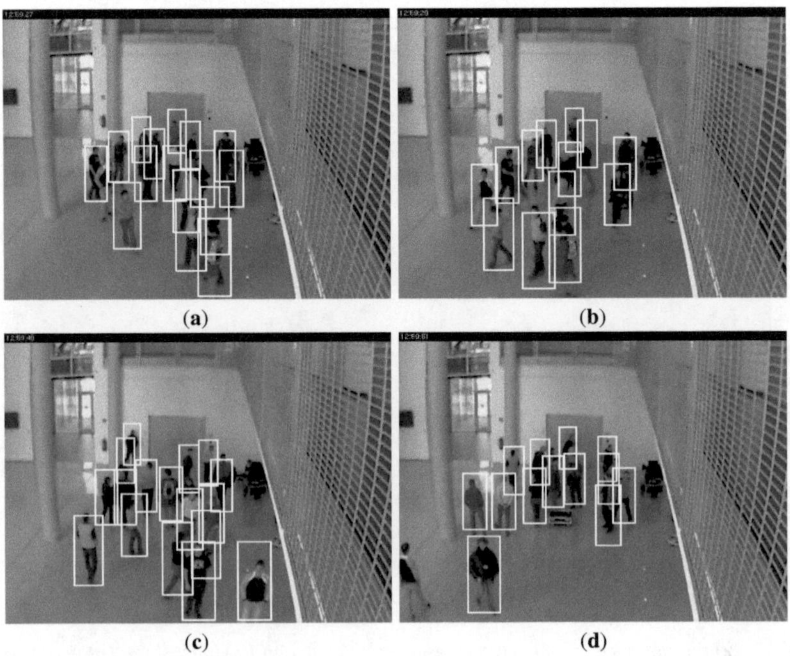

Fig. 6.14 Conservative learning framework on a more complex scene

Figure 6.12(b) depicts the evaluation results of this experiment. The main improvement is achieved within the first 100 frames (false positives in the background). The precision (and therefore the F-measure) can be clearly increased. Please note that the obtained performance of approximately 60% detection rate might look quite low, but this is actually quite good considering the given ground truth. If we exclude all persons that are not at least 50% visible we get a detection rate of approximately 79%.

The noisy behavior of the curves can be explained by the nature of online learning and the way we perform the evaluation. We have a fixed test set, therefore it may happen that particular cases occurring in the test set are not occurring in the training sequence for some time. Thus, the classifier will not perform well on these particular data, i.e., if this happens to parts of the background that are visible most of the time. Examples of the finally obtained classifier that was trained with 1200 frames are shown in Fig. 6.13.

(a) (b) (c) (d)

Fig. 6.15 Conservative learning framework applied for learning a car detector

6.3.4 Further Detection Results

To demonstrate that the proposed framework even works on more complex scenarios we show detection results on a *crowdy scene*. Therefore, a classifier was trained online from 1000 frames. This classifier was then evaluated on an independent test sequence of 600 frames. Examples of detections are shown in Fig. 6.14.

Finally, we show that the proposed framework is quite general and is not limited to train a person model. Therefore, Fig. 6.15 shows results of a car detector applied in a tunnel that was trained using online conservative learning.

6.4 Summary and Conclusions

We have presented an online learning framework that avoids hand labeling of training data. This framework has been used on challenging person detection tasks. We have demonstrated that online learning obtains comparable results to off-line learning. We have also shown that we can adapt an already trained person detector to a quite different setup (i.e., by reducing the number of false positives). While the *CoffeeCam* sequence was obtained with a high-quality camera looking down a corridor, the *Caviar* sequence was obtained with a consumer video camera mounted almost horizontally. Moreover, the sequences contain a considerable amount of compression noise. Nevertheless, the online framework was able to adapt to this quite different scenario. The proposed framework is quite general. It can be used to learn completely different objects (e.g., cars) and can be extended in several ways. More modules (generative and discriminative classifiers) operating on different modalities will further increase the robustness and generality of the system. In particular, we will add a tracking algorithm to obtain a wider variety of positive samples (which in turn should increase the detection rate).

References

1. S. Agarwal, A. Awan, and D. Roth. Learning to detect objects in images via a sparse, part-based representation. *IEEE Transaction on Pattern Analysis and Machine Intelligence*, 26(11):1475–1490, 2004.
2. H. Breu, J. Gil, D. Kirkpatrick, and M. Werman. Linear time euclidean distance transform algorithms. *IEEE Transaction on Pattern Analysis and Machine Intelligence*, 17(5):529–533, 1995.
3. S. Chandrasekaran, B. S. Manjunath, Y. Wang, J. Winkeler, and H. Zhang. An eigenspace update algorithm for image analysis. *Graphical Models and Image Processing*, 59(5): 321–332, 1997.
4. S.-C. S. Cheung and C. Kamath. Robust techniques for background subtraction in urban traffic video. In *Proceedings of SPIE Visual Communications and Image Processing*, pages 881–892, 2004.

5. N. Dalal and B. Triggs. Histograms of oriented gradients for human detection. In *Proceedings IEEE Conference Computer Vision and Pattern Recognition*, volume 1, pages 886–893, 2005.
6. F. de la Torre and M. J. Black. A framework for robust subspace learning. *International Journal of Computer Vision*, 54(1):117–142, 2003.
7. A. Demiriz, K. R. Bennett, and J. Shawe-Taylor. Linear programming boosting via column generation. *Machine Learning*, 46:225–254, 2002.
8. Y. Freund and R. E. Shapire. A decision-theoretic generalization of on-line learning and an application to boosting. *Journal of Computer and System Sciences*, 55:119–139, 1997.
9. H. Grabner and H. Bischof. On-line boosting and vision. In *Proceedings of IEEE Conference on Computer Vision and Pattern Recognition*, volume I, pages 260–267, 2006.
10. D. Hall, J. Nascimento, P. C. Ribeiroand, E. Andrade, P. Moreno, Sebastien Pesneland Thor List, R. Emonent, Robert B. Fisher, Jose Santos-Victor, and James L. Crowley. Comparison of target detection algorithms using adaptive background models. In *Proceedings of IEEE International Workshop on Visual Surveillance and Performance Evaluation of Tracking and Surveillance*, pages 113–120, 2005.
11. P. Hall, D. Marshall, and R. Martin. Merging and splitting eigenspace models. *IEEE Transaction on Pattern Analysis and Machine Intelligence*, 22(9):1042–1049, 2000.
12. D. Koller, J. Weber, T. Huang, J. Malik, G. Ogasawara, B. Rao, and S. Russell. Towards robust automatic traffic scene analysis in real-time. In *Proceedings of International Conference on Pattern Recognition*, volume I, pages 126–131, 1994.
13. A. Leonardis and H. Bischof. Robust recognition using eigenimages. *Computer Vision and Image Understanding*, 78(1):99–118, 2000.
14. K. Levi and Y. Weiss. Learning object detection from a small number of examples: The importance of good features. In *Proceedings IEEE Conference Computer Vision and Pattern Recognition*, pages 53–60, 2004.
15. A. Levin, P. Viola, and Y. Freund. Unsupervised improvement of visual detectors using co-training. In *Proceedings of IEEE International Conference on Computer Vision*, volume I, pages 626–633, 2003.
16. A. Levy and M. Lindenbaum. Sequential karhunen-loeve basis extraction and its application to images. *IEEE Transation on Image Processing*, 9(8):1371–1374, 2000.
17. Y. Li. On incremental and robust subspace learning. *Pattern Recognition*, 37(7):1509–1518, 2004.
18. N. Littlestone. Learning quickly when irrelevant attributes abound. *Machine Learning*, 2: 285–318, 1987.
19. B. Lo and S. A. Velastin. Automatic congestion detection system for underground platforms. In *Proceedings of IEEE International Symposium on Intelligent Multimedia , Video and Speech Processing*, pages 158–161, 2001.
20. N. J. B . McFarlane and C. P. Schofield. Segmentation and tracking of piglets. *Machine Vision and Applications*, 8(3):187–193, 1995.
21. V. Nair and J. J. Clark. An unsupervised, online learning framework for moving object detection. In *Proceedings of IEEE Conference on Computer Vision and Pattern Recognition*, volume II, pages 317–324, 2004.
22. T. Ojala, M. Pietikäinen, and T. Mäenpää. Multiresolution gray-scale and rotation invariant texture classification with local binary patterns. *IEEE Transactions on Pattern Analysis and Machine Intelligence*, 24(7):971–987, 2002.
23. A. Opelt, M. Fussenegger, A. Pinz, and P. Auer. Weak hypotheses and boosting for generic object detection and recognition. In *Proceedings of European Conference on Computer Vision*, volume II, pages 71–84, 2004.
24. N. Oza and S. Russell. Experimental comparisons of online and batch versions of bagging and boosting. In *Proceedings ACM SIGKDD International Conference on Knowledge Discovery and Data Mining*, 2001.
25. N. Oza and S. Russell. Online bagging and boosting. In *Proceedings Artificial Intelligence and Statistics*, pages 105–112, 2001.
26. J.-H. Park and Y.-K. Choi. On-line learning for active pattern recognition. *IEEE Signal Processing Letters*, 3(11):301–303, 1996.

27. M. Piccardi. Background subtraction techniques: a review. In *Proceedings of IEEE International Conference on Systems, Man and Cybernetics*, volume 4, pages 3099–3104, 2004.
28. F. Porikli. Integral histogram: A fast way to extract histograms in cartesian spaces. In *Proceedings IEEE Conference Computer Vision and Pattern Recognition*, volume 1, pages 829–836, 2005.
29. P. M. Roth and H. Bischof. On-line learning a person model from video data. In *Video Proceedings of IEEE Conference on Computer Vision and Pattern Recognition*, 2006.
30. P. M. Roth, H. Grabner, D. Skočaj, H. Bischof, and A. Leonardis. Conservative visual learning for object detection with minimal hand labeling effort. In *Proceedings of DAGM Symposium*, volume 3663 of *LNCS*, pages 293–300. Springer, New York, 2005.
31. P. M. Roth, H. Grabner, D. Skočaj, H. Bischof, and A. Leonardis. On-line conservative learning for person detection. In *Proceedings of IEEE International Workshop on Visual Surveillance and Performance Evaluation of Tracking and Surveillance*, pages 223–230, 2005.
32. H. Rowley, S. Baluja, and T. Kanade. Neural network-based face detection. *IEEE Transactions on Pattern Analysis and Machine Intelligence*, 20(1):23–38, 1998.
33. D. Skočaj, H. Bischof, and A. Leonardis. A robust PCA algorithm for building representations from panoramic images. In *Proceedings of European Conference on Computer Vision*, volume IV, pages 761–775, 2002.
34. D. Skočaj and A. Leonardis. Weighted and robust incremental method for subspace learning. In *Proceedings of IEEE International Conference on Computer Vision*, volume II, pages 1494–1501, 2003.
35. K.-K. Sung and T. Poggio. Example-based learning for view-based face detection. *IEEE Transaction on Pattern Analysis and Machine Intelligence*, 20(1):39–51, 1998.
36. K. Tieu and P. Viola. Boosting image retrieval. In *Proceedings IEEE Conference Computer Vision and Pattern Recognition*, pages 228–235, 2000.
37. V. N. Vapnik. *The Nature of Statistical Learning Theory*. Springer, New York, 1995.
38. P. Viola and M. J. Jones. Rapid object detection using a boosted cascade of simple features. In *Proceedings of IEEE Conference on Computer Vision and Pattern Recognition*, volume II, pages 511–518, 2001.
39. P. Viola, M. J. Jones, and D. Snow. Detecting pedestrians using patterns of motion and appearance. In *Proceedings of IEEE International Conference on Computer Vision*, volume II, pages 734–741, 2003.

Chapter 7
Machine Learning Techniques for Face Analysis

Roberto Valenti, Nicu Sebe, Theo Gevers, and Ira Cohen

Abstract In recent years there has been a growing interest in improving all aspects of the interaction between humans and computers with the clear goal of achieving a natural interaction, similar to the way human–human interaction takes place. The most expressive way humans display emotions is through facial expressions. Humans detect and interpret faces and facial expressions in a scene with little or no effort. Still, development of an automated system that accomplishes this task is rather difficult. There are several related problems: detection of an image segment as a face, extraction of the facial expression information, and classification of the expression (e.g., in emotion categories). A system that performs these operations accurately and in real time would be a major step forward in achieving a human-like interaction between the man and machine. In this chapter, we present several machine learning algorithms applied to face analysis and stress the importance of learning the structure of Bayesian network classifiers when they are applied to face and facial expression analysis.

Roberto Valenti
Faculty of Science, University of Amsterdam, Amsterdam, The Netherlands, e-mail: `rvalenti@science.uva.nl`

Nicu Sebe
Faculty of Science, University of Amsterdam, Amsterdam, The Netherlands, e-mail: `nicu@science.uva.nl`

Theo Gevers
Faculty of Science, University of Amsterdam, Amsterdam, The Netherlands, e-mail: `gevers@science.uva.nl`

Ira Cohen
HP Labs, Palo Alto, CA, USA, e-mail: `iracohen@hp.com`

7.1 Introduction

Information systems are ubiquitous in all human endeavors including scientific, medical, military, transportation, and consumer. Individual users use them for learning, searching for information (including data mining), doing research (including visual computing), and authoring. Multiple users (groups of users, and groups of groups of users) use them for communication and collaboration. And either single or multiple users use them for entertainment. An information system consists of two components: computer (data/knowledge base and information processing engine) and humans. It is the intelligent interaction between the two that we are addressing in this chapter.

Automatic face analysis has attracted increasing interest in the research community mainly due to its many useful applications. A system involving such an analysis assumes that the face can be accurately detected and tracked, the facial features can be precisely identified, and that the facial expressions, if any, can be precisely classified and interpreted. For doing this, in the following, we present in detail the three essential components of our automatic system for human–computer interaction: face detection, facial feature detection, and facial emotion recognition. This chapter presents our real time facial expression recognition system [10] which uses a facial features detector and a model-based non-rigid face tracking algorithm to extract motion features that serve as input to a Bayesian network classifier used for recognizing the different facial expressions. Parts of this system have been developed in collaboration with our colleagues from the Beckman Institute, University of Illinois at Urbana-Champaign, USA. We present here the components of the system and give reference to the publications that contain extensive details on the individual components [9, 40].

7.2 Background

7.2.1 Face Detection

Images containing face are essential to intelligent vision-based human–computer interaction. The rapidly expanding research in face processing is based on the premise that information about user's identity, state, and intend can be extracted from images and that computers can react accordingly, e.g., by observing a person's facial expression. Given an arbitrary image, the goal of face detection is to automatically locate a human face in an image or video, if it is present. Face detection in a general setting is a challenging problem for various reasons. The first set of reasons are inherent: there are many types of faces, with different colors, texture, sizes, etc. In addition, the face is a non-rigid object which can change its appearance. The second set of reasons are environmental: changing lighting, rotations, translations, and scales of the faces in natural images.

To solve the problem of face detection, two main approaches can be taken. The first is a model-based approach, where a description of what is a human face is used for detection. The second is an appearance-based approach, where we learn what faces are directly from their appearance in images. In this work, we focus on the latter.

There have been numerous appearance-based approaches. We list a few from recent years and refer to the reviews of Yang et al. [46] and Hjelmas and Low [23] for further details. Rowley et al. [37] used neural networks to detect faces in images by training from a corpus of face and non-face images. Colmenarez and Huang [11] used maximum entropic discrimination between faces and non-faces to perform maximum likelihood classification, which was used for a real time face tracking system. Yang et al. [47] used SNoW-based classifiers to learn the face and non-face discrimination boundary on natural face images. Wang et al. [44] learned a minimum spanning weighted tree for learning pairwise dependencies graphs of facial pixels, followed by a discriminant projection to reduce complexity. Viola and Jones [43] used boosting and a cascade of classifiers for face detection.

Very relevant to our work is the research of Schneiderman [38] who learns a sparse structure of statistical dependencies for several object classes including faces. While analyzing such dependencies can reveal useful information, we go beyond the scope of Schneiderman's work and present a framework that not only learns the structure of a face but also allows the use of unlabeled data in classification.

Face detection provides interesting challenges to the underlying pattern classification and learning techniques. When a raw or filtered image is considered as input to a pattern classifier, the dimension of the space is extremely large (i.e., the number of pixels in normalized training images). The classes of face and non-face images are decidedly characterized by multimodal distribution functions and effective decision boundaries are likely to be non-linear in the image space. To be effective, the classifiers must be able to extrapolate from a modest number of training samples.

7.2.2 Facial Feature Detection

Various approaches to facial feature detection exist in the literature. Although many of the methods have been shown to achieve good results, they mainly focus on finding the location of some facial features (e.g., eyes and mouth corners) in restricted environments (e.g., constant lighting, simple background, etc.). Since we want to obtain a complex and accurate system of feature annotation, these methods are not suitable for us.

In recent years deformable model-based approaches for image interpretation have been proven very successful, especially in images containing objects with large variability such as faces. These approaches are more appropriate for our specific case since they make use of a template (e.g., the shape of an object). Among the early deformable template models is the active contour model by Kass et al. [26] in which a correlation structure between shape markers is used to constrain local changes. Cootes et al. [14] proposed a generalized extension, namely active shape models (ASM), where deformation variability is learned using a training set. Active appearance models (AAM) were later proposed in [12] and they are closely related to the simultaneous formulation of active blobs [39] and morphble models [24]. AAM can be seen as an extension of ASM which includes the appearance information of an object.

While active appearance models have been shown to be very successful, they suffer from important drawbacks such as background handling and initialization. Previous work tried to solve the latter by using an object detector to provide an acceptable model initialization. In Sect. 7.5.2, we bring this concept one step further and we reduce the existing AAM problems by considering the initialization information as a part of the active appearance model.

7.2.3 Emotion Recognition Research

Ekman and Friesen [17] developed the facial action coding system (FACS) to code facial expressions where movements on the face are described by a set of action units (AUs). Each AU has some related muscular basis. This system of coding facial expressions is done manually by following a set of prescribed rules. The inputs are still images of facial expressions, often at the peak of the expression. This process is very time-consuming.

Ekman's work inspired many researchers to analyze facial expressions by means of image and video processing. By tracking facial features and measuring the amount of facial movement, they attempt to categorize different facial expressions. Recent work on facial expression analysis and recognition has used these 'basic expressions' or a subset of them. The two recent surveys in the area [19, 35] provide an in-depth review of many of the research done in automatic facial expression recognition in recent years.

The work in computer-assisted quantification of facial expressions did not start until the 1990s. Black and Yacoob [2] used local parameterized models of image motion to recover non-rigid motion. Once recovered, these parameters were used as inputs to a rule-based classifier to recognize the six basic facial expressions. Essa and Pentland [18] used an optical flow region-based method to recognize expressions. Oliver et al. [32] used lower face tracking to extract mouth shape features and used them as inputs to an HMM-based facial expression recognition system (recognizing neutral, happy, sad, and an open mouth). Chen [5] used a suite of static classifiers to recognize facial expressions, reporting on both person-dependent and person-independent results. Cohen et al. [10] describe classification schemes for facial expression recognition in two types of settings: dynamic and static classification. In the static setting, the authors learn the structure of Bayesian networks classifiers using as input 12 motion units given by a face tracking system for each frame in a video. For the dynamic setting, they used a multilevel HMM classifier that combines the temporal information and allows not only to perform the classification of a video segment to the corresponding facial expression, as in the previous works on HMM-based classifiers, but also to automatically segment an arbitrary long sequence to the different expression segments without resorting to heuristic methods of segmentation.

These methods are similar in that they first extract some features from the images, then these features are used as inputs into a classification system, and the outcome is one of the preselected emotion categories. They differ mainly in the features extracted from the video images and in the classifiers used to distinguish between the different emotions.

7.3 Learning Classifiers for Human–Computer Interaction

Many pattern recognition and human–computer interaction applications require the design of classifiers. Classification is the task of systematic arrangement in groups or categories according to some set of observations, e.g., classifying images to those containing human faces and those that do not or classifying individual pixels as being skin or non-skin. Classification is a natural part of daily human activity and is performed on a routine basis. One of the tasks in machine learning has been to give the computer the ability to perform classification in different problems. In machine classification, a classifier is constructed which takes as input a set of observations (such as images in the face detection problem) and outputs a prediction of the class *label* (e.g., face or no face). The mechanism which performs this operation is the *classifier*.

We are interested in probabilistic classifiers, in which the observations and class are treated as random variables, and a classification rule is derived using probabilistic arguments (e.g., if the probability of an image being a face given that we observed two eyes, nose, and mouth in the image is higher than some threshold, classify the image as a face). We consider two aspects. First, most of the research mentioned in the previous section tried to classify each observable independent from each of the others. We want to take a different approach: can we learn the dependencies (the structure) between the observables (e.g., the pixels in an image patch)? Can we use this structure for classification? To achieve this we use Bayesian networks. Bayesian networks can represent joint distributions in an intuitive and efficient way; as such, Bayesian networks are naturally suited for classification. Second, we are interested in using a framework that allows for the usage of labeled and unlabeled data (also called semi-supervised learning). The motivation for semi-supervised learning stems from the fact that labeled data are typically much harder to obtain compared to unlabeled data. For example, in facial expression recognition it is easy to collect videos of people displaying emotions, but it is very tedious and difficult to label the video to the corresponding expressions. Bayesian networks are very well suited for this task: they can be learned with labeled and unlabeled data using maximum likelihood estimation.

Is there value to unlabeled data in supervised learning of classifiers? This fundamental question has been increasingly discussed in recent years, with a general optimistic view that unlabeled data hold great value. Due to an increasing number of applications and algorithms that successfully use unlabeled data [1, 31, 41] and magnified by theoretical issues over the value of unlabeled data in certain cases [4, 33], semi-supervised learning is seen optimistically as a learning paradigm that can relieve the practitioner from the need to collect many expensive labeled training data. However, several disparate empirical evidences in the literature suggest that there are situations in which the addition of unlabeled data to a pool of labeled data causes degradation of the classifier's performance [1, 31, 41], in contrast to improvement of performance when adding more labeled data. Intrigued by these discrepancies, we performed extensive experiments, reported in [9]. Our experiments suggested that performance degradation can occur when the assumed classifier's model is incorrect. Such situations are quite common, as one rarely knows whether the assumed model is an accurate description of the underlying true data generating distribution. More details are given below (for the sake of consistency we keep the same notations as the one introduced in [9]).

The goal is to classify an incoming vector of observables $\mathbf{X}$. Each instantiation of $\mathbf{X}$ is a *sample*. There exists a *class variable* C; the values of C are the *classes*. Let $P(C,\mathbf{X})$ be the *true* joint distribution of the class and features from which any sample of some (or all) of the variables from the set $\{C,\mathbf{X}\}$ is drawn, and let $p(C,\mathbf{X})$ be the density distribution associated with it. We want to build *classifiers* that receive a sample $\mathbf{x}$ and output either one of the values of C.

Probabilities of $(C,\mathbf{X})$ are estimated from data and then are fed into the optimal classification rule. Also, a parametrical model $p(C,\mathbf{X}|\theta)$ is adopted. An estimate of θ is denoted by $\hat{\theta}$ and we denote throughout by $\hat{\theta}^*$ the asymptotic value of $\hat{\theta}$. If the distribution $p(C,\mathbf{X})$ belongs to the family $p(C,\mathbf{X}|\theta)$, we say the 'model is correct'; otherwise, we say the 'model is incorrect'. We use 'estimation bias' loosely to mean the expected difference between $p(C,\mathbf{X})$ and the estimated $p\left(C,\mathbf{X}|\hat{\theta}\right)$.

The analysis presented in [9] and summarized here is based on the work of White [45] on the properties of maximum likelihood estimators without assuming model correctness. White [45] showed that under suitable regularity conditions, maximum likelihood estimators converge to a parameter set θ^* that minimizes the Kullback–Leibler (KL) distance between the assumed family of distributions, $p(Y|\theta)$, and the true distribution, $p(Y)$. White [45] also shows that the estimator is asymptotically normal, i.e., $\sqrt{N}(\hat{\theta}_N - \theta^*) \sim \mathcal{N}(0, C_Y(\theta))$ as N (the number of samples) goes to infinity. $C_Y(\theta)$ is a covariance matrix equal to $A_Y(\theta)^{-1}B_Y(\theta)A_Y(\theta)^{-1}$, evaluated at θ^*, where $A_Y(\theta)$ and $B_Y(\theta)$ are matrices whose (i,j)th element ($i,j = 1,...,d$, where d is the number of parameters) is given by

$$A_Y(\theta) = E\left[\partial^2 \log p(Y|\theta)\,/\partial\theta_i\theta_j\right],$$
$$B_Y(\theta) = E[(\partial \log p(Y|\theta)\,/\partial\theta_i)(\partial \log p(Y|\theta)\,/\partial\theta_j)].$$

Using these definitions, in [9] the following theorem was introduced:

Theorem 7.1. *Consider supervised learning where samples are randomly labeled with probability λ. Adopt the regularity conditions in Theorems 3.1, 3.2, 3.3 from [45], with Y replaced by $(C,\mathbf{X})$ and by $\mathbf{X}$, and also assume identifiability for the marginal distributions of $\mathbf{X}$. Then the value of θ^*, the limiting value of maximum likelihood estimates, is*

$$\arg\max_{\theta} \left(\lambda E[\log p(C,\mathbf{X}|\theta)] + (1-\lambda)E[\log p(\mathbf{X}|\theta)]\right), \quad (7.1)$$

where the expectations are with respect to $p(C,\mathbf{X})$. Additionally, $\sqrt{N}(\hat{\theta}_N - \theta^*) \sim \mathcal{N}(0, C_\lambda(\theta))$ as $N \to \infty$, where $C_\lambda(\theta)$ is given by

$$\begin{aligned} C_\lambda(\theta) &= A_\lambda(\theta)^{-1} B_\lambda(\theta) A_\lambda(\theta)^{-1} \text{ with,} \quad (7.2)\\ A_\lambda(\theta) &= \left(\lambda A_{(C,\mathbf{X})}(\theta) + (1-\lambda) A_{\mathbf{X}}(\theta)\right) \text{ and} \\ B_\lambda(\theta) &= \left(\lambda B_{(C,\mathbf{X})}(\theta) + (1-\lambda) B_{\mathbf{X}}(\theta)\right), \end{aligned}$$

evaluated at θ^*.

For a proof of this theorem we direct the interested reader to [9]. Here we restrict only to a few observations. Expression (7.1) indicates that semi-supervised learning can be viewed asymptotically as a 'convex' combination of supervised and unsupervised learning. As such, the objective function for semi-supervised learning is a combination of the objective function for supervised learning ($E[\log p(C,\mathbf{X}|\theta)]$) and the objective function for unsupervised learning ($E[\log p(\mathbf{X}|\theta)]$).

Denote by θ^*_λ the value of θ that maximizes Expression (7.1) for a given λ. Then, θ^*_1 is the asymptotic estimate of θ for *supervised* learning, denoted by θ^*_l. Likewise, θ^*_0 is the asymptotic estimate of θ for *unsupervised* learning, denoted by θ^*_u.

The asymptotic covariance matrix is positive definite as $B_Y(\theta)$ is positive definite, $A_Y(\theta)$ is symmetric for any Y, and

$$\theta A(\theta)^{-1} B_Y(\theta) A(\theta)^{-1} \theta^T = w(\theta) B_Y(\theta) w(\theta)^T > 0,$$

where $w(\theta) = \theta A_Y(\theta)^{-1}$. We see that asymptotically, an increase in N, the number of labeled and unlabeled samples, will lead to a reduction in the variance of $\hat{\theta}$. Such a guarantee can perhaps be the basis for the optimistic view that unlabeled data should always be used to improve classification accuracy. In [9] it was shown that this observation holds when the model is correct, and that when the model is incorrect this observation might not always hold.

7.3.1 Model Is Correct

Suppose first that the family of distributions $P(C,\mathbf{X}|\theta)$ contains the distribution $P(C,\mathbf{X})$; that is, $P(C,\mathbf{X}|\theta_\top) = P(C,\mathbf{X})$ for some $\theta_\top$. Under this condition, the maximum likelihood estimator is consistent; thus, $\theta^*_l = \theta^*_u = \theta_\top$ given identifiability. Thus, $\theta^*_\lambda = \theta_\top$ for any $0 \le \lambda \le 1$.

Additionally, using White's results [45], $A(\theta_\lambda^*) = -B(\theta_\lambda^*) = \mathbf{I}(\theta_\lambda^*)$, where $\mathbf{I}()$ denotes the Fisher information matrix. Thus, the Fisher information matrix can be written as

$$\mathbf{I}(\theta) = \lambda \mathbf{I}_l(\theta) + (1-\lambda)\mathbf{I}_u(\theta), \tag{7.3}$$

which matches the derivations made by Zhang and Oles [48]. The significance of Expression (7.3) is that it allows the use of the Cramer–Rao lower bound (CRLB) on the covariance of a consistent estimator:

$$\mathrm{Cov}(\hat{\theta}_N) \geq \frac{1}{N}(\mathbf{I}(\theta))^{-1}, \tag{7.4}$$

where N is the number of data (both labeled and unlabeled) and $\mathrm{Cov}(\hat{\theta}_N)$ is the estimator's covariance matrix with N samples.

Consider the Taylor expansion of the classification error around $\theta_\top$, as suggested by Shahshahani and Landgrebe [41], linking the decrease in variance associated with unlabeled data to a decrease in classification error, and assume the existence of necessary derivatives:

$$\mathbf{e}(\hat{\theta}) \approx \mathbf{e}_\mathrm{B} + \left.\frac{\partial \mathbf{e}(\theta)}{\partial \theta}\right|_{\theta_\top} (\hat{\theta}-\theta_\top) + \frac{1}{2}\mathrm{tr}\left(\left.\frac{\partial^2 \mathbf{e}(\theta)}{\partial \theta^2}\right|_{\theta_\top} (\hat{\theta}-\theta_\top)(\hat{\theta}-\theta_\top)^T\right). \tag{7.5}$$

Take expected values on both sides. Asymptotically the expected value of the second term in the expansion is zero, as maximum likelihood estimators are asymptotically unbiased when the model is correct. Shahshahani and Landgrebe [41] thus argue that

$$E\left[\mathbf{e}(\hat{\theta})\right] \approx \mathbf{e}_\mathrm{B} + (1/2)\mathrm{tr}\left((\partial^2 \mathbf{e}(\theta)/\partial\theta^2)|_{\theta_\top}\mathrm{Cov}(\hat{\theta})\right),$$

where $\mathbf{e}_\mathrm{B} = \mathbf{e}(\theta_\top)$ is the Bayes error rate. They also show that if $\mathrm{Cov}(\theta') \geq \mathrm{Cov}(\theta'')$ for some θ' and θ'', then the second term in the approximation is larger for θ' than for θ''. Because $\mathbf{I}_u(\theta)$ is always positive definite, $\mathbf{I}_l(\theta) \leq \mathbf{I}(\theta)$. Thus, using the Cramer–Rao lower bound (Expression (7.4)) the covariance with labeled and unlabeled data is smaller than the covariance with just labeled data, leading to the conclusion that *unlabeled data must cause a reduction in classification error when the model is correct.* It should be noted that this argument holds as the number of records goes to infinity and is an approximation for finite values.

7.3.2 Model Is Incorrect

A more realistic scenario described in detail in [9] is when the distribution $P(C,\mathbf{X})$ does not belong to the family of distributions $P(C,\mathbf{X}|\theta)$. In view of Theorem 7.1, it is clear that unlabeled data can have the deleterious effect observed occasionally in the literature. Suppose that $\theta_u^* \neq \theta_l^*$ and that $\mathbf{e}(\theta_u^*) > \mathbf{e}(\theta_l^*)$ (for the difficulties in estimating $\mathbf{e}(\theta_u^*)$ and a solution for this please see [9]). If a large number of labeled samples is observed, the classification error is approximated by $\mathbf{e}(\theta_l^*)$. If we then have more samples, most of which are unlabeled, we eventually reach a point where the classification error approaches $\mathbf{e}(\theta_u^*)$. So, the net result is that we started with classification error close to $\mathbf{e}(\theta_l^*)$, and by adding a large number of unlabeled samples, classification performance degraded (see again [9] for more details). The basic fact here is that estimation and classification bias are affected differently by different values of λ. Hence, a necessary condition for this kind of performance degradation is that $\mathbf{e}(\theta_u^*) \neq \mathbf{e}(\theta_l^*)$; a sufficient condition is that $\mathbf{e}(\theta_u^*) > \mathbf{e}(\theta_l^*)$.

The focus on asymptotics is adequate as we want to eliminate phenomena that can vary from data set to data set. If $\mathbf{e}(\theta_l^*)$ is smaller than $\mathbf{e}(\theta_u^*)$, then a large enough labeled data set can be dwarfed by a much larger unlabeled data set—the classification error using the whole data set can be larger than the classification error using the labeled data only.

7.3.3 Discussion

Despite the shortcomings of semi-supervised learning presented in the previous sections, we do not discourage its use. Understanding the causes of performance degradation with unlabeled data motivates the exploration of new methods attempting to use positively the available unlabeled data. Incorrect modeling assumptions in Bayesian networks culminate mainly as discrepancies in the graph structure, signifying incorrect independence assumptions among variables. To eliminate the increased bias caused by the addition of unlabeled data we can try simple solutions, such as model switching (Sect. 7.4.2) or attempt to learn better structures. We describe likelihood-based structure learning methods (Sect. 7.4.3) and a possible alternative: classification-driven structure learning (Sect. 7.4.4). In cases where relatively mild changes in structure still suffer from performance degradation from unlabeled data, there are different approaches that can be taken: discard the unlabeled data, give them a different weight (Sect. 7.4.5), or use the alternative of actively labeling some of the unlabeled data (Sect. 7.4.6).

To summarize, the main conclusions that can be derived from our analysis are the following:

- Labeled and unlabeled data contribute to a reduction in variance in semi-supervised learning under maximum likelihood estimation. *This is true regardless of whether the model is correct or not.*
- If the model is correct, the maximum likelihood estimator is unbiased and both labeled and unlabeled data contribute to a reduction in classification error by reducing variance.
- If the model is incorrect, there may be different asymptotic estimation biases for different values of λ (the ratio between the number of labeled and unlabeled data). Asymptotic classification error may also be different for different values of λ. An increase in the number of unlabeled samples may lead to a larger bias from the true distribution and a larger classification error.

In the next section, we discuss several possible solutions for the problem of performance degradation in the framework of Bayesian network classifiers.

7.4 Learning the Structure of Bayesian Network Classifiers

The conclusion of the previous section indicates the importance of obtaining the correct structure when using unlabeled data in learning a classifier. If the correct structure is obtained, unlabeled data improve the classifier; otherwise, unlabeled data can actually degrade performance. Somewhat surprisingly, the option of searching for better structures was not proposed by researchers that previously witnessed the performance degradation. Apparently, performance degradation was attributed to unpredictable, stochastic disturbances in modeling assumptions, and not to mistakes in the underlying structure—something that can be detected and fixed.

7.4.1 Bayesian Networks

Bayesian networks [36] are tools for modeling and classification. A Bayesian network (BN) is composed of a directed acyclic graph in which every node is associated with a variable X_i and a conditional distribution $p(X_i|\Pi_i)$, where Π_i denotes the parents of X_i in the graph. The joint probability distribution is factored to the collection of conditional probability distributions of each node in the graph as

$$p(X_1,...,X_n) = \prod_{i=1}^{n} p(X_i|\Pi_i) . \tag{7.6}$$

The directed acyclic graph is the *structure*, and the distributions $p(X_i|\Pi_i)$ represent the *parameters* of the network. We say that the assumed structure for a network, S', is *correct* when it is possible to find a distribution, $p(C,\mathbf{X}|S')$, that matches the distribution that generates data, $p(C,\mathbf{X})$; otherwise, the structure is *incorrect*. In the above notations, $\mathbf{X}$ is an incoming vector of features. The classifier receives a record $\mathbf{x}$ and generates a label $\hat{c}(\mathbf{x})$. An optimal classification rule can be obtained from the exact distribution $p(C,\mathbf{X})$ which represents the a posteriori probability of the class given the features.

Maximum likelihood estimation is one of the main methods to learn the parameters of the network. When there are missing data in the training set, the expectation maximization (EM) algorithm [15] can be used to maximize the likelihood.

As a direct consequence of the analysis in Sect. 7.3, a Bayesian network that has the correct structure and the correct parameters is also optimal for classification because the a posteriori distribution of the class variable is accurately represented (see [9] for a detailed analysis on this issue). As pointed out in [9] and [8] to solve the problem of performance degradation in BNs, there is a need to carefully analyze the structure of the BN classifier used in the classification.

7.4.2 Switching Between Simple Models

One attempt to overcome the performance degradation from unlabeled data could be to switch models as soon as degradation is detected. Suppose that we learn a classifier with labeled data only and we observe a degradation in performance when the classifier is learned with labeled and unlabeled data. We can switch to a more complex structure at that point. An interesting idea is to start with a Naive Bayes classifier in which the features are assumed independent given the class. If performance degrades with unlabeled data, switch to a different type of Bayesian network classifier, namely the tree-augmented Naive Bayes classifier (TAN) [21].

In the TAN classifier structure the class node has no parents and each feature has the class node and at most one other feature as parents, such that the result is a tree structure for the features. Learning the most likely TAN structure has an efficient and exact solution [21] using a modified Chow–Liu algorithm [7]. Learning the TAN classifiers when there are unlabeled data requires a modification of the original algorithm to what we named the EM-TAN algorithm [10].

If the correct structure can be represented using a TAN structure, this approach will indeed work. However, even the TAN structure is only a small set of all possible structures. Moreover, as the examples in the experimental section show, switching from NB to TAN does not guarantee that the performance degradation will not occur.

Very relevant is the research of Baluja [1]. The author uses labeled and unlabeled data in a probabilistic classifier framework to detect the orientation of a face. In his results, he obtained excellent classification results, but there were cases where unlabeled data degraded performance. As a consequence, he decided to switch from a Naive Bayes approach to more complex models. Following this intuitive direction, we explain Baluja's observations and provide a solution to the problem: structure learning.

7.4.3 Beyond Simple Models

A different approach to overcome performance degradation is to learn the structure of the Bayesian network without restrictions other than the generative one.[1] There are a number of such algorithms in the literature (among them [3, 6, 20]). Nearly all structure learning algorithms use the 'likelihood-based' approach. The goal is to find structures that best fit the data (with perhaps a prior distribution over different

[1] A Bayesian network classifier is a *generative* classifier when the class variable is an ancestor (e.g., parent) of some (or all) features.

structures). Since more complicated structures have higher likelihood scores, penalizing terms are added to avoid overfitting to the data, e.g., the minimum description length (MDL) term. The difficulty of structure search is the size of the space of possible structures. With finite amounts of data, algorithms that search through the space of structures maximizing the likelihood can lead to poor classifiers because the a posteriori probability of the class variable could have a small effect on the score [21]. Therefore, a network with a higher score is not necessarily a better classifier. Friedman et al. [21] suggest changing the scoring function to focus only on the posterior probability of the class variable, but show that it is not computationally feasible.

The drawbacks of likelihood-based structure learning algorithms could be magnified when learning with unlabeled data; the posterior probability of the class has a smaller effect during the search, while the marginal of the features would dominate. Therefore, we decided to take a different approach presented in the next section.

7.4.4 Classification-Driven Stochastic Structure Search

As pointed out in [8] one elegant solution is to find the structure that minimizes the probability of classification error directly. To do so the classification-driven stochastic search algorithm (SSS) was proposed in [9]. The basic idea of this approach is that, since one is interested in finding a structure that performs well as a classifier, it is natural to design an algorithm that uses classification error as the guide for structure learning. For completeness we summarize the main observation here and we direct the interested reader to [8] for a complete analysis.

One important observation is that unlabeled data can indicate incorrect structure through degradation of classification performance. Additionally, we also saw previously that classification performance improves with the correct structure. As a consequence, a structure with higher classification accuracy over another indicates an improvement towards finding the optimal classifier.

To learn structure using classification error, it is necessary to adopt a strategy for efficiently searching through the space of all structures while avoiding local maxima. As there is no simple closed-form expression that relates structure with classification error, it is difficult to design a gradient descent algorithm or a similar iterative method which would in any case be prone to find local minima due to the size of the search space.

In [8] the following measure was proposed to be maximized:

Definition 7.1. The *inverse error measure* for structure S' is

$$inv_e(S') = \frac{\frac{1}{p_{S'}(\hat{c}(X) \neq C)}}{\sum_S \frac{1}{p_S(\hat{c}(\mathbf{X}) \neq C)}}, \tag{7.7}$$

where the summation is over the space of possible structures and $p_S(\hat{c}(\mathbf{X}) \neq C)$ is the probability of error of the best classifier learned with structure S.

Metropolis–Hastings sampling [30] can be used to generate samples from the inverse error measure, without the need to compute it for all possible structures. For constructing the Metropolis–Hastings sampling, a neighborhood of a structure is defined as the set of directed acyclic graphs to which we can transit in the next step. Transition is done using a predefined set of possible changes to the structure; at each transition a change consists of a single edge addition, removal, or reversal. In [8] the acceptance probability of a candidate structure, S_{new}, to replace a previous structure, S_t is defined as follows:

$$\min\left(1, \left(\frac{inv_e(S^{\mathrm{new}})}{inv_e(S^t)}\right)^{1/T} \frac{q(S^t|S^{\mathrm{new}})}{q(S^{\mathrm{new}}|S^t)}\right) = \min\left(1, \left(\frac{p^t_{\mathrm{error}}}{p^{\mathrm{new}}_{\mathrm{error}}}\right)^{1/T} \frac{N_t}{N_{\mathrm{new}}}\right), \qquad (7.8)$$

where $q(S'|S)$ is the transition probability from S to S' and N_t and N_{new} are the sizes of the neighborhoods of S_t and S_{new}, respectively; this choice corresponds to equal probability of transition to each member in the neighborhood of a structure. This choice of neighborhood and transition probability creates a Markov chain which is aperiodic and irreducible, thus satisfying the Markov chain Monte Carlo (MCMC) conditions [27].

The parameter T is used as a temperature factor in the acceptance probability. As such, T close to 1 would allow acceptance of more structures with higher probability of error than previous structures. T close to 0 mostly allows acceptance of structures that improve probability of error. A fixed T amounts to changing the distribution being sampled by the MCMC, while a decreasing T is a simulated annealing run, aimed at finding the maximum of the inverse error measures. The rate of decrease of the temperature determines the rate of convergence. Asymptotically in the number of data, a logarithmic decrease of T guarantees convergence to a global maximum with probability that tends to 1 [22].

The SSS algorithm, with a logarithmic cooling schedule T, can find a structure that is close to minimum probability of error. The estimate of the classification error of a given structure is obtained by using the labeled training data. Therefore, to avoid overfitting, a multiplicative penalty term is required. This penalty term, derived from the Vapnik–Chervonenkis (VC) bound on the empirical classification error, penalizes complex classifiers thus keeping the balance between bias and variance (for more details we refer the reader to [9]).

7.4.5 *Should Unlabeled Be Weighed Differently?*

An interesting strategy, suggested by Nigam et al. [31], is to change the weight of the unlabeled data (reducing their effect on the likelihood). The basic idea in Nigam et al.'s estimators is to produce a modified log-likelihood that is of the form.

$$\lambda' L_l(\theta) + (1-\lambda')L_u(\theta), \qquad (7.9)$$

where $L_l(\theta)$ and $L_u(\theta)$ are the likelihoods of the labeled and unlabeled data, respectively. For a sequence of λ', maximize the modified log-likelihood functions to obtain $\hat{\theta}_{\lambda'}$ ($\hat{\theta}$ denotes an estimate of θ), and choose the best one with respect to cross-validation or testing. This estimator is simply modifying the ratio of labeled to unlabeled samples for any fixed λ'. Note that this estimator can only make sense under the assumption that the model is incorrect. Otherwise, both terms in Expression (7.9) lead to unbiased estimators of θ.

Our experiments in [8] suggest that there is then no reason to impose different weights on the data, and much less reason to search for the best weight, when the differences are solely in the rate of reduction of variance. Presumably, there are a few labeled samples available and a large number of unlabeled samples; why should we increase the importance of the labeled samples, giving more weight to a term that will contribute more heavily to the variance?

7.4.6 Active Learning

All the methods presented above consider a ‘passive’ use of unlabeled data. A different approach is known as active learning, in which an oracle is queried as to the label of some of the unlabeled data. Such an approach increases the size of the labeled data set, reduces the classifier’s variance, and thus reduces the classification error. There are different ways to choose which unlabeled data to query. The straightforward approach is to choose a sample randomly. This approach ensures that the data distribution $p(C,\mathbf{X})$ is unchanged, a desirable property when estimating generative classifiers. However, the random sample approach typically requires many more samples to achieve the same performance as methods that choose to label data close to the decision boundary. We note that, for generative classifiers, the latter approach changes the data distribution therefore leading to estimation bias. Nevertheless, McCallum and Nigam [29] used active learning with generative models with success. They proposed to first actively query some of the labeled data followed by estimation of the model’s parameters with the remainder of the unlabeled data.

We performed extensive experiments in [8]. Here we present only the main conclusions. With correctly specified generative models and a large pool of unlabeled data, ‘passive’ use of the unlabeled data is typically sufficient to achieve good performance. Active learning can help reduce the chances of numerical errors (improve EM starting point, for example) and help in the estimation of classification error. With incorrectly specified generative models, active learning is very profitable in quickly reducing the error, while adding the remainder of unlabeled data might not be desirable.

7.4.7 Summary

The idea of structure search is particularly promising when unlabeled data are present. It seems that simple heuristic methods, such as the solution proposed by Nigam et al. [31] of weighing down the unlabeled data, are not the best strategies for unlabeled data. We suggest that structure search, and in particular stochastic structure search, holds the most promise for handling large amount of unlabeled data and relatively scarce labeled data for classification. We also believe that the success of structure search methods for classification increases significantly the breadth of applications of Bayesian networks.

In a nutshell, when faced with the option of learning with labeled and unlabeled data, our discussion suggests adopting the following path. Start with Naive Bayes and TAN classifiers, learn with only labeled data and test whether the model is correct by learning with the unlabeled data, using EM and EM-TAN. If the result is not satisfactory, then SSS can be used to attempt to further improve performance with enough computational resources. If none of the methods using the unlabeled data improve performance over the supervised TAN (or Naive Bayes), active learning can be used, as long as there are resources to label some samples.

7.5 Experiments

For the experiments, we used our real time facial expression recognition system [10]. This is composed of a face detector which is used as an input to a facial feature detection module. Using the extracted facial features, a face tracking algorithm outputs a vector of motion features of certain regions of the face. The features are used as inputs to a Bayesian network classifier.

The face tracking we use in our system is based on a system developed by Tao and Huang [42] called the piecewise Bézier volume deformation (PBVD) tracker. The face tracker uses a model-based approach where an explicit 3D wireframe model of the face is constructed. A generic face model is then warped to fit the detected facial features. The face model consists of 16 surface patches embedded in Bézier volumes. The surface patches defined in this way are guaranteed to be continuous and smooth. The shape of the mesh can be changed by changing the locations of the control points in the Bézier volume. A snapshot of the system with the face tracking and the corresponding recognition result is shown in Fig. 7.1.

In Sect. 7.5.1, we start by investigating the use Bayesian network classifiers learned with labeled and unlabeled data for face detection. We present our results on two standard databases and show good results even if we use a very small set of labeled data. Subsequently, in Sect. 7.5.2, we present our facial feature detection module which uses the input given from the face detector and outputs the location of relevant facial features. Finally, in Sect. 7.5.3, we discuss the facial expression recognition results obtained by incorporating the facial feature detected inside the PBVD tracker.

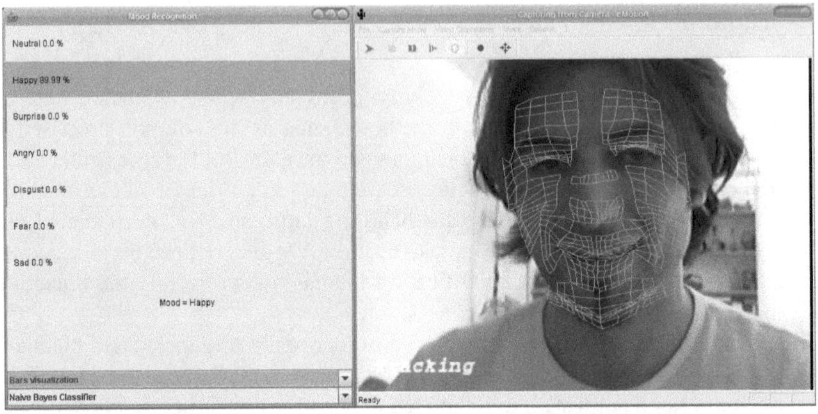

Fig. 7.1 A snapshot of our real time facial expression recognition system. On the left side is a wireframe model overlaid on a face being tracked. On the right side the correct expression, Happy, is detected (the bars show the relative probability of Happy compared to the other expressions). The subject shown is from the Cohn–Kanade database

7.5.1 Face Detection Experiments

In our face detection experiments we propose to use Bayesian network classifiers with the image pixels of a predefined window size as the features in the Bayesian network. Among the different works, those of Colmenarez and Huang [11] and Wang et al. [44] are more related to the Bayesian network classification methods for face detection. Both learn some 'structure' between the facial pixels and combine them to a probabilistic classification rule. Both use the entropy between the different pixels to learn pairwise dependencies.

Our approach in detecting faces is an appearance-based approach, where the intensity of image pixels serve as the features for the classifier. In a natural image, faces can appear at different scales, rotations, and location. For learning and defining the Bayesian network classifiers, we must look at fixed size windows and learn how a face appears in such windows, where we assume that the face appears in most of the window's pixels.

The goal of the classifier is to determine if the pixels in a fixed size window are those of a face or non-face. While faces are a well-defined concept, and have a relatively regular appearance, it is harder to characterize non-faces. We therefore model the pixel intensities as discrete random variables, as it would be impossible to define a parametric probability distribution function (pdf) for non-face images. For 8-bit representation of pixel intensity, each pixel has 256 values. Clearly, if all these values are used for the classifier, the number of parameters of the joint distribution is too large for learning dependencies between the pixels (as is the case of TAN classifiers). Therefore, there is a need to reduce the number of values representing pixel intensity. Colmenarez and Huang [11] used four values per pixel using fixed and equal bin sizes. We use non-uniform discretization using the class conditional entropy as the mean to bin the 256 values to a smaller number. We use the MLC++ software for that purpose as is described in [16].

Note that our methodology can be extended to other face detection methods which use different features. The complexity of our method is $O(n)$, where n is the number of features (pixels in our case) considered in each image window.

We test the different approaches described in Sect. 7.4, with both labeled and unlabeled data. For training the classifier we used a data set consisting of 2,429 faces and 10,000 non-faces obtained from the MIT CBCL Face database #1.[2] Examples of face images from the database are presented in Fig. 7.2. Each face image is cropped and resampled to a 19×19 window, thus we have a classifier with 361 features. We also randomly rotate and translate the face images to create a training set of 10,000 face images. In addition we have available 10,000 non-face images. We leave out 1,000 images (faces and non-faces) for testing and train the Bayesian network classifiers on the remaining 19,000. In all the experiments we learn a Naive Bayes, TAN, and a general generative Bayesian network classifier, the latter using the SSS algorithm.

In Table 7.1 we summarize the results obtained for different algorithms and in the presence of increasing number of unlabeled data. We fixed the false alarm to 1, 5, and 10% and we computed the detection rates. We first learn using all the training data being labeled (that is 19,000 labeled images). The classifier learned with the SSS algorithm outperforms both TAN and NB classifiers, and all perform quite well, achieving high detection rates with a low rate of false alarm. Next we remove the labels of some of the training data and train the classifiers. In the first case, we remove the labels of 97.5% of the training data (leaving only 475 labeled images). We see that the NB classifier using both labeled and unlabeled data performs very poorly. The TAN based only on the 475 labeled images and the TAN based on the

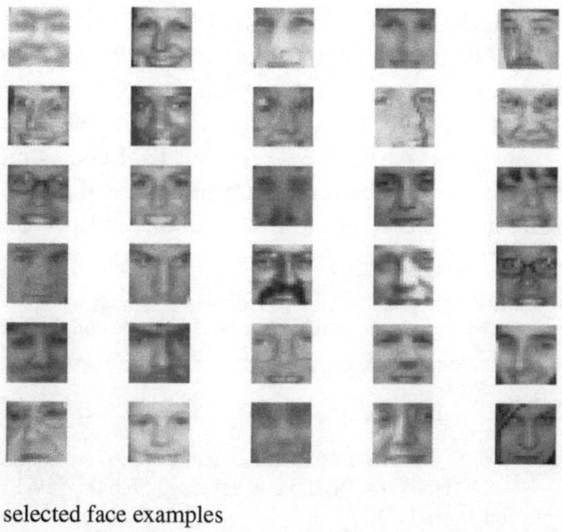

Fig. 7.2 Randomly selected face examples

[2] http://www.ai.mit.edu/projects/cbcl

Table 7.1 Detection rates (%) for various numbers of false positives

Detector \ False positives		1%	5%	10%
NB	19,000 labeled	74.31	89.21	92.72
	475 labeled	68.37	86.55	89.45
	475 labeled + 18,525 unlabeled	66.05	85.73	86.98
	250 labeled	65.59	84.13	87.67
	250 labeled + 18,750 unlabeled	65.15	83.81	86.07
TAN	19,000 labeled	91.82	96.42	99.11
	475 labeled	86.59	90.84	94.67
	475 labeled + 18,525 unlabeled	85.77	90.87	94.21
	250 labeled	75.37	87.97	92.56
	250 labeled + 18,750 unlabeled	77.19	89.08	91.42
SSS	19,000 labeled	90.27	98.26	99.87
	475 labeled + 18,525 unlabeled	88.66	96.89	98.77
	250 labeled + 18,750 unlabeled	86.64	95.29	97.93
SVM	19,000 labeled	87.78	93.84	94.14
	475 labeled	82.61	89.66	91.12
	250 labeled	77.64	87.17	89.16

labeled and unlabeled images are close in performance, thus there was no significant degradation of performance when adding the unlabeled data. When only 250 labeled data are used (the labels of about 98.7% of the training data were removed), NB with both labeled and unlabeled data performs poorly, while SSS outperforms the other classifiers with no great reduction of performance compared to the previous cases. For benchmarking, we also implemented a SVM classifier (we used the implementation of Osuna et al. [34]). Note that this classifier starts off very good, but does not improve performance.

In summary, note that the detection rates for NB are lower than the ones obtained for the other detectors. Overall, the results obtained with SSS are the best. We see that even in the most difficult cases, there was sufficient amount of unlabeled data to achieve almost the same performance as with a large-sized labeled data set.

We also tested our system on the CMU test set [37] consisting of 130 images with a total of 507 frontal faces. The results are summarized in Table 7.2. Note that we obtained comparable results with the results obtained by Viola and Jones [43]

Table 7.2 Detection rates (%) for various numbers of false positives on the CMU test set

Detector \ False positives		10%	20%
SSS	19,000 labeled	91.7	92.84
	475 labeled + 18,525 unlabeled	89.67	91.03
	250 labeled + 18,750 unlabeled	86.64	89.17
Viola and Jones [43]		92.1	93.2
Rowley et al. [37]		–	89.2

Fig. 7.3 Output of the system on some images of the CMU test using the SSS classifier learned with 19,000 labeled data. MFs represents the number of missed faces and FDs is the number of false detections

and better than the results of Rowley et al. [37]. Examples of the detection results on some of the images of the CMU test are presented in Fig. 7.3. We noticed similar failure modes as Viola and Jones [43]. Since, the face detector was trained only on frontal faces our system failed to detect faces if they have a significant rotation out of the plane (toward a profile view). The detector also has problems with the images in which the faces appear dark and the background is relatively light. Inevitably, we also detect false positive especially in some texture regions.

7.5.2 Facial Feature Detection

In this section, we introduce a novel way to unify the knowledge of a face detector inside an active appearance model [12], using what we call a 'virtual structuring element', which limits the possible settings of the AAM in an appearance-driven manner. We propose this visual artifact as a good solution for the background linking problems and respective generalization problems of basic AAMs.

The main idea of using an AAM approach is to learn the possible variations of facial features exclusively on a probabilistic and statistical basis of the existing observations (i.e., which relation holds in all the previously seen instances of facial features). This can be defined as a combination of shapes and appearances.

At the basis of AAM search is the idea to treat the fitting procedure of a combined shape–appearance model as an optimization problem in trying to minimize the difference vector between the image **I** and the generated model **M** of shape and appearance: $\delta\mathbf{I} = \mathbf{I} - \mathbf{M}$.

Cootes et al. [12] observed that each search corresponds to a similar class of problems where the initial and the final model parameters are the same. This class can be learned offline (when we create the model) saving high-dimensional computations during the search phase.

Learning the class of problems means that we have to assume a relation **R** between the current error image $\delta\mathbf{I}$ and the needed adjustments in the model parameters m. The common assumption is to use a linear relation: $\delta m = \mathbf{R}\delta\mathbf{I}$. Despite the fact that more accurate models were proposed [28], the assumption of linearity was shown to be sufficiently accurate to obtain good results. To find **R** we can conduct a series of experiments on the training set, where the optimal parameters m are known. Each experiment consists of displacing a set of parameters by a known amount and in measuring the difference between the generated model and the image under it. Note that when we displace the model from its optimal position and calculate the error image $\delta\mathbf{I}$, the image will surely contain parts of the background.

What remains to be discussed is an iterative optimization procedure that uses the found predictions. The first step is to initialize the mean model in an initial position and the parameters within the reach of the parameter prediction range (which depends on the perturbation used during training). Iteratively, a sample of the image under initialization is taken and compared with the model instance. The differences between the two appearances are used to predict the set of parameters that would perhaps improve the similarity. In case a prediction fails to improve the similarity, it is possible to damp or amplify the prediction several times and maintain the one with the best result. For an overview of some possible variations to the original

(**a**) Unseen face (**b**) Initialization (**c**) Converged model

Fig. 7.4 Results of an AAM search on an unseen face

AAM algorithm refer to [13]. An example of the AAM search is shown in Fig. 7.4 where a model is fitted to a previously unseen face.

One of the main drawbacks of the AAM is coming from its very basic concept: when the algorithm learns how to solve the optimization offline, the perturbation applied to the model inevitably takes parts of the background into account. This means that instead of learning how to generally solve the class of problems, the algorithm actually learns how to solve it only for the same or similar background. This makes AMMs domain-specific, that is, the AAM trained for a shape in a predefined environment has difficulties when used on the same shape immersed in a different environment. Since we always need to perturbate the model and to take into account the background, an often used idea is to constrain the shape deformation within predefined boundaries. Note that a shape constraint does not adjust the deformation, but will only limit it when it is found to be invalid.

To overcome these deficiencies of AAMs, we propose a novel method to visually integrate the information obtained by a face detector inside the AAM. This method is based on the observation that an object with a specific and recognizable feature would ease the successful alignment of its model. As the face detector we can choose between the one proposed by Viola and Jones [43] and the one presented in Sect. 7.5.1.

Since faces have many highly relevant features, erroneously located ones could lead the optimization process to converge to local minima. The novel idea is to add a virtual artifact in each of the appearances in the training and the test sets that would inherently prohibit some deformations. We call this artifact a *virtual structuring element* (or VSE) since it adds structure in the data that was not available otherwise. In our specific case, this element adds visual information about the position of the face. If we assume that the face detector successfully detects a face, we can use that information to build this artifact.

After experimenting with different VSEs, we propose the following guideline to choose a good VSE. We should choose a VSE that: (1) is big enough to steer the optimization process; (2) does not create additional uncertainty by covering relevant features (e.g., the eyes or nose); (3) scales accordingly to the dimension of the detected face; and (4) completely or partially removes the high-variance areas in the model (e.g., background) with uniform ones.

Fig. 7.5 The effect of a virtual structuring element to the annotation, appearance, and variance (white indicates a larger variance)

In the used VSE, a black frame with width equal to 20% of the size of the detected face is built around the face itself. Besides the regular markers that capture the facial features (see Fig. 7.5 and [10] for details) four new markers are added in the corners to stretch the convex hull of the shape to take in consideration the structuring element. Around each of those four points, a black circle with the radius of one third of the size of the face is added. The resulting annotation, shape, and appearance variance are displayed in Fig. 7.5. Note that in the variance map the initialization variance of the face detector is automatically included in the model (i.e., the thick white border delimitating the VSE).

This virtual structuring element visually passes information between the face detection and the AAM. We show in the experiments that VSE helps the basic AAMs in the model generalization and fitting performances.

Two data sets were used during the evaluation: (1) a part of the Cohn–Kanade [25] data set consisting of 53 male and female subjects, showing neutral frontal faces in a controlled environment; (2) the Unilever data set consisting of 50 females, showing natural poses in an outdoor uncontrolled environment. The idea is to investigate the influence of the VSE when the background is unchanged (Cohn–Kanade) and when more difficult conditions are present (Unilever).

We evaluate two specific annotations, one named 'relevant' (Fig. 7.6(a)) describing the facial features that are relevant for the facial expression classifiers including the face contours that are needed for face tracking, and the other one named 'inside'

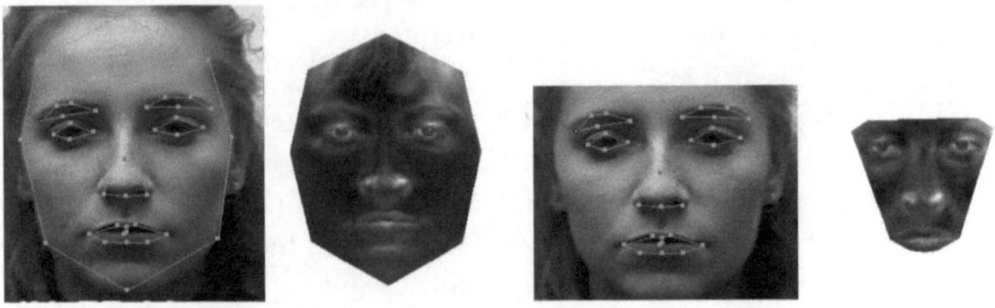

(**a**) Relevant (**b**) Inside

Fig. 7.6 The annotations and their respective variance maps over the data sets

(Fig. 7.6(b)) describing the facial features without the face contours. Note that the 'inside' model is surrounded only by face area (so not by background) so its variance is lower and the model is more robust. To assess the performance of the AAM we initialize the mean model (i.e., the mean shape with the mean appearance) shifted in the Cartesian plane with a predefined amount. This simulates some extremes in the initialization error obtained by the face detector.

The common approach to assess performance of AAM is to compare the results to a ground truth (i.e., the annotations in the training set). The following measures are used: *point to point error* is the Euclidean distance between each point of the true shape and the corresponding fitted shape; *point to curve error* is the Euclidean distance between a fitted shape point and the closest point on the linear spline obtained from the true shape points; and *Mahalanobis distance* defined as

$$D^2 = \sum_{i=1}^{t} \frac{m_i^2}{\lambda_i}, \tag{7.10}$$

where m_i represents the AAM parameters and λ_i their respective principal components.

We perform two types of experiments. In the person-independent case we perform a leave-one-out cross-validation. For the second experiment, the generalized AAM test, we merge the two data sets and we create a model which includes all the different lighting conditions, backgrounds, subject features, and annotations (together with their respective errors). The goal of this experiment is to test whether the generalization problems of AAMs could be solved just by using a greater amount of training data.

Table 7.3 shows the results obtained for the two data sets in the person-independent experiment. Important to notice that the results obtained with Cohn–Kanade data sets are in most of the cases better than the one obtained with the Unilever data set. This has to do with the fact that in the Unilever data set, the effect of the uncontrolled lighting condition and background change is more relevant and the model fitting is more difficult. However, in both cases one can see that the use of VSE significantly improved the results. Another important aspect is that the use of VSE is more effective in the case of Unilever database and this is because the VSE is reducing the background influence to a larger

Table 7.3 Mean and standard error in the person-independent test for the two data sets

	Cohn–Kanade			Unilever		
	Point–Point	Point–Curve	Mahalanobis	Point–Point	Point–Curve	Mahalanobis
Relev.	16.72(5.53)	9.09 (3.36)	47.93 (4.90)	54.84(10.58)	29.82 (6.22)	79.41 (6.66)
Relev. VSE	6.73(0.21)	4.34 (0.15)	26.46 (1.57)	10.14(2.07)	6.53 (1.30)	24.75 (3.57)
Inside	9.53(3.48)	6.19 (2.47)	39.55 (3.66)	25.98(7.29)	17.69 (5.16)	38.20 (4.52)
Inside VSE	5.85(0.24)	3.76 (0.13)	27.14 (1.77)	8.99(1.90)	6.37 (1.46)	23.45 (2.81)

extent. Interesting to note is that, while the use of a VSE does not excessively improve the accuracy of the 'inside' model, the use of VSE on the 'relevant' model drastically improves its accuracy making it even better than the basic 'inside' model. This result is surprising since in the 'relevant' model parts of the markers are covered by the VSE (i.e., the forehead and chin markers) we expected the final model to inherently generate some errors. Instead, it seems that the inner parts of the face might steer the outer markers to the optimal position. This could only mean that there is a proportional relation between the facial contours and the inside features, which is a very interesting and unexpected property.

In the generalized AAM experiment (see Table 7.4), we notice that the results are generally worse when compared with the person-independent results on the 'controlled' Cohn–Kanade data set, but better when compared with the same experiment on the 'uncontrolled' Unilever data set. Also in this case the VSE implementation shows very good improvements over the basic AAM implementation. What is important to note is that the VSE implementation brings the results of the generalized AAM very close to the data set-specific results, improving the generalization of basic AAM.

While the 'relevant VSE' model is better than the normal 'inside' model, the 'inside VSE' is the model of choice to obtain the best overall results on facial features detection. In our specific task, we can use the 'inside VSE' model to obtain the best results but we will additionally need some heuristics to correctly position the other markers which are not included in the model. These missing markers are relevant for robust face tracking and implicit for facial expression classification so their accurate positioning is very important. Since in the case of 'inside VSE' model these markers are not detected explicitly, we indicate the 'relevant VSE' model as the best choice for our purposes.

To better illustrate the effect of using a VSE, Fig. 7.7 shows an example of the difference in the results when using a 'relevant' model and a 'relevant VSE' model. While the first failed to correctly converge, the second result is optimal for inner facial features. Empirically, VSE models showed to always overlap to the correct annotation, avoiding the mistakes generated by unsuccessful alignments like the one in Fig. 7.7(a).

Table 7.4 Mean and standard error for generalized AAM

	Generalized AAM		
	Point–Point	Point–Curve	Mahalanobis
Relevant	21.05(0.79)	8.45 (0.27)	116.22(3.57)
Relevant VSE	8.50(0.20)	5.38 (0.12)	51.11(0.91)
Inside	8.11(0.21)	4.77 (0.10)	85.22(1.98)
Inside VSE	7.22(0.17)	4.65 (0.09)	52.84(0.96)

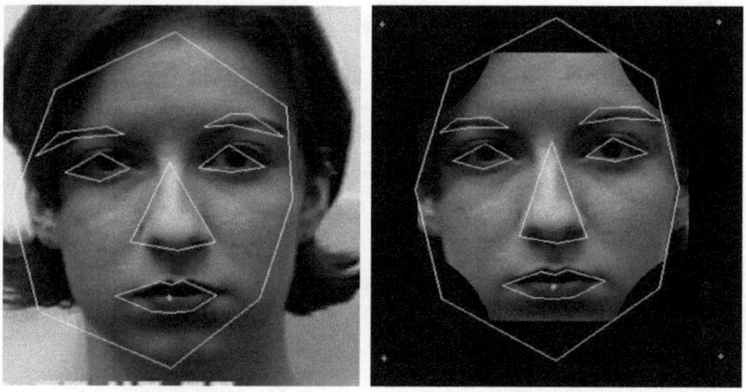

(**a**) Relevant (**b**)Relevant VSE

Fig. 7.7 An example of the difference in the results between a 'relevant' and a 'relevant VSE' model

7.5.3 Facial Expression Recognition Experiments

As mentioned previously, our system uses a generic face model consisting of 16 surface patches embedded in Bézier volumes which is warped to fit the detected facial features. This model is used for tracking the detected facial features. The recovered motions are represented in terms of magnitudes of some predefined motion of the facial features. Each feature motion corresponds to a simple deformation on the face, defined in terms of the Bézier volume control parameters. We refer to these motion vectors as motion units (MUs). Note that they are similar but not equivalent to Ekman's AUs [17], and are numeric in nature, representing not only the activation of a facial region, but also the direction and intensity of the motion. The 12 MUs used in the face tracker are shown in Fig. 7.8. The MUs are used as the features for the Bayesian network classifiers learned with labeled and unlabeled data.

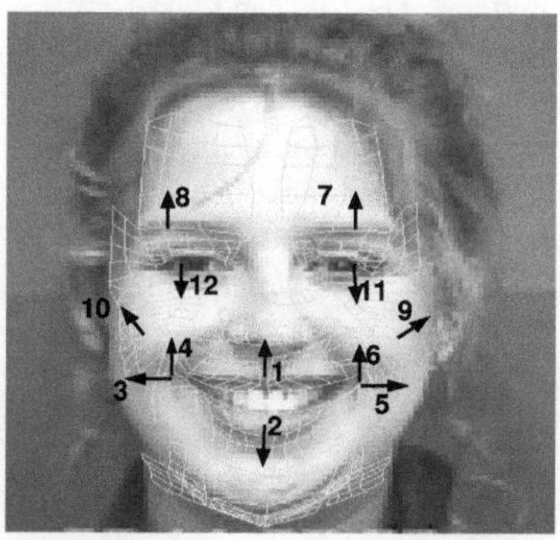

Fig. 7.8 The facial motion measurements

Table 7.5 The experimental setup and the classification results for facial expression recognition with labeled data (L) and labeled + unlabeled data (LUL). Accuracy is shown with the corresponding 95% confidence interval

Dataset	Train		Test	NB-L	NB-LUL	TAN-L	TAN-LUL	SSS-LUL
	# Lab.	# Unlab.						
Chen–Huang	300	11,982	3,555	71.25±0.75%	58.54±0.81%	72.45±0.74%	62.87±0.79%	74.99±0.71%
Cohn–Kanade	200	2,980	1,000	72.50±1.40%	69.10±1.44%	72.90±1.39%	69.30±1.44%	74.80±1.36%

There are seven categories of facial expressions corresponding to *neutral, joy, surprise, anger, disgust, sad,* and *fear*. For testing we use two databases, in which all the data are labeled. We removed the labels of most of the training data and learned the classifiers with the different approaches discussed in Sect. 7.4.

The first database was collected by Chen and Huang [5] and is a database of subjects that were instructed to display facial expressions corresponding to the six types of emotions. All the tests of the algorithms are performed on a set of five people, each one displaying six sequences of each one of the six emotions, starting and ending at the Neutral expression. The video sampling rate was 30 Hz, and a typical emotion sequence is about 70 samples long (~2 s). The second database is the Cohn–Kanade database [25] introduced in the previous section. For each subject there is at most one sequence per expression with an average of eight frames for each expression.

We measure the accuracy with respect to the classification result of each frame, where each frame in the video sequence was manually labeled to one of the expressions (including Neutral). The results are shown in Table 7.5, showing classification accuracy with 95% confidence intervals. We see that the classifier trained with the SSS algorithm improves classification performance to about 75% for both data sets. Model switching from Naive Bayes to TAN does not significantly improve the performance; apparently, the increase in the likelihood of the data does not cause a decrease in the classification error. In both the NB and TAN cases, we see a performance degradation as the unlabeled data are added to the smaller labeled data set (TAN-L and NB-L compared to TAN-LUL and NB-LUL). An interesting fact arises from learning the same classifiers with all the data being labeled (i.e., the original database without removal of any labels). Now, SSS achieves about 83% accuracy, compared to the 75% achieved with the unlabeled data. Had we had more unlabeled data, it might have been possible to achieve similar performance as with the fully labeled database. This result points to the fact that labeled data are more valuable than unlabeled data (see [4] for a detailed analysis).

7.6 Conclusion

In this work we presented a complete system that aims at human–computer interaction applications. We considered several instances of Bayesian networks and we showed that learning the structure of Bayesian networks classifiers enables learning good classifiers with a small labeled set and a large unlabeled set.

Our discussion of semi-supervised learning for Bayesian networks suggests the following path: when faced with the option of learning Bayesian networks with labeled and unlabeled data, start with Naive Bayes and TAN classifiers, learn with only labeled data and test whether the model is correct by learning with the unlabeled data. If the result is not satisfactory, then SSS can be used to attempt to further improve performance with enough computational resources. If none of the methods using the unlabeled data improve performance over the supervised TAN (or Naive Bayes), either discard the unlabeled data or try to label more data, using active learning for example.

In closing, it is possible to view some of the components of this work independent of each other. The theoretical results of Sect. 7.3 do not depend on the choice of probabilistic classifier and can be used as a guide to other classifiers. Structure learning of Bayesian networks is not a topic motivated solely by the use of unlabeled data. The three applications we considered could be solved using classifiers other than Bayesian networks. However, this work should be viewed as a combination of all three components; (1) the theory showing the limitations of unlabeled data is used to motivate (2) the design of algorithms to search for better performing structures of Bayesian networks and finally, (3) the successful applications to an human–computer interaction problem we are interested in solving by learning with labeled and unlabeled data.

Acknowledgements We would like to thank Marcelo Cirelo, Fabio Cozman, Ashutosh Garg, and Thomas Huang for their suggestions, discussions, and critical comments. This work was supported by the Muscle NoE and MIAUCE European projects.

References

1. S. Baluja. Probabilistic modelling for face orientation discrimination: Learning from labeled and unlabeled data. In *Neural Information and Processing Systems*, pages 854—860, 1998.
2. M.J. Black and Y. Yacoob. Tracking and recognizing rigid and non-rigid facial motions using local parametric models of image motion. In *Proceedings of the International Conference Computer Vision*, pages 374–381, 1995.
3. M. Brand. An entropic estimator for structure discovery. In *Neural Information and Processing Systems*, pages 723–729, 1998.
4. V. Castelli. *The relative value of labeled and unlabeled samples in pattern recognition.* PhD thesis, Stanford, 1994.
5. L.S. Chen. *Joint processing of audio-visual information for the recognition of emotional expressions in human–computer interaction.* PhD thesis, University of Illinois at Urbana-Champaign, 2000.
6. J. Cheng, R. Greiner, J. Kelly, D.A. Bell, and W. Liu. Learning Bayesian networks from data:An information-theory based approach. *The Artificial Intelligence Journal*, 137:43–90, 2002.
7. C.K. Chow and C.N. Liu. Approximating discrete probability distribution with dependence trees. *IEEE Transactions on Information Theory*, 14:462–467, 1968.
8. I. Cohen. *Semi-supervised learning of classifiers with application to human computer interaction.* PhD thesis, University of Illinois at Urbana-Champaign, 2003.

9. I. Cohen, F. Cozman, N. Sebe, M. Cirello, and T.S. Huang. Semi-supervised learning of classifiers: Theory, algorithms, and their applications to human–computer interaction. *IEEE Transactions on Pattern Analysis and Machine Intelligence*, 26(12):1553–1567, 2004.
10. I. Cohen, N. Sebe, A. Garg, L. Chen, and T.S. Huang. Facial expression recognition from video sequences: Temporal and static modelling. *Computer Vision and Image Understanding*, 91(1–2):160–187, 2003.
11. A.J. Colmenarez and T.S. Huang. Face detection with information based maximum discrimination. In *IEEE Conference on Computer Vision and Pattern Recogntion*, pages 782–787, 1997.
12. T. Cootes, G. Edwards, and C. Taylor. Active appearance models. *Pattern Analysis and Machine Intelligence*, 23(6):681–685, 2001.
13. T. Cootes and P. Kittipanya-ngam. Comparing variations on the active appearance model algorithm. In *BMVC*, pages 837–846., 2002.
14. T. Cootes, C. Taylor, D. Cooper, and J. Graham. Active shape models – Their training and application. *Computer Vision and Image Understanding*, 61(1):38–59, 1995.
15. A.P. Dempster, N.M. Laird, and D.B. Rubin. Maximum likelihood from incomplete data via the EM algorithm. *Journal of the Royal Statistical Society, Series B*, 39(1):1–38, 1977.
16. J. Dougherty, R. Kohavi, and M. Sahami. Supervised and unsupervised discretization of continuous features. In *International Conference on Machine Learning*, pages 194–202, 1995.
17. P. Ekman and W.V. Friesen. *Facial Action Coding System: Investigator's Guide*. Consulting Psychologists Press, Palo Alto, CA, 1978.
18. I.A. Essa and A.P. Pentland. Coding, analysis, interpretation, and recognition of facial expressions. *IEEE Transactions on Pattern Analysis and Machine Intelligence*, 19(7):757–763, 1997.
19. B. Fasel and J. Luettin. Automatic facial expression analysis: A survey. *Pattern Recognition*, 36:259–275, 2003.
20. N. Friedman. The Bayesian structural EM algorithm. In *Proceedings of the Conference on Uncertainty in Artificial Intelligence*, pages 129–138, 1998.
21. N. Friedman, D. Geiger, and M. Goldszmidt. Bayesian network classifiers. *Machine Learning*, 29(2):131–163, 1997.
22. B. Hajek. Cooling schedules for optimal annealing. *Mathematics of Operational Research*, 13:311–329, 1988.
23. E. Hjelmas and B.K. Low. Face detection: A survey. *Computer Vision and Image Understanding*, 83:236–274, 2003.
24. M. Jones and T. Poggio. Multidimensional morphable models. In *ICCV*, pages 683–688, 1998.
25. T. Kanade, J.F. Cohn, and Y. Tian. Comprehesive database for facial expression analysis. In *International Conference on Automatic Face and Gesture Recognition*, pages 46–53, 2000.
26. M. Kass, A. Witkin, and D. Terzopoulos. Snakes: Active contour models. *International Journal of Computer Vision*, 1(4):321–331, 1987.
27. D. Madigan and J. York. Bayesian graphical models for discrete data. *International Statistical Review*, 63:215–232, 1995.
28. I. Matthews and S. Baker. Active appearance models revisited. *International Journal of Computer Vision*, 60(2):135–164, 2004.
29. A.K. McCallum and K. Nigam. Employing EM in pool-based active learning for text classification. In *International Conference on Machine Learning*, pages 350–358, 1998.
30. N. Metropolis, A.W. Rosenbluth, M.N. Rosenbluth, A.H. Teller, and E. Teller. Equation of state calculation by fast computing machines. *Journal of Chemical Physics*, 21:1087–1092, 1953.
31. K. Nigam, A. McCallum, S. Thrun, and T. Mitchell. Text classification from labeled and unlabeled documents using EM. *Machine Learning*, 39:103–134, 2000.
32. N. Oliver, A. Pentland, and F. Bérard. LAFTER: A real-time face and lips tracker with facial expression recognition. *Pattern Recognition*, 33:1369–1382, 2000.
33. T.J. O'Neill. Normal discrimination with unclassified obseravations. *Journal of the American Statistical Association*, 73(364):821–826, 1978.

34. E. Osuna, R. Freund, and F. Girosi. Training support vector machines: An application to face detection. In *Proceedings of IEEE Conference on Computer Vision and Pattern Recognition*, pages 130–136, 1997.
35. M. Pantic and L.J.M. Rothkrantz. Automatic analysis of facial expressions: The state of the art. *IEEE Transactions on Pattern Analysis and Machine Intelligence*, 22(12):1424–1445, 2000.
36. J. Pearl. *Probabilistic Reasoning in Intelligent Systems: Networks of Plausible Inference*. Morgan Kaufmann, San Francisco, CA, 1988.
37. H. Rowley, S. Baluja, and T. Kanade. Neural network-based face detection. *IEEE Transactions on Pattern Analysis and Machine Intelligence*, 20(1):23–38, 1998.
38. H. Schneiderman. Learning a restricted Bayesian network for object detection. In *CVPR*, pages 639–646, 2004.
39. S. Sclaroff and J. Isidoro. Active blobs. In *ICCV*, 1998.
40. N. Sebe, I. Cohen, F.G. Cozman, and T.S. Huang. Learning probabilistic classifiers for human–computer interaction applications. *ACM Multimedia Systems*, 10(6):484–498, 2005.
41. B. Shahshahani and D. Landgrebe. Effect of unlabeled samples in reducing the small sample size problem and mitigating the Hughes phenomenon. *IEEE Transactions on Geoscience and Remote Sensing*, 32(5):1087–1095, 1994.
42. H. Tao and T.S. Huang. Connected vibrations: A modal analysis approach to non-rigid motion tracking. In *IEEE Conference on Computer Vision and Pattern Recognition*, pages 735–740, 1998.
43. P. Viola and M.J. Jones. Robust real-time object detection. *International Journal of Computer Vision*, 57(2), 2004.
44. R.R. Wang, T.S. Huang, and J. Zhong. Generative and discriminative face modeling for detection. In *Automatic Face and Gesture recognition*, 2002.
45. H. White. Maximum likelihood estimation of misspecified models. *Econometrica*, 50(1): 1–25, 1982.
46. M.-H. Yang, D. Kriegman, and N. Ahuja. Detecting faces in images: A survey. *IEEE Transactions on Pattern Analysis and Machine Intelligence*, 24(1):34–58, 2002.
47. M.-H. Yang, D. Roth, and N. Ahuja. SNoW based face detector. In *Neural Information Processing Systems*, pages 855–861, 2000.
48. T. Zhang and F. Oles. A probability analysis on the value of unlabeled data for classification problems. In *International Conference on Machine Learning*, 2000.

Chapter 8
Mental Search in Image Databases: Implicit Versus Explicit Content Query

Simon P. Wilson, Julien Fauqueur, and Nozha Boujemaa

Abstract In comparison with the classic query-by-example paradigm, the "mental image search" paradigm lifts the strong assumption that the user has a relevant example at hand to start the search. In this chapter, we review different methods that implement this paradigm, originating from both the content-based image retrieval and the object recognition fields. In particular, we present two complementary methods. The first one allows the user to reach the target mental image by relevance feedback, using a Bayesian inference. The second one lets the user specify the mental image visual composition from an automatically generated visual thesaurus of segmented regions. In this scenario, the user formulates the query with an *explicit* representation of the image content, as opposed to the first scenario which accommodates an *implicit* representation. In terms of usage, we will show that the second approach is particularly suitable when the mental image has a well-defined visual composition. On the other hand, the Bayesian approach can handle more "semantic" queries, such as emotions for which the visual characterization is more implicit.

8.1 Introduction

Mental image search is a paradigm for content-based image retrieval (CBIR) in which an image is searched for in a database without the benefit of an example image to give to the system. The user only has a "mental image" in his/her head. In this chapter we discuss mental image search, how it relates to other types of image search and describe two different but complementary approaches to it.

Simon P. Wilson
Trinity College Dublin, Dublin, Ireland, e-mail: swilson@tcd.ie

Julien Fauqueur
University of Cambridge, Cambridge, UK, e-mail: jf330@cam.ac.uk

Nozha Boujemaa
Projet IMEDIA, INRIA, Le Chesnav Cedex, France, e-mail: Nozha.Boujemaa@inria.fr

The chapter is structured as follows. In Sect. 8.2, we describe some of the different CBIR paradigms and discuss how they differ from mental image search. Section 8.3 develops a Bayesian learning approach to mental image search that is illustrated with an example of searching for images by emotional content. Section 8.4 develops a more visually explicit approach called the visual thesaurus. A short conclusion compares and contrasts the two approaches.

8.2 "Mental Image Search" Versus Other Search Paradigms

When searching for images, the range and variety of visual queries can be vast. On one side, we have the user who may express an extremely wide range of visual queries relating to visual appearance, objects or emotions. On the other side, we have an image database which may provide an enormous amount of visual information (e.g. a heterogeneous database of thousands or millions of images).

The option of keyword search can be very efficient to address specific semantic queries, but its great weakness is that there must be an a priori text annotation of the images. As well as the tedious task of image annotation, annotations limit significantly the information the user can access, because all the visual latent information which is contained in images but not annotated is lost.

It is intuitive to search text documents with keywords, and by extension it is intuitive to search image documents with visual data. One of the difficulties of CBIR/visual search is finding the means to facilitate the user to express his/her needs visually. The usual way to handle visual queries is through the query-by-example paradigm in which the user provides a starting example image and the engine retrieves all images which have a similar visual content. Providing an image sketch is a related idea (called query by sketch) that inputs to the system visual information on the query.

Keyword search and image example are the two common ways in which a user can provide a system with information on the image being searched for. The assumption of the mental image search paradigm is that neither of these is present; alternative means of inputting information about the query image are needed. Some of the alternatives are discussed in [2], which defined the term mental image search. The two extensive survey reports [7, 10] introduced the idea of "target search" for mental image search; there is a target image in the database, for which the user has a mental image, and the objective is to find it by providing the search system with information about it. The idea of a target image is useful in what we describe.

The two approaches discussed here have several differences. The first uses the idea of relevance feedback; there is an interaction between the user and the system that allows the system to refine its search for the target. The form of this interaction is that the system displays images from the database and the user selects one that is most relevant to the query. This interaction is treated as data that can be used by a Bayesian inference algorithm to gain information about the query. The second has no such interaction but allows the user to formulate the visual content of the image from a large set of image pieces that have been extracted from the database beforehand.

We view the most important conceptual difference between these two algorithms to be how they model the visual content of the mental image. For the Bayesian relevance feedback approach, the mental image is *implicitly* defined through images that exist in the database; the user never attempts to express the mental image directly. The thesaurus approach, on the other hand, *explicitly* describes the visual content of the mental image because the user is forced to construct it.

The notion of the mental image is more of a concept for the Bayesian approach while it concerns low-level visual composition for the thesaurus.

8.3 Implicit Content Query: Mental Image Search Using Bayesian Inference

In this section we describe one approach to mental image search: a Bayesian learning method that starts from an initial random set of images from the database and uses relevance feedback information to learn about what the user is looking for. The work described builds on the relevance feedback mechanism of the PicHunter Bayesian retrieval system of Cox et al. [3]. We extend this scenario in mental image search to where a user is looking for several images from a semantic class; in this section we take examples from a database of art paintings. We call this mental image category search.

One significance of this work is that it learns about aspects of the retrieval process; Bayesian statistical methods are suited to this. The question then arises: can this information from previous runs of the system aid the user, by reducing the length of the search process for subsequent similar queries? This is tested on queries with highly subjective meaning, in this case searches for categories of images in a database of paintings with different emotional content.

8.3.1 Bayesian Inference for CBIR

Our approach is an extension of PicHunter [3], although there are some other examples of Bayesian CBIR: for example [8] uses a query-shifting approach to relevance feedback, motivated by Bayesian decision theory, while [11] attempts to infer a distribution of the features of the class of relevant images using Bayesian inference.

We consider a database of K images. We assume that the mental image is represented by a specific "target" image T. Our goal is to infer T and this is accomplished by the usual iterative relevance feedback process; to start, a set of K_D randomly selected images from the database are displayed, from which the user picks the most relevant. The system uses this information to select another image set, from which the user picks one again, and so on. We define D_i to be the set of displayed images at the ith iteration of this process and $A_i \in D_i$ to be the image picked. We define $H_t = \{D_1, A_1, D_2, A_2, \ldots, D_t, A_t\}$ to be the history of displayed images and user actions up to the tth iteration.

The learning algorithm is based around the model for the probability of which image a user picks from D_i given the target image T:

$$\mathbb{P}(A_i \mid D_i, T, \sigma) = \frac{\exp\left(-d(A_i, T)/\sigma\right)}{\sum_{A \in D_i} \exp\left(-d(A, T)/\sigma\right)}, A_i \in D_i, \tag{8.1}$$

where σ is a precision parameter and d is Euclidean distance measure in the set of normalized image features. This form for $\mathbb{P}$ comes from [3] although others are possible; the key is that the probability of A_i decreases as $d(A_i, T)$ increases. The parameter σ is a scale factor and is a measure of the validity of the feature space for the query at hand. A small value of σ implies that images in D_i that are closest to T will be picked with high probability. That is to say, the feature space is able to describe the user's search well; close images in feature space correspond to closeness in the query space. On the other hand, a large value of σ implies that each image in D_i is picked with a more or less equal probability, regardless of its distance to T, so the feature space does not describe well the relevance of one image to another.

We generalize this model by partitioning the feature vector into three sub-vectors that represent global colour, texture and segmentation features, and define $F \in \{\text{GC, TX, SG}\}$ to be the sub-vector used in the relevance feedback process. This allows a simple form of feature weighting and is similar to probabilistic feature relevance techniques; see [9]. The probability of selecting A_i from D_i is now considered as a function of T, σ and F:

$$\mathbb{P}(A_i \mid D_i, T, \sigma, F) = \frac{\exp\left(-d_F(A_i, T)/\sigma\right)}{\sum_{A \in D_i} \exp\left(-d_F(A, T)/\sigma\right)}, A_i \in D_i, \tag{8.2}$$

where d_F is the distance measure over the sub-space spanned by the feature sub-vector F.

Bayesian learning quantifies what is known about (T, σ, F) through the posterior distribution given H_t. By Bayes law and assuming independence of each iteration of the relevance feedback

$$\mathbb{P}(T, \sigma, F \mid H_t) \propto \left(\prod_{i=1}^{t} \mathbb{P}(A_i \mid D_i, T, \sigma, F)\right) \mathbb{P}(T, \sigma, F), \tag{8.3}$$

where $\mathbb{P}(T, \sigma, F)$ is the prior distribution that we assume is independent and uniform: $\mathbb{P}(T) = K^{-1}$, $\mathbb{P}(\sigma) = \sigma_{\max}^{-1}, 0 \leq \sigma \leq \sigma_{\max}$ and $\mathbb{P}(F) = 1/3, F \in \{\text{GC,TX,SG}\}$. Other choices, for example for T to be based on information from text annotation or sketches, are possible.

Of immediate interest is the marginal posterior distribution of T:

$$\mathbb{P}(T \mid H_t) = \int_0^{\sigma_{\max}} \sum_{F \in \{\mathrm{GC,TX,SG}\}} \mathbb{P}(T_i, \sigma, F \mid H_t)\, \mathrm{d}\sigma. \tag{8.4}$$

This posterior distribution of T represents the system's state of knowledge about relevant images and has taken account of what we have learned about the performance of the feature space – through σ – and which sub-vectors are most important for the relevance feedback through F. Once $\mathbb{P}(T \mid H_t)$ is computed then D_{t+1}, taken to be the N_D most probable images from $\mathbb{P}(T \mid H_t)$, is displayed, although other strategies are possible. This process repeats until the user encounters a satisfactory image.

One important type of mental image search is the search for images that invoke different emotional reactions. To illustrate mental image search we apply the method of Sect. 8.3.1 to such a search. We applied this system to mental image search on a database of 1066 paintings supplied to us by the Bridgeman Art Library, London. This is a challenging database for a CBIR system, containing a very wide variety of content.

Table 8.1 shows the results from 30 independent searches in three categories: romantic, violent and sad paintings, along with the number of iterations of the relevance feedback required to achieve the result. Uniform prior distributions and a randomly selected initial display set were used and nine images used per display set. The results shown are those images that the user first deemed to represent such images from the set being then displayed.

8.3.2 Mental Image Category Search

For category search, where the goal is to build up a set of images, the system is repeatedly used from random initial display sets. The question is how to use information from previous searches to aid the next search. Because the images that were retrieved are highly dependent on the initial display set, information on T in the posterior distribution may not be a good representation of all images in the database that are relevant to the query, and its use may lead to images that have already been discovered. However, σ and F should be more robust to the initial display set, since they model properties of the feature space and user search strategy.

This observation may be modelled by letting the prior for T in the current search be independent of data from previous searches, while the prior for σ and F is dependent. Specifically, suppose that M successful searches for particular content have taken place in the past. Data from unsuccessful searches are ignored. Let

$$H_{t_m}(m) = \{D_1(m), A_1(m), \ldots, D_{t_m}(m), A_{t_m}(m)\}$$

Table 8.1 Examples of the result of running the Bayesian CBIR system on the Bridgeman Art Library database, where the search was being conducted for romantic, violent and sad images, respectively. The images as well as the number of iterations conducted to find each image are recorded. The mean and standard deviation of the number of iterations over 10 examples per query are also shown

Romantic		Violent		Sad	
Images	Iterations	Images	Iterations	Images	Iterations
	2		5		5
	6		4		7
	16		2		12
	4		10		12
	3		4		3
Mean	7.4	Mean	5.9	Mean	6.6
Std. Dev.	6.5	Std. Dev.	3.5	Std. Dev.	3.1

be the data from the mth search, and let $\mathscr{H}_M = \{H_{t_1}(1), \ldots, H_{t_M}(M)\}$ denote data from all M previous searches. For the current query, the prior distribution for (T, σ, F) is

$$\mathbb{P}(T, \sigma, F \mid \mathscr{H}_M) = \mathbb{P}(T)\mathbb{P}(\sigma, F \mid T, \mathscr{H}_M),$$

where, by Bayes' law,

$$\mathbb{P}(\sigma, F \mid T, \mathscr{H}_M) \propto \left[\prod_{m=1}^{M} \left(\prod_{k=1}^{tm} \mathbb{P}(A_k(m) \mid D_k(m), T, \sigma, F) \right) \right] \mathbb{P}(\sigma, F). \tag{8.5}$$

At iteration t of the current search, having observed data H_t, the posterior distribution of (T, σ, F) is then

$$\mathbb{P}(T,\sigma,F\,|\,\mathscr{H}_M,H_t) \propto \left(\prod_{k=1}^{t}\mathbb{P}(A_k\,|\,D_k,T,\sigma,F)\right)\mathbb{P}(T)\,\mathbb{P}(\sigma,F\,|\,T,\mathscr{H}_M). \quad (8.6)$$

The next display set is chosen as before to be the N_D most probable images from $\mathbb{P}(T\,|\,\mathscr{H}_M,H_t)$. This is repeated until a relevant image is found. The data from this latest search are then added to $\mathscr{H}_M$ to form $\mathscr{H}_{M+1}$ and another search may be done using the prior $\mathbb{P}(T)\,\mathbb{P}(\sigma,F\,|\,T,\mathscr{H}_{M+1})$.

8.3.3 Evaluation

The aim of this section is to test whether there is information in the posterior distribution, from previous searches for the same emotional content, that shortens the retrieval process.

The experiment consisted of two stages, called A and B. Stage A repeats the example of Table 8.1. The system is used to search for the three different queries used in Table 8.1 – romantic, violent and sad images. The user is asked to conduct 10 successful searches for each query, with each search using the initial uniform prior on (T,σ,F). The number of iterations taken to find a relevant image is recorded.

Stage B consists of repeating the procedure of Stage A but using as the prior for (σ,F) the posterior distribution from the initial ten searches in that class, examples of which are seen in Table 8.1, i.e. $\mathbb{P}(T)\,\mathbb{P}(\sigma,F\,|\,\mathscr{H}_{10})$. The number of iterations taken by the user until a relevant image was found was again recorded. Several months lay between the searches of Stage A and Stage B, to try to eliminate any effect of user experience in Stage A affecting the results of Stage B. Of interest is whether the searches in Stage B, based on the information from Stage A, are completed any quicker. We show the results of Stage B in Table 8.2.

The first observation from these tables is that a variety of images have been obtained for each query, and sometimes the same image is re-discovered. However, the set of recovered images forms a good representation of the semantic class that are spread out over the feature space. For the three queries, the mean number of iterations to find an image in Stage B was less than in Stage A; this difference was statistically significant in the case of romantic queries.

Finally, we also look at the posterior distributions of F and σ for each query given Stage A results and given both Stages A and B, e.g. $\mathbb{P}(F,\sigma\,|\,\mathscr{H}_{10})$ and $\mathbb{P}(F,\sigma\,|\,\mathscr{H}_{20})$. Figure 8.1 shows the results for romantic queries on the left and sad queries on the right. Each pair of posterior distributions place probability mass over the same values of σ and F, accounting for the fact that the posterior given Stage A data is flatter as it is conditional on fewer data. This shows that there is some consistency in the values of F and σ for these queries. For romantic queries, large weight was placed on texture features while for the sad queries the largest posterior weight was assigned to segmentation features.

Table 8.2 Some results for Stage B of the experiment for user 2, given that Stage A is the result of Table 8.1. The images as well as the number of iterations conducted to find each image are recorded. The mean and standard deviation of the number of iterations over 10 examples per query are also shown

User 2 – Stage B					
Romantic		Violent		Sad	
Images	Iterations	Images	Iterations	Images	Iterations
	2		2		2
	6		3		9
	3		4		5
	2		5		6
	2		11		2
Mean Std. Dev.	4.1 2.3	Mean Std. Dev.	4.9 3.2	Mean Std. Dev.	4.7 2.3

8.3.4 Remarks

In this section we have demonstrated a Bayesian approach to mental image search that starts from a display of a random subset of images of a database. The approach then updates the display in light of relevance feedback on the mental image from the user. If the same query arises again, for example in a category search, the system tries to use information gained from previous runs to improve the retrieval process in subsequent runs. The approach takes account of the observation that direct

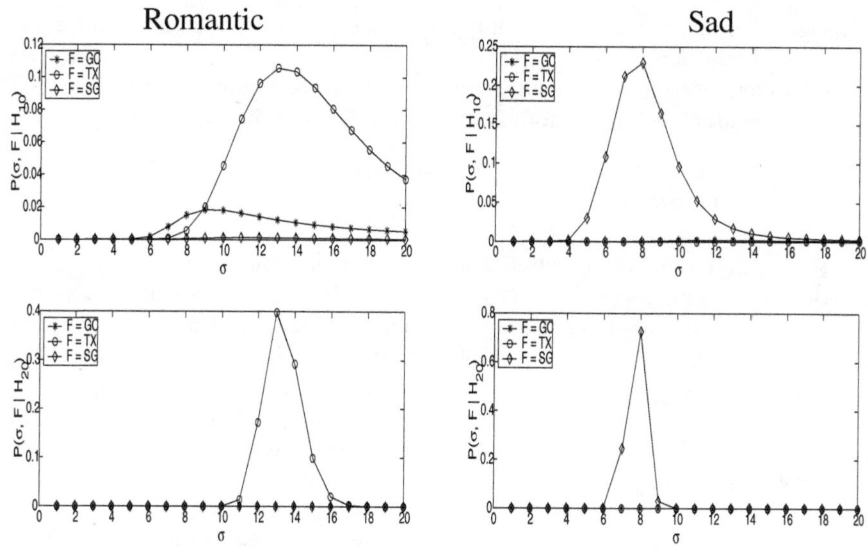

Fig. 8.1 The posterior distribution of F and σ given the data for searches for romantic images (left) and sad images (right). At the top is the posterior given data from Stage A and the bottom is the posterior given data from Stages A and B

information about which images are relevant from previous searches is not appropriate as it eliminates the possibility of finding new clusters of relevant images.

The tentative conclusion that we draw is that there is evidence of some decrease in the number of iterations to retrieve similar images when we use information from previous similar searches. However, this is not true for all the queries that we looked at and is based on a small sample. The information used from one run to another – the posterior distribution of σ and F – is a function of the query and should also be of the user, but crucially appears to be consistent for a particular user and query, as evidenced by the consistency of the posterior distribution before and after Stage B in Fig. 8.1.

8.4 Explicit Content Query: Mental Image Search by Visual Composition Formulation

With the previous Bayesian approach, the user just has to say whether the displayed images are relevant (or not?) to the target image and the system reweighs the feature vector accordingly to refine the search. This approach is suitable for highly semantic queries (as illustrated previously with the "emotion scenarios").

However, this approach requires some interaction from the user which may become tedious after a few iterations, especially when starting from a random selection of images. For more visually explicit types of searches, it is possible to reduce this amount of interaction as shown here.

In this section we present the "Boolean composition of region categories" method by which the user can directly formulate the visual content he/she is looking for.

8.4.1 System Summary

The idea relies on the unsupervised generation of a summary of the visual components of the images present in the database. It is a thesaurus of segmented regions. By interacting with the thesaurus, the user directly selects which types of regions should be present or absent in the mental image. In the same way as one searches text documents using keywords, the user can search images by specifying the image region content using a Boolean query. The indexing scheme relies on inverted tables, which efficiently support Boolean queries and allow for instantaneous search time. The idea is, in the absence of a starting example, to provide a set of initial relevant images (possibly followed by a query-by-example search step).

We will now describe the creation process of the visual thesaurus.

8.4.2 Visual Thesaurus Construction

Region detection. The first step is the detection of coarse salient regions in each image of the database. This is achieved automatically by an unsupervised segmentation algorithm (see [4]) based on the clustering of a descriptor which captures the local colour diversity.

Region description. For each detected region, we characterize its content by a visual descriptor. Here we chose a simple average colour descriptor for a proof of concept of the approach, but other region descriptors may also be used.

Region grouping into categories. Then regions are grouped into categories by unsupervised clustering of their descriptors. The goal is to produce a summary of all present regions in the database by grouping the visually similar ones. The number of clusters should depend on the heterogeneity of the region appearance in the database. We use the competitive agglomeration (CA) clustering algorithm [6], which is a version of fuzzy C-means [1] which automatically estimates the optimal number of clusters given a clustering granularity. The fuzzy nature of this technique accommodates the clustering ambiguities arising from the distribution of the natural image descriptors. We use the Mahalanobis distance to capture ellipsoidal clusters, which are more general than the spherical ones when using the Euclidean distance. After convergence, the CA algorithm produces P clusters, the region–cluster associations and P prototypes $\{p_1, ..., p_P\}$ associated with each cluster (a prototype corresponds to the average descriptor value for the cluster). In the region description space, we define the closest region to each cluster prototype as a "representative region" r_i. The P clusters of visually similar regions as referred to as the "region categories" and are labelled $\{C_1, ..., C_P\}$. Figure 8.2 shows an example of a category which contains similar green regions.

Fig. 8.2 Example of a region category: this one contains mostly green regions. Note that those categories can be made homogeneous according to other visual attributes by using other descriptors (e.g. texture, shape)

Definition of the Visual Thesaurus. The set $\{r_1, ..., r_P\}$ defines the region photometric thesaurus. It can be seen as a visualization of the region clustering outcome to show the user the visual components which are present in the database. The user interacts with the thesaurus to specify the "types of regions" which are present or absent in the target image. More details about the construction of this thesaurus can be found in [5]. Figure 8.4 shows an example of thesaurus for the Corel database we used.

8.4.3 Symbolic Indexing, Boolean Search and Range Query Mechanism

In this section, we explain how region category labels are used for indexing the region composition in images, in a similar fashion to text retrieval.

Image indexing. The indexing scheme is based on the region category labels only. In the same way as one browses an index page in a book to find a page with a specific keyword, our search method relies on the use of inverted tables to find images with a particular type of region. Borrowed from text retrieval [12], inverted tables are a powerful and simple way to index documents and search them using Boolean queries. The first indexing table, $IC(C)$, is a mapping from $\{C_1, ..., C_P\}$ to $\{1, ..., K\}$, where K is the number of images in the database. Given a selected region category C in the thesaurus, this table gives direct access to the set $IC(C)$ of relevant images. Then we create a second table $CI(I)$ as the reverse association between images and categories.

Range query mechanism. Regions which belong to the same category are considered visually similar in our indexing scheme. When the user selects a category, his/her idea of the type of region may be more or less precise. In order to allow him/her to expand the similarity search in the region description space, we implement a *range query mechanism* which relies on the neighbour categories. At the indexing stage, we define a *neighbour category* of a category C_q with prototype p_q as a category C_j whose prototype p_j satisfies $d(C_q,C_j) = \| p_q - p_j \|_{L^2} \leq \gamma$, for a given range radius threshold γ. We call $N^\gamma(C_q)$ the set of neighbour categories of a category C_q and define it as

$$N^\gamma(C) = \{C' \in N(C) \mid d(C,C') \leq \gamma\}. \tag{8.7}$$

By convention, we decide that a category C belongs to $N^\gamma(C)$ as a neighbour of itself at distance zero. The range radius γ is selected by the user at retrieval phase depending on the required precision of search: for a given category A, increasing γ will increase the number of categories that are integrated in the search. Note that neighbour categories are defined from distances between category prototypes, so their integration in the search process defines a range query mechanism in the description space.

In addition to $IC(C)$ and $CI(I)$, we define a third indexing table $N(C)$ in order to implement this range query mechanism. For each category C_q, the ordered set $N(C_q)$ contains the list of distances to all categories sorted by increasing order of distance values, i.e. $N(C_q) = ((C_j, d(C_q,C_j)), \forall j = 1,...,P)$ with $d(C_q,C_j) \leq d(C_q,C_{j+1})$, where $d(C_q,C_j) = \| p_q - p_j \|_{L^2}$. Figure 8.3 illustrates an example of this table for the Corel database for which we have 9,995 images and 91 categories.

As a summary, the indexing scheme relies on the three indexing tables $N(C)$, $CI(I)$ and $IC(C)$ which provide associations between images and categories and between categories and their neighbour categories.

Processing a Boolean query. By selecting types of regions which should be present or absent in the mental image, the user can formulate Boolean queries on the image content such as "Find images composed of a region of this type and this type but with no region of this type".[1] More formally, we suppose the user selects in the thesaurus a set $\{C_{pq_1},...,C_{pq_M}\}$ of types of regions which should be present and a set $\{C_{nq_1},...,C_{nq_R}\}$ of types of regions which should be absent. We also let the user select the range radius γ (see above) to specify the search precision.

For each given query category C (each $\{C_{pq_i}\}$ and each $\{C_{nq_j}\}$), the first Boolean operation consists in retrieving all images which have a region in C or any of its neighbour category $N^\gamma(C)$. This image set is given by the expression: $S_N^\gamma(C) = \bigcup_{C' \in N^\gamma(C)} IC(C')$. Then the set S_Q of images which have a region in C_{pq_1} or its neighbours *and* ... a region in C_{pq_M} or its neighbours is: $S_Q = \bigcap_{i=1}^{M} S_N^\gamma(C_{pq_i})$.

categories	neighbour category and prototype distances
0	(0,0.00) (12,11.13) ... (9,164.47)
:	
90	(90,0.00) (89,15.06) ... (3,153.87)

Fig. 8.3 Table $N(C)$ associates neighbour categories to categories. Each row corresponds to one category. For each category its corresponding neighbour categories are stored as pairs (category label, prototype distance to the category)

[1] A "type of region" refers to a visual patch in the user's mind and, in our framework, corresponds to a region category or to a set of neighbour (i.e. similar) region categories.

For the specification of absence of regions we introduce the set S_{NQ} of images which have a region in C_{nq_1} or its neighbours *and* ... a region in C_{nq_R} or its neighbours: $S_{NQ} = \bigcap_{i=1}^{R} S_N^{\gamma}(C_{nq_i})$.

The final set S_{result} of relevant images, i.e. which have regions in the different PQCs and which do not have regions in the NQCs, is expressed as the set subtraction of S_Q and S_{NQ}: $S_{\text{result}} = S_Q \setminus S_{NQ}$. We invite the reader to refer to [5] for a description of the work flow to determine S_{result}.

We draw the reader's attention to the fact that the retrieval process involves no distance computation. It is simply based on accesses to the three indexing tables.

8.4.4 Results

In this section, we present an example of use of this system on a generic Corel photostock database with 9,995 images which contain landscapes, portraits, objects, flowers, cars, animals, kitchens, food, etc.

After image segmentation 50,220 regions are automatically extracted from the 9,995 images. Clustering the 50,220 region mean colour descriptors produces 91 categories with the CA algorithm. Category populations range from 112 regions to 2,048 regions. Figure 8.2 illustrates ones of these categories. As expected, CA generates categories which are homogeneous with respect to the region mean colour descriptor. The perceptual difference between regions within a category is due to regions which have similar mean colour with different textures. Using additional region descriptors would guarantee a stronger visual homogeneity within categories.

Figure 8.4 shows the visual thesaurus that was automatically generated for this database. It is composed of the 91 category region representatives which provide an overview of the available types of regions in the Corel database.

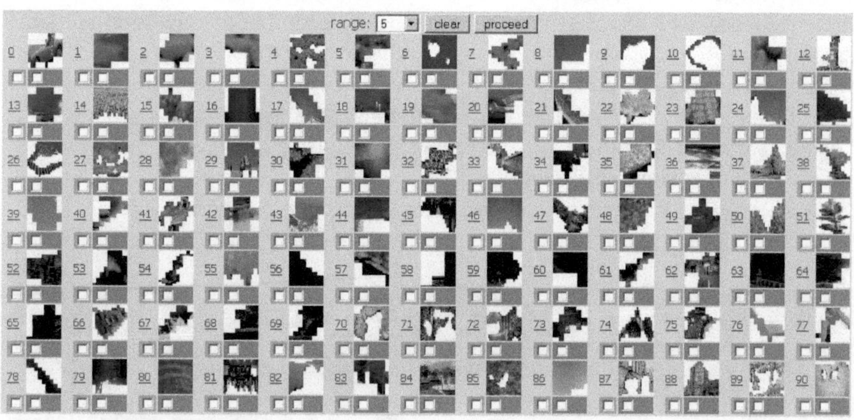

Fig. 8.4 Query interface for the Corel database: 91 region categories constitute the visual thesaurus. Each region category can be clicked to specify a type of region which should be absent or present in the query

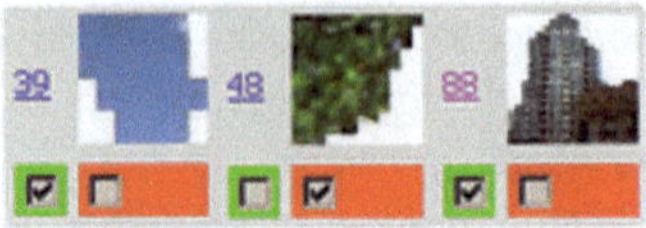

Fig. 8.5 Query formulated by the user to find cityscapes ("images with a blue region and a grey region and no green region")

The thesaurus is the query interface of the system. By clicking red and green boxes, the user can specify the corresponding types of regions that should be present or absent in the mental image. The user can set the appropriate range radius value in the "range" box (see Fig. 8.4).

Let us now consider a query composition scenario. To find cityscapes in a photo agency context, the user may want to search images with a building, some sky and no vegetation. In the region photometric thesaurus this query can be expressed by the following composition: "grey region *and* blue region *and no* green region" (see Fig. 8.5). Given the range value, the system determines the possible neighbours of each query category (grey, blue, green) and translates the query into a Boolean composition query (Fig. 8.6).

Figure 8.7 shows the set of relevant images retrieved for this query. Note that retrieved images are displayed in *random order*. From a visual point of view all retrieved images are relevant to the Boolean query since they do contain a grey and a blue region and no green region. From the semantic point of view retrieved images contain not only many cityscapes but also ruins and monuments. False positives correspond to scenes which match the visual composition but are semantically irrelevant such as a painting or seascapes. Such false positives can be easily rejected by using extra features such as texture and spatial information. This mental search is successful since it filtered out the database to provide a starting set of cityscape images. If a more precise search on cityscapes is needed the retrieved images constitute a satisfactory starting point for a query-by-example or relevance feedback search.

Figure 8.8 shows the images which were rejected for this query due to the presence of a green region (in addition to a blue and grey region). It is interesting to observe that almost all these rejected images depict natural landscapes which are semantically opposite to the query for "cityscapes".

Another search scenario on a TV news broadcast is illustrated in [5].

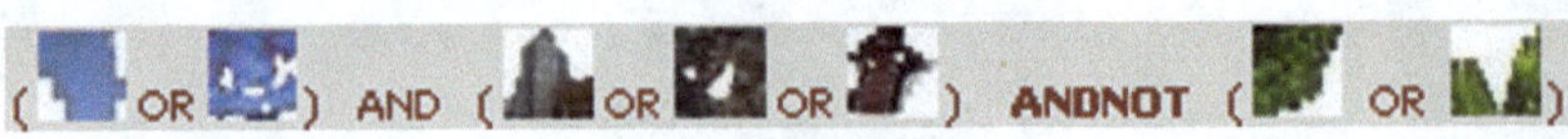

Fig. 8.6 Query expanded by the system including the neighbour categories

Fig. 8.7 Result for the above "cityscapes" query. Retrieved images do satisfy the composition constraint. Semantically, many of those images correspond to cityscapes. If the user needs to find more of those cityscape images, he/she can select one of them to perform a further query-by-example search

Fig. 8.8 For information, the system also displays the rejected images which also contain a green region. Interestingly, those rejected image correspond to "landscapes"

8.4.5 Summary

In this section, we have presented a second approach which implements the mental image search paradigm. The primary goal here is to retrieve quickly a first set of relevant images which have a visual composition of regions specified by the user, in the absence of a starting example. Although sophisticated queries can be expressed, very little interaction is required from the user. Result images can then be used as starting examples in a classic QbE scenario to retrieve more specific images from the database.

For more details about this approach, we invite the reader to refer to [5].

8.5 Conclusions

One obvious question is: can the two methods be combined? The answer is yes, but we feel that this rather misses the point about the two approaches. We could for example use the visual thesaurus to define a prior distribution on the target image, based on a measure of distance of an image from the query specified in the thesaurus.

This would replace the current uniform prior for image in the Bayesian relevance feedback algorithm. Then the relevance feedback would begin, first by selecting images from the database randomly according to the new prior and then updating by Bayes law.

The strength of the visual thesaurus approach is that it does not require extensive (and time-consuming) feedback, something that is lost in a joint algorithm.

References

1. J. C. Bezdek. *Pattern Recognition with Fuzzy Objective Functions*. Plenum, New York, 1981.
2. N. Boujemaa, J. Fauqueur, and V. Gouet. What's beyond query by example? In L. Shapiro, H. P. Kriegel, R. Veltkamp (ed.), *Trends and Advances in Content-Based Image and Video Retrieval*. LNCS, Springer Verlag, New York, 2004.
3. I. J. Cox, M. L. Miller, T. P. Minka, T. V. Papathomas, and P. N. Yianilos. The Bayesian image retrieval system, PicHunter: Theory, implementation and psychophysical experiments. *IEEE Transactions on Image Processing*, 9:20–37, 2000.
4. J. Fauqueur and N. Boujemaa. Region-based image retrieval: Fast coarse segmentation and fine color description. *Journal of Visual Languages and Computing (JVLC), Special Issue on Visual Information Systems*, 15(1):69–95, 2004.
5. J. Fauqueur and N. Boujemaa. Mental image search by Boolean composition of region categories. *Multimedia Tools and Applications*, 31(1):95–117, 2006.
6. H. Frigui and R. Krishnapuram. Clustering by competitive agglomeration. *Pattern Recognition*, 30(7):1109–1119, 1997.
7. Th. Gevers and A.W.M. Smeulders. Content-based image retrieval: An overview. In G. Medioni and S. B. Kang (Eds.), Emerging Topics in Computer Vision. Prentice Hall, New Jersey, pages 333–384, 2004.
8. G. Giacinto and F. Roli. Bayesian relevance feedback for content-based image retrieval. *Pattern Recognition*, 37:1499–1508, 2004.
9. J. Peng, B. Bhanu, and S. Qing. Probabilistic feature relevance learning for content-based image retrieval. *Computer Vision and Image Understanding*, 65:150–164, 1999.
10. A Smeulders, M. Worring, S. Santini, A. Gupta, and R. Jain. Content based image retrieval at the end of the early years. *IEEE Transactions on Pattern Analysis and Machine Intelligence (PAMI)*, 22(12):1349–1380, 2000.
11. J.-L. Tao and Y.-P. Hung. A Bayesian method for content-based image retrieval by use of relevance feedback. In S.-K. Chang, Z. Chen, and S.-Y. Lee, editors, *VISUAL 2002, Lecture Notes in Computer Science*, volume 2314, pages 76–87. Springer, Berlin, 2002.
12. I. H. Witten, A. Moffat, and T. C. Bell. *Managing Gigabytes: Compressing and Indexing Documents and Images*. Van Nostrand Reinhold, New York, 1994.

Chapter 9
Combining Textual and Visual Information for Semantic Labeling of Images and Videos*

Pınar Duygulu, Muhammet Baştan, and Derya Ozkan

Abstract Semantic labeling of large volumes of image and video archives is difficult, if not impossible, with the traditional methods due to the huge amount of human effort required for manual labeling used in a supervised setting. Recently, semi-supervised techniques which make use of annotated image and video collections are proposed as an alternative to reduce the human effort. In this direction, different techniques, which are mostly adapted from information retrieval literature, are applied to learn the unknown one-to-one associations between visual structures and semantic descriptions. When the links are learned, the range of application areas is wide including better retrieval and automatic annotation of images and videos, labeling of image regions as a way of large-scale object recognition and association of names with faces as a way of large-scale face recognition. In this chapter, after reviewing and discussing a variety of related studies, we present two methods in detail, namely, the so called "translation approach" which translates the visual structures to semantic descriptors using the idea of statistical machine translation techniques, and another approach which finds the densest component of a graph corresponding to the largest group of similar visual structures associated with a semantic description.

Pınar Duygulu
Bilkent University, Ankara, Turkey, e-mail: `duygulu@cs.bilkent.edu.tr`

Muhammet Baştan
Bilkent University, Ankara, Turkey, e-mail: `bastan@cs.bilkent.edu.tr`

Derya Ozkan
Bilkent University, Ankara, Turkey, e-mail: `deryao@cs.bilkent.edu.tr`

* This research is partially supported by TÜBİTAK Career grant number 104E065 and grant number 104E077.

9.1 Introduction

There is an increasing demand to efficiently and effectively access large volumes of image and video collections. This demand leads to many systems to be introduced for indexing, searching and browsing of multimedia data. However, as observed in user studies [4], most of the systems do not satisfy the user requirements. The main bottleneck in building large-scale and realistic systems is the disconnection between the low-level representation of the data and the high-level semantics, which is usually referred as the "semantic gap" problem [41].

Early work on image retrieval systems were based on text input, in which the images are annotated by text and then text-based methods are used for retrieval [15]. However, two major difficulties are encountered with text-based approaches: first, manual annotation, which is a necessary step for these approaches, is labor-intensive and becomes impractical when the collection is large. Second, keyword annotations are subjective; the same image/video may be annotated differently by different annotators.

In order to overcome these difficulties, content-based image retrieval (CBIR) was proposed in the early 1990s. In CBIR systems, instead of text-based annotations, images are indexed, searched or browsed by their visual features, such as color, texture or shape (see [21, 41, 42] for recent surveys in this area). However, such systems are usually based only on low-level features and therefore cannot capture semantics.

In some studies, the semantics derived from the text is incorporated with the visual appearance to make use of both textual and visual information. The Blobworld system [13] uses a simple conjuction of keywords and image features to search for the images. In [14], Cascia et al. unify the textual and visual statistics in a single indexing vector for retrieval of web documents. Similarly, Zhao and Grosky [47] use both textual keywords and image features to discover the latent semantic structure of web documents. In the work of Chen et al. [17], image and text features are used together to iteratively narrow the search space for browsing and retrieval of web documents. Benitez and Chang [7] combine the textual and visual descriptors in the annotated image collections for clustering and further for sense disambiguation. Srihari et al. [43] use the textual captions for the interpretation of the corresponding photographs, especially for face identification applications.

Although these methods provide better access to image and video collections, the "semantic gap" problem still exists. Many systems are proposed to bridge the semantic gap in the form of recognition of objects and scenes. However, most of these systems require supervised input for labeling and therefore cannot be adapted to large-scale problems.

Recently, methods that try to bridge the semantic gap by learning the associations between textual and visual information from "loosely labeled data" are proposed as an alternative to supervised methods. Loosely labeled data means that the labels are not provided for individual items but for a collection of items. For example, in an annotated image collection, textual descriptions are available for the entire image but not for individual regions (Fig. 9.1). Loosely labeled data are available in large

Fig. 9.1 Examples of annotated images. **Top:** Corel data set. **Bottom:** TRECVID news videos data set

quantities in a variety of data sets (such as stock photographs annotated with keywords, museum images associated with metadata or descriptions, news photographs with captions and news videos with associated speech transcripts) and are easy to obtain. Therefore, the manual effort required for manual labeling can largely be reduced by the use of data sets providing loosely labeled data. However, these data sets do not provide one-to-one correspondences, and therefore it requires different learning techniques to learn the links between semantics derived from text and visual appearance.

In this direction, a variety of approaches are proposed in recent studies for semantic labeling of images and videos using the loosely labeled data sets. Although, each study brings its own solution to the problem, most of them rely on the joint occurrences of the textual and visual information, and use statistical approaches which are mostly adapted from text retrieval literature. The range of applications which make use of such approaches is large: improved search and browsing capabilities, automatic annotation of images, region naming as a possible direction to recognize large number of objects, face naming as a way of recognizing large number of people in different conditions, semantic alignment of video sequences with speech transcripts, etc.

In the following, first we will discuss some of these studies by focusing on the automatic image annotation problem. Then, we will present two methods in detail, namely the so-called "translation model" with a focus on the association of videos and speech transcript text and a graph model for solving the face naming problem.

9.2 Semantic Labeling of Images

The need for the labels in semantic retrieval of images and videos and the difficulty and the subjectivity of manual labeling make automatic image annotation, which is the process of automatically assigning textual descriptions to images, a desired and attractive task resulting in many studies to be proposed recently for automatic annotation of images.

In some studies, semi-automatic strategies are used for annotation of images. Wenyin et al. [45] use the query keywords which receive positive feedback from the user as possible annotations to the retrieved images. Similarly, Izquierdo and Dorado [24] propose a semi-automatic image annotation strategy by first manually labeling a set of images and then extracting candidate keywords for annotating a new image from its most similar images after a frequent pattern mining process.

For automatic indexing of pictures, Li and Wang [29] propose a method which models image concepts by 2D multi-resolution hidden Markov models and then labels an image with the concepts best fit the content. In [34], Monay et al. extend latent semantic analysis (LSA) models proposed for text and combine keywords with visual terms in a single vector representation. Singular value decomposition (SVD) is used to reduce the dimensionality and annotation is then performed by propagating the annotations of the most similar images in the corpus to the un-annotated image in the projected space.

In some other studies image annotation is viewed as a classification problem where the goal is to classify the entire image or the parts of the image into one of the categories corresponding to annotation keywords. In [32], Maron and Ratan propose multiple-instance learning as a way of classifying the images and use labeled images as bags of examples. Classifiers are built for each concept separately and an image is taken as positive if it contains a concept (e.g., waterfall) somewhere in the image and negative if it does not. Using a similar multiple-instance learning formulation Argillander et al. [3] propose a maximum entropy based approach and build multiple binary classifiers to annotate the images. Carneiro and Vasconcelos [12] formulate the problem as M-ary classification where each of the semantic concepts of interest defines an image class and the classes directly compete at the annotation time. Feng et al. [19] develop a co-training framework to bootstrap the process of annotating large WWW image collections by exploiting both the visual contents and their associated HTML text and build text-based and visual-based classifiers using probabilistic SVM.

Recently, probabilistic approaches which model the joint distribution of words and image components are proposed. The first model proposed by Mori et al. [36] uses a fixed size grid representation and obtains visual clusters by vector quantization of features extracted from these grids. The joint distribution of visual clusters and words are then learned using the co-occurrence statistics.

In [18] it is shown that learning the associations between visual elements and words can be attacked as a problem of translating visual elements into words. The visual elements, called as *blobs*, are constructed by vector quantization of the features extracted from the regions obtained using a segmentation algorithm. Given a set of training images, the problem is turned into creating a probability table that translates blobs to words. The probability table is learned using a method adapted from the statistical machine translation literature [10]. Different models are experimented in [44].

Pan et al. [38] use a similar blob representation and construct word-by-document and blob-by-document matrices. They discover the correlations between blobs and words based on the co-occurrence counts and also on the cosine similarity of the occurrence patterns – the documents including those items. For improvement, words and blobs are weighted inversely proportional to their occurrences and singular value decomposition (SVD) is applied over the matrices to suppress the noise.

In [25], Jeon et al. adapt the relevance-based language models to annotation problem and introduce cross-media relevance model. The images are represented in terms of both words and blobs. Given an image the probability of a word/blob is found as the ratio of the occurrence of the word/blob in the image to the total count of the word/blob in the training set. Then, they use the training set of annotated images to estimate the joint probability of observing a word with a set of blobs in the same image.

In the following studies of the same authors, the use of discrete blob representation is replaced with the direct modeling of continuous features, and two new models are proposed: continuous relevance model [28] and multiple Bernoulli relevance model [20].

The model proposed by Barnard et al. [5, 6] is a generative hierarchical aspect model inspired from text-based studies. The model combines the aspect model with a soft clustering model. Images and corresponding words are generated by nodes arranged in a tree structure. Image regions represented by continuous features are modeled using a Gaussian distribution, and words are modeled using a multinomial distribution.

Blei and Jordan [9] propose Corr-LDA (correspondence latent Dirichlet allocation) model which finds conditional relationships between latent variable representations of sets of image regions and sets of words. The model first generates the region descriptions and then the caption words.

In [35], Monay and Gatica-Perez modify the probabilistic latent space models to give higher importance to the semantic concepts and propose linked pair of PLSA models. They first constrain the definition of the latent space by focusing on textual features, and then learn visual variations conditioned on the space learned from text.

Carbonetto et al. [11] consider the spatial context and estimate the probability of an image blob being aligned to a particular word depending on the word assignments of its neighboring blobs using markov random fields.

Other approaches proposed for image annotation include maximum entropy based model [26], a method based on hidden Markov model [23], a graph-based approach [39] and active learning method [27].

Automatic annotation of images is important since considerable amount of work for manual annotation of images can be eliminated and semantic retrieval of large number of images can be achieved. However, most of the methods mentioned above do not learn the direct links between image regions and words, but rather models the joint occurrences of regions and words through an image.

In the following, we give the details of the so-called "translation method", which is first proposed in [18], as a way of learning the links between image regions and words. Then, we discuss the generalization of this method for the problem of learning the links between textual and visual elements in videos.

9.3 Translation Approach

In the annotated image and video collections, the images are annotated with a few keywords which describe the images. However, one-to-one correspondences between image regions and words are unknown. That is, for an annotation keyword, we know that the visual concept associated with the keyword is likely to be in the image, but we do not know which part of the image corresponds to that word (Fig. 9.2).

The correspondence problem is very similar to the correspondence problem faced in statistical machine translation literature (Fig. 9.2). There is one form of data (image structures or words in one language) and we want to transform it into another form of data (annotation keywords or words in another language). Learning a lexicon (a device that can predict one representation given the other representation) from large data sets (referred as aligned bitext) is a standard problem in the statistical machine translation literature [10]. Aligned bitexts consist of many small blocks of text in both languages, corresponding to each other at paragraph or sentence level, but not at the word level. Using the aligned bitexts the problem of lexicon learning is transformed into the problem of finding the correspondences between words of different languages, which can then be tackled by machine learning methods.

Due to the similarity of the problems, correspondence problem between image structures and keywords can be attacked as a problem of translating visual features into words, as first proposed in [18]. Given a set of training images, the problem is to create a probability table that associates words and visual elements. This translation table can then be used to find the corresponding words for the given test images (*auto-annotation*) or to label the image components with words as a novel approach to recognition (*region labeling*).

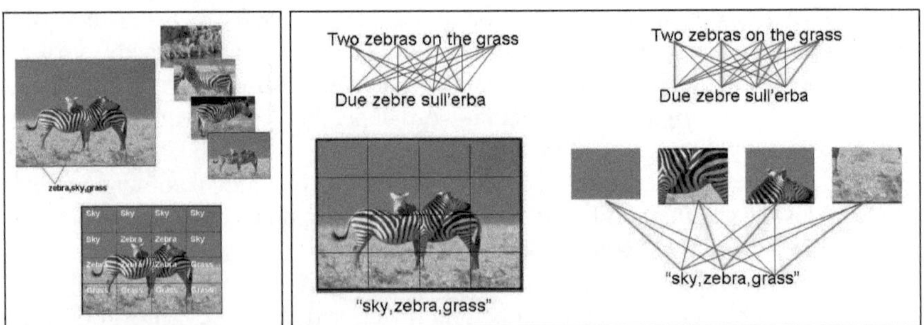

(a) (b)

Fig. 9.2 (**a**) The correspondence problem between image regions and words. The words `zebra`, `grass` and `sky` are associated with the image, but the word-to-region correspondences are unknown. If there are other images, the correct correspondences can be learned and used to automatically label each region in the image with correct words or to auto-annotate a given image. (**b**) The analogy with the statistical machine translation. We want to transform one form of data (image regions or words in one language) to another form of data (concepts or words in another language)

In the following sections, we will first present the details of the approach to learn the correspondences between image regions and words and then show how it can be generalized to other problems.

9.3.1 Learning Correspondences Between Words and Regions

Brown et al. [10] propose a set of models for statistical machine translation. These models aim to maximize the conditional probability density $P(\mathbf{f} \mid \mathbf{e})$, which is called as the likelihood of translation $(\mathbf{f}, \mathbf{e})$, where $\mathbf{f}$ is a set of French words and $\mathbf{e}$ is a set of English words.

In machine translation, a lexicon links a set of discrete objects (words in one language) to another set of discrete objects (words in the other language). In our case, the data consist of visual elements associated with words. The *words* are in discrete form. In order to exploit the analogy with machine translation, the visual data, represented as a set of feature vectors, also need to be broken up into discrete items. For this purpose, the features are grouped by vector quantization techniques such as k-means and the labels of the classes, which we call as *blobs*, are used as the discrete items for the visual data. Then, an aligned bitext, consisting of the blobs and the words for each image, is obtained and used to construct a probability table linking blobs with words.

In our case, the goal is to maximize $P(\mathbf{w} \mid \mathbf{b})$, where $\mathbf{b}$ is a set of blobs and $\mathbf{w}$ is a set of words. Each word is aligned with the blobs in the image. The alignments (referred as $\mathbf{a}$) provide a correspondence between each word and all the blobs. The model requires the sum over all possible assignments for each pair of aligned sentences, so that $P(\mathbf{w} \mid \mathbf{b})$ can be written in terms of the conditional probability density $P(\mathbf{w}, \mathbf{a} \mid \mathbf{b})$ as

$$P(\mathbf{w} \mid \mathbf{b}) = \sum_{\mathbf{a}} P(\mathbf{w}, \mathbf{a} \mid \mathbf{b}). \tag{9.1}$$

The simplest model (Model 1) assumes that all connections for each French position are equally likely. This model is adapted to translate blobs into words, since there is no order relation among the blobs or words in the data [44]. In Model 1 it is assumed that each word is aligned exactly with a single blob. If the image has l blobs and m words, the alignment is determined by specifying the values of a_j such that if the jth word is connected to the ith blob, then $a_j = i$, and if it is not connected to any blob $a_j = 0$. Assuming a uniform alignment probability (each alignment is equally probable), given a blob the joint likelihood of a word and an alignment is then written as

$$P(\mathbf{w}, \mathbf{a} \mid \mathbf{b}) = \frac{\varepsilon}{(l+1)^m} \prod_{j=1}^{m} t(w_j \mid b_{a_j}), \tag{9.2}$$

where $t(w_j \mid b_{a_j})$ is the translation probability of the word w_j given the blob b_{a_j}, and ε is a fixed small number.

The alignment is determined by specifying the values of a_j for j from 1 to m each of which can take a value from 0 to l. Then, $P(\mathbf{w} \mid \mathbf{b})$ can be written as

$$P(\mathbf{w} \mid \mathbf{b}) = \frac{\varepsilon}{(l+1)^m} \sum_{a_1=0}^{l} \cdots \sum_{a_m=0}^{l} \prod_{j=1}^{m} t(w_j \mid b_{a_j}). \tag{9.3}$$

Our goal is to maximize $P(\mathbf{w} \mid \mathbf{b})$ subject to the constraint that for each b

$$\sum_{w} t(w \mid b) = 1. \tag{9.4}$$

This maximization problem can be solved with the expectation maximization (EM) formulation [10, 18]. In this study, we use the Giza++ tool [37] – which is a part of the statistical machine translation toolkit developed during summer 1999 at CLSP at Johns Hopkins University – to learn the probabilities. Note that, we use the direct translation model throughout the study.

The learned association probabilities are kept in a translation probability table, and then used to predict words for the test data.

Region naming refers to predicting the labels for the regions, which is clearly recognition. For region naming, given a blob b corresponding to the region, the word w with the highest probability ($P(w \mid b)$) is chosen and used to label the region.

In order to automatically annotate the images, the word posterior probabilities for the entire image are obtained by marginalizing the word posterior probabilities of all the blobs in the image as

$$P(w|I_b) = 1/|I_b| \sum_{b \in I_b} P(w|b), \tag{9.5}$$

where b is a blob, I_b is the set of all blobs of the image and w is a word. Then, the word posterior probabilities are normalized. The first N words with the highest posterior probabilities are used as the annotation words.

9.3.2 Linking Visual Elements to Words in News Videos

Being an important source, broadcast news videos are acknowledged by NIST as a challenging data set and used for TRECVID Video Retrieval and Evaluation track since 2003.[2] For retrieving the relevant information from the news videos, it is common to use speech transcript or closed caption text and perform text-based queries. However, there are cases where text is not available or full of errors. Also, text is aligned with the shots only temporally and therefore the retrieved shots may not be related to the visual content (we refer this problem as "video association problem"). For example, when we retrieve the shots where a keyword is spoken in the transcript we may come up with visually non-relevant shots where an anchor/reporter is introducing or wrapping up a story (Fig. 9.3). An alternative is to use the annotation

[2] http://www-nlpir.nist.gov/projects/trecvid.

... (1) so today it was an energized president **CLINTON** who formally presented his one point seven three trillion dollar budget to the congress and told them there'd be money left over first of the white house a.b.c's sam donaldson (2) ready this (3) morning here at the whitehouse and why not (4) next year's projected budget deficit zero where they've presidential shelf and tell *this* (5) *budget marks the hand of an era and ended decades of deficits that have shackled our economy paralyzed our politics and held our people back* ..... (6) [empty] (7) [empty] (8) administration officials say this balanced budget are the results of the president's sound policies he's critics say it's merely a matter of benefiting from a strong economy that other forces are driving for the matter why it couldn't come at a better time just another upward push for mr **CLINTON**'s new sudden sky high job approval rating peter thanks very ...

Fig. 9.3 Key-frames and corresponding speech transcripts for a sample sequence of shots for a story related to Clinton. Italic text shows Clinton's speech, and capitalized letters show when Clinton's name appears in the transcript. Note that, Clinton's name is mentioned when an anchorperson or reporter is speaking, but not when he is in the picture

words, but due to the huge amount of human effort required for manual annotation it is not practical.

As can be seen, a correspondence problem similar to the one faced in annotated image collections occurs in video data. There are sets of video frames and transcripts extracted from the audio speech narrative, but the semantic correspondences between them are not fixed because they may not be co-occurring in time. If there is no direct association between text and video frames, a query based on text may produce incorrect visual results. For example, in most news videos (see Fig. 9.3) the anchorperson talks about an event, place or person, but the images related to the event, place or person appear later in the video. Therefore, a query based only on text related to a person, place or event, and showing the frames at the matching narrative, may yield incorrect frames of the anchorperson as the result.

Some of the studies discussed above are applied to video data, and the limited amount of manually annotated key-frames are used to annotate the other key-frames [23, 44]. However, since the vocabulary is limited and annotations are full of errors, the usability of such data is narrow. On the other hand, speech transcript text is available for all the videos and provides an unrestricted vocabulary. However, since the frames are aligned with the speech transcript text temporally the semantic relationships are lost.

Now, we discuss how the methods proposed for linking images with words can be generalized to solve the video association problem and to recognize the objects and scenes in news videos, and present an extended version of machine translation method proposed in [18] adapted for this problem.

9.3.3 Translation Approach to Solve Video Association Problem

An annotated image consists of two parts: set of regions and set of words. As discussed above, the goal was to learn the links between the elements of these two sets.

A similar relationship occurs in video data. There are a set of video frames and a set of words extracted from the speech recognition text. While, the temporal alignment do not relate these two sets semantically, there is a unit which provides a semantic association: a video story. Therefore, we can redefine the video association problem as finding the links between frames and words of a video story.

The translation approach to learn the associations between image regions and annotation words is then modified to solve the video association problem. Each story is taken as the basic unit, and the problem is turned into finding the associations between the key-frames and the speech transcript words of the story segments. To make the analogy with the association problem between image regions and annotation keywords, the stories correspond to images, the key-frames correspond to image regions and speech transcript text corresponds to annotation keywords. The features extracted from the key-frames are vector quantized to represent each image with labels which are again called blobs. Then, the translation tables are constructed similar to the one constructed for annotated images. The associations can then be used either to associate the key-frames with the correct words or for predicting words for the entire stories.

9.3.4 Experiments on News Videos Data Set

In the following, we will present the results of experiments of using the translation approach on news videos data set. First, the translation approach proposed for learning region to word correspondences will be directly adapted to solve the correspondences between regions of the key-frames and the corresponding annotation words. Then, speech transcript words will be associated with key-frames of the stories to solve the video association problem.

9.3.4.1 Translation Using Manual Annotations

In the experiments we use the TRECVID 2004 corpus provided by NIST which contains over 150 hours of CNN and ABC broadcast news videos. The shot boundaries and the key-frames extracted from each shot are provided by NIST. One hundred and fourteen videos are manually annotated with a collaborative effort of the TRECVID participants with a few keywords [30]. The annotations are usually incomplete.

In total, 614 words are used for annotation, most of which have very low frequencies, and with spelling and format errors. After correcting the errors and removing the least frequent words, we pruned the vocabulary down to 62 words. We only use the annotations for the key-frames, and therefore eliminate the videos where the annotations are provided for the frames which are not key-frames, resulting in 92 videos with 17177 images, 10164 used for training and 7013 for testing.

The key-frames are divided into 5×7 rectangular grids and each grid is represented with various color (RGB, HSV mean-std) and texture (Canny edge orientation histograms, Gabor filter outputs) features. These features are then vector quantized using k-means to obtain the visual terms (blobs). The correspondences between the blobs and manual annotation words are learned in the form of a probability table using *Giza++* [37]. The translation probabilities in the probability table are used for auto-annotation, region labeling and ranked retrieval.

Figures 9.4 and 9.5 shows some region-labeling and auto-annotation examples. When predicted annotation words are compared with the actual annotation words, for each image, the average annotation prediction performance is around 30%. Since the manual annotations are incomplete (for example in the third example of Fig. 9.5, although *sky* is in the picture and predicted it is not in the manual annotations), the actual annotation prediction performance should be higher than 30%.

Figure 9.6 shows query results for some words (with the highest rank). By visually inspecting the top 10 images retrieved for 62 words, the mean average precision (MAP) is determined to be 63%. MAP is 89% for the best (with highest precision) 30 words and 99% for the best 15 words. The results show that when the annotations are not available the proposed system can effectively be used for ranked retrieval.

9.3.4.2 Translation Using Speech Transcripts in Story Segments

The automatic speech recognition (ASR) transcripts for TRECVID2004 corpus are provided by LIMSI and are aligned with the shots on time basis [22]. The speech transcripts (ASR) are in the free text form and need preprocessing. Therefore, we applied tagging, stemming and stop word elimination and used only the nouns having frequencies more than 300 as our final vocabulary resulting in 251 words.

The story boundaries provided by NIST are used. We remove the stories associated with less than four words, and use the remaining 2503 stories consisting of 31450 key-frames for training and 2900 stories consisting of 31464 key-frames for testing. The number of words corresponding to the stories vary between 4 and 105, and the average number of words per story is 15.

The key-frames are represented by blobs obtained by vector quantization of HSV, RGB color histograms, Canny edge orientation histograms and bags of SIFT keypoints extracted from entire images. The correspondences between the blobs and speech transcript words in each story segment are learned in the form of a probability table again using Giza++.

Fig. 9.4 Examples for blob-to-word matches

studio-setting female-news-person male-news-subject graphics person — female-news-person studio-setting people male-face graphics person scene-text	people basketball — people graphics basketball female-news-person scene-text male-news-subject studio-setting	water-body boat — sky graphics water-body building boat person male-news-person	sky building road car graphics — road man-made-object people sky building car man-made-scene

Fig. 9.5 Auto-annotation examples. The manual annotations are shown at the top, and the predicted words, top seven words with the highest probability, are shown at the bottom

The translation probabilities are used for predicting words for the individual shots (Fig. 9.7) and for the stories (Fig. 9.8). The results show that especially for the stories related to weather, sports or economy, which frequently appear in the broadcast news, the system can predict the correct words. Note that the system can predict words which are better than the original speech transcript words. This characteristic is important for a better retrieval.

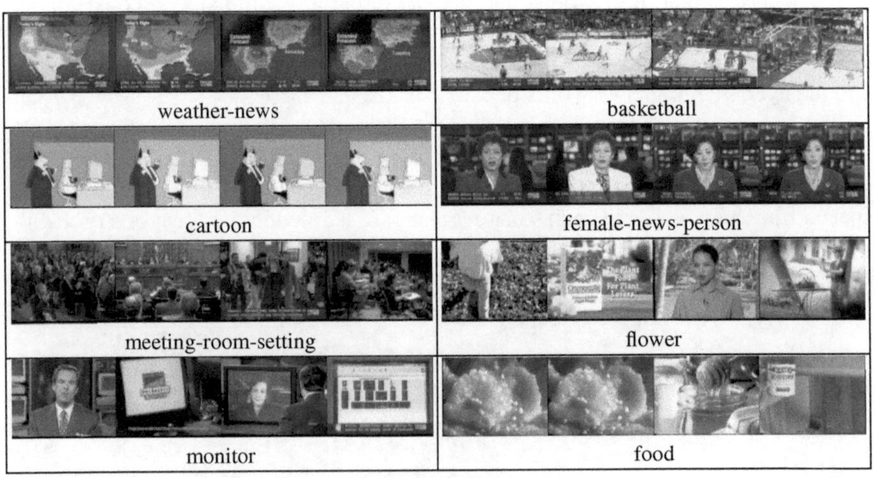

Fig. 9.6 Ranked query results for some words using manual annotations in learning the correspondences

Fig. 9.7 Top three words predicted for some shots using ASR

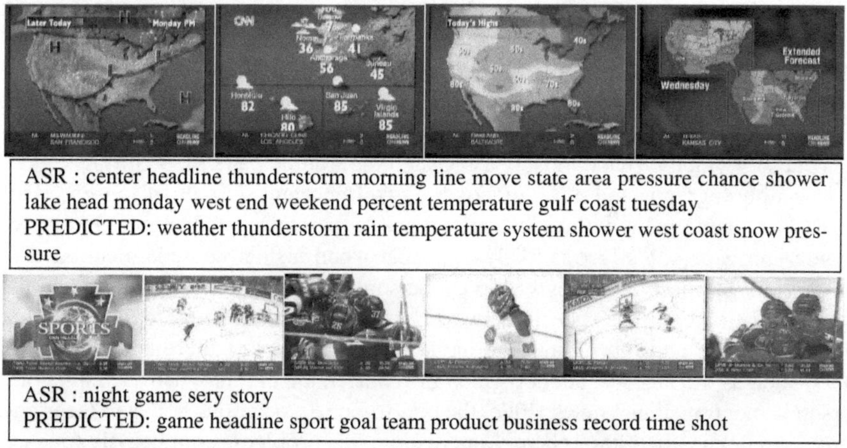

Fig. 9.8 For sample stories corresponding ASR outputs and top 10 words predicted

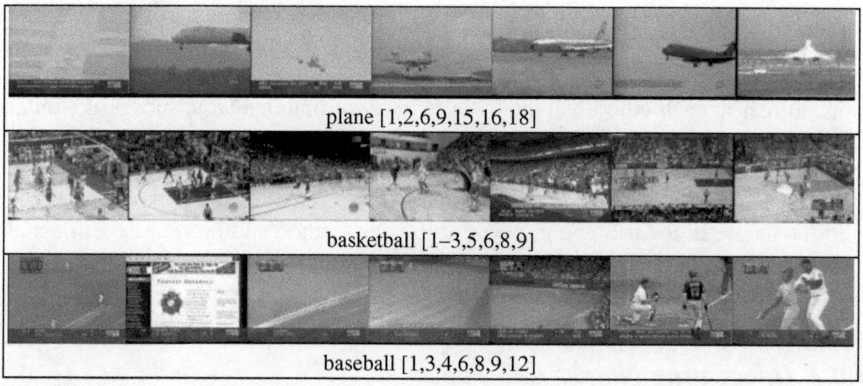

Fig. 9.9 Ranked story-based query results for ASR. Numbers in square brackets show the rank of retrieval

Story-based query results in Fig. 9.9 show that the proposed system is able to detect the associations between the words (objects) and scenes. In these examples, the shots within each story are ranked according to the marginalized word posterior probabilities, and the shots matching the query word with highest probability are retrieved; a final ranking is done among all shots retrieved from all stories and all videos and final ranked query results are returned to the user.

9.4 Naming Faces in News

While face recognition is a long-standing problem, recognition of large number of faces is still a challenge especially for uncontrolled environments with a variety of illumination and pose changes, and in the case of occlusion [48]. Recently, following a similar direction in image/video annotation, alternative solutions are proposed for retrieval and recognition of faces by associating the name and face information using the data sets which provide loosely associated face–name pairs [8, 40, 46].

News photographs and news videos are two important sources of loosely associated face–name pairs with their heavy focus on person-related stories. In a news photograph, a person's face is likely to appear when his/her name is mentioned in the caption. Similarly, in a news video, a person appears more frequently when the name is mentioned in the speech. However, since there could be more than one face in a photograph and more than one name in the caption, or there could be other faces, such as the faces of anchorperson or reporter, appearing when the name of a person is mentioned in a news video, the one-to-one associations between faces and names are unknown. Although very interesting, this problem is not heavily touched as in the case of image annotation, mostly due to the difficulty of representing face similarities.

In this chapter, we present an approach to name the faces by introducing a graph-based approach. The approach lies on two observations: (i) a person's face is more likely to appear when his/her name is mentioned in the related text together with many non-face images and faces of other people, and (ii) the faces of that person are more similar to each other than the other faces, resulting in a large subset of similar faces in a set of faces extracted around the name. Based on these observations, we represent the similarity of faces appearing around a name in a graph structure and then find the densest subgraph corresponding to the faces of the person associated with the name. In the following, we describe our method and present experimental results on news photo and video collections.

9.4.1 Integrating Names and Faces

The first step in our method is to integrate the face and the name information. We use the name information mainly to limit the search space, since a person is likely to appear in news photos/videos when his/her name is mentioned. Using this assumption, first we detect the faces and use name information to further reduce these set of faces. In the news photographs, we reduce the set for a queried person only to the photographs that include the name of that person in the associated caption and having one or more faces. However, in news videos this assumption does not usually hold and choosing the shot where the name is mentioned may yield incorrect results since there is mostly a time shift between the visual appearance of a person and his/her name. In order to handle this alignment problem, we also choose one preceding and two succeeding shots along with the shot in which the name of the queried person is mentioned.

We assume that the integration of name and face information removes most of the unrelated faces, and in the subset of remaining faces, although there might be faces corresponding to other people in the story, or some non-face images due to the errors of the face detection method, the faces of the query name are likely to be the most frequently appearing ones than any other person. Also, we assume that even if the expressions or poses vary, different appearances of the face of the same person tend to be more similar to each other than to the faces of others.

In the following, we will present a method which seeks for the most similar faces in the reduced face set corresponding to the faces of the person whose name is mentioned, first starting with the discussion of how the similarity of the faces can be represented.

9.4.2 Finding Similarity of Faces

Finding the most similar faces in the subset of faces associated with the name requires a good representation for faces and a similarity measure which is invariant to view point, pose and illumination changes. Based on their successful results in matching images for object and scene recognition, we represent the similarity of faces using the interest points extracted from the detected face areas. Lowe's SIFT operator [31] is used for extracting the interest points.

The similarity of two faces are computed based on the interest points that are matched. To find the matching interest points on two faces, each point on one face is compared with all the points on the other face and the points with the least Euclidean distance are selected. Since this method produces many matching points including the wrong ones, two constraints are applied to obtain only the correct matches, namely the geometrical constraint and the unique match constraint.

Geometrical constraint expects the matching points to appear around similar positions on the face when the normalized positions are considered. The matches whose interest points do not fall in close positions on the face are eliminated. Unique match constraint ensures that each point matches to only a single point by eliminating multiple matches to one point and also by removing one-way matches. Example of matches after applying these constraints are shown in Fig. 9.10.

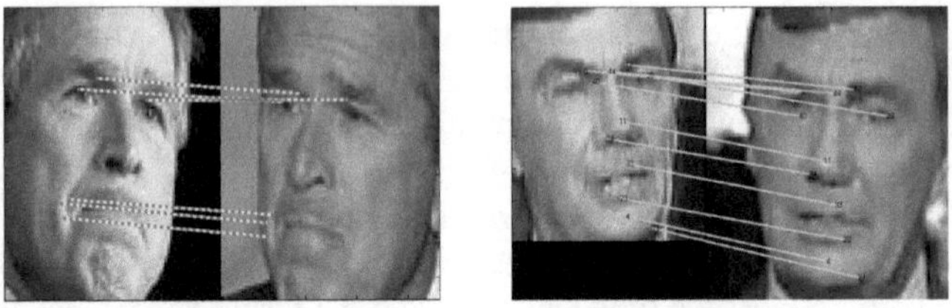

Fig. 9.10 Sample matching points for two faces from news photographs on the left and news videos on the right. Note that, even for faces with different size, pose or expressions the method successfully finds the corresponding points

After applying the constraints, the distance between the two faces is defined as the average distance of all matching points between these two faces. A similarity graph for all the faces in the search space is then constructed using these distances.

9.4.3 Finding the Densest Component in the Similarity Graph

In the similarity graph, faces represent the nodes and the distances between the faces represent the edge weights. We assume that in this graph the nodes of a particular person will be close to each other (highly connected) and distant from the other nodes (weakly connected). Hence, the problem of finding the largest subset of similar faces can be transformed into finding the densest subgraph (component) in the similarity graph. To find the densest component, we adapt the method proposed by Charikar [16] where the density of subset S of a graph G is defined as

$$f(S) = \frac{|E(S)|}{|S|},$$

in which $E(S) = \{i, j \in E : i \in S, j \in S\}$ and E is the set of all edges in G and $E(S)$ is the set of edges induced by subset S. The subset S that has maximum $f(S)$ is defined as the densest component.

Initially, the algorithm presented in [16] starts from the entire graph and in each step, the vertex of minimum degree is removed from the set S. The $f(S)$ value is also computed for each step. The algorithm continues until the set S is empty. Finally, the subset S with maximum $f(S)$ value is returned as the densest component of the graph.

In order to apply the above algorithm to the constructed dissimilarity graph, we need to convert it into a binary form, in which 0 indicates no edge and 1 indicates an edge between the two nodes. This conversion is carried out by applying a threshold on the distance between the nodes. For instance, if 0.5 is used as the threshold value, then edges in the similarity graph having higher values than 0.5 are assigned as 0, and the others as 1. In other words, the threshold can be thought of as an indicator of two nodes being near-by and/or remote.

9.4.4 Experiments

We applied the proposed method on two data sets: news photographs and news videos. News photographs data set consists of a total of 30,281 detected faces from half a million captioned news images collected from Yahoo! News on the Web, which is constructed by Berg et al. [8]. Each image in this set is associated with a set of names. In the experiments, the top 23 people, whose name appears with the highest frequencies (more than 200 times) are used.

Average precision value for the baseline method is 48%, which assumes that all the faces appearing around the name is correct. With the proposed, method we achieved 68% recall and 71% precision values on average. The method can achieve up to 84% recall and 100% precision for some people. We had initially assumed that after associating names, true faces of the queried person appear more than any other person in the search space. However, when this is not the case, the algorithm gives bad retrieval results. For example, there is a total of 913 images associated with name *Saddam Hussein*, but only 74 of them are true *Saddam Hussein* images while 179 of them are *George Bush* images. Some sample images retrieved for three people are shown in Fig. 9.11.

News videos data set is the broadcast news videos provided by NIST for TRECVID video retrieval evaluation competition 2004. It consists of 229 movies (30 minutes each) from ABC and CNN news. The shot boundaries and the key-frames are provided by NIST. Speech transcripts extracted by LIMSI [22] are used to obtain the associated text for each shot. The face detection algorithm provided by Mikolajcyzk [33] is used to extract faces from key-frames. Due to high noise levels and low image resolution quality, the face detector produces many false alarms.

For the experiments, we chose five people, namely Bill Clinton, Benjamin Netanyahu, Sam Donaldson, Saddam Hussein and Boris Yeltsin. In the speech transcript text, their names appear 991, 51, 100, 149 and 78 times, respectively.

When the shots including the query name are selected as explained above, the faces of the anchorpeople appear more frequently making our assumption that the most frequent face will correspond to the query name wrong. Hence, before

Fig. 9.11 Sample images retrieved for three person queries in the experiments on news photographs. Each row corresponds to samples for George Bush, Hans Blix and Colin Powell queries respectively

Table 9.1 Numbers in the table indicate the number of correct images retrieved/total number of images retrieved for the query name

Query name	Clinton	Netanyahu	Sam Donaldson	Saddam	Yeltsin
Text-only	160/2457	6/114	102/330	14/332	19/157
Anchor removed	150/1765	5/74	81/200	14/227	17/122
Method applied	109/1047	4/32	67/67	9/110	10/57

applying the proposed method, we detect the anchorpeople and remove them from the selected shots by applying the densest component based method to each news video separately. The idea is based on the fact that the anchorpeople are usually the most frequently appearing people in one broadcast news video. If we construct a similarity graph for the faces in a news video, the densest component in this graph will correspond to the faces of the anchorperson. We ran the algorithm on 229 videos in our test set and obtained average recall and precision values as 90 and 85%, respectively.

We have recorded the number of true faces of the query name and total number of images retrieved as in Table 9.1. The first column of the table refers to total number of true images retrieved vs. total number of true images retrieved by using only the

Fig. 9.12 Sample images retrieved for five person queries in the experiments on news videos. Each row corresponds to samples for Clinton, Netanyahu, Sam Donaldson, Saddam and Yeltsin queries, respectively

speech transcripts – selecting the shots within interval [1, 2]. The numbers after removing the detected anchorpeople by the algorithm from the text-only results are given in the second column. And the last column is for applying the algorithm to this set, from which the anchorpeople are removed. Some sample images retrieved for each person are shown in Fig. 9.12.

As can be seen from the results, we keep most of the correct faces (especially after anchorperson removal) and reject many of the incorrect faces. Hence the number of images presented to the user is decreased. Also, our improvement in precision values are relatively high. Average precision of only text-based results, which was 11.8% is increased to 15% after ancherperson removal, and to 29.7% after applying the proposed algorithm.

9.5 Conclusion and Discussion

In this chapter, we present recent approaches for semantic labeling of images and videos and describe two methods in detail: (i) translation approach for solving the correspondences between visual and textual elements and (ii) naming faces using a graph-based method. The results promise that the use of loosely labeled data sets allows to learn large number of semantic labels and results in novel applications.

References

1. Giza++. http://www.fjoch.com/GIZA++.html.
2. Trec vieo retrieval evaluation. http://www-nlpir.nist.gov/projects/trecvid.
3. J. Argillander, G. Iyengar, and H. Nock. Semantic annotation of multimedia using maximum entropy models. In *IEEE International Conference on Acoustics, Speech, and Signal Processing (ICASSP 2005)*, Philadelphia, PA, USA, March 18–23 2005.
4. L. H. Armitage and P.G.B. Enser. Analysis of user need in image archives. *Journal of Information Science*, 23(4):287–299, 1997.
5. K. Barnard, P. Duygulu, N. de Freitas, D.A. Forsyth, D. Blei, and M. Jordan. Matching words and pictures. *Journal of Machine Learning Research*, 3:1107–1135, 2003.
6. K. Barnard and D.A. Forsyth. Learning the semantics of words and pictures. In *International Conference on Computer Vision*, pages 408–415, 2001.
7. A. B. Benitez and S.-F. Chang. Semantic knowledge construction from annotated image collections. In *IEEE International Conference On Multimedia and Expo (ICME-2002)*, Lausanne, Switzerland, August 2002.
8. T. Berg, A.C. Berg, J. Edwards, and D.A. Forsyth. Who is in the picture. In *Neural Information Processing Systems (NIPS)*, 2004.
9. D.M. Blei and M.I. Jordan. Modeling annotated data. In *26th Annual International ACM SIGIR Conference*, pages 127–134, Toronto, Canada, July 28 – August 1 2003.
10. P.F. Brown, S.A. Della Pietra, V.J. Della Pietra, and R.L. Mercer. The mathematics of statistical machine translation: Parameter estimation. *Computational Linguistics*, 19(2):263–311, 1993.
11. P. Carbonetto, N. de Freitas, and K. Barnard. A statistical model for general contextual object recognition. In *Eight European Conference on Computer Vision (ECCV)*, Prague, Czech Republic, May 11–14 2004.

12. G. Carneiro and N. Vasconcelos. Formulating semantic image annotation as a supervised learning problem. In *Proceedings of IEEE Conference on Computer Vision and Pattern Recognition*, San Diego, 2005.
13. C. Carson, S. Belongie, H. Greenspan, and J. Malik. Blobworld: Image segmentation using expectation-maximization and its application to image querying. *IEEE Transactions on Pattern Analysis and Machine Intelligence*, 24(8):1026–1038, August 2002.
14. M.L. Cascia, S. Sethi, and S. Sclaroff. Combining textual and visual cues for content-based image retrieval on the world wide web. In *Proceedings of the IEEE Workshop on Content-Based Access of Image and Video Libraries*, Santa Barbara CA USA, June 1998.
15. S. Chang and A. Hsu. Image information systems: Where do we go from here? *IEEE Trans. on Knowledge and Data Enginnering*, 4(5):431–442, October 1992.
16. M. Charikar. Greedy approximation algorithms for finding dense components in a graph. In *APPROX '00: Proceedings of the 3rd International Workshop on Approximation Algorithms for Combinatorial Optimization*, London, UK, 2000.
17. F. Chen, U. Gargi, L. Niles, and H. Schuetze. Multi-modal browsing of images in web documents. In *Proceedings of SPIE Document Recognition and Retrieval VI*, 1999.
18. P. Duygulu, K. Barnard, N.d. Freitas, and D.A. Forsyth. Object recognition as machine translation: learning a lexicon for a fixed image vocabulary. In *Seventh European Conference on Computer Vision (ECCV)*, volume 4, pages 97–112, Copenhagen Denmark, May 27 – June 2 2002.
19. H. Feng, R. Shi, and T.-S. Chua. A bootstrapping framework for annotating and retrieving www images. In *Proceedings of the 12th annual ACM international conference on Multimedia*, pages 960–967, New York, NY, USA, 2004.
20. S.L. Feng, R. Manmatha, and V. Lavrenko. Multiple bernoulli relevance models for image and video annotation. In *the Proceedings of the International Conference on Pattern Recognition (CVPR 2004)*, volume 2, pages 1002–1009, 2004.
21. D.A. Forsyth and J. Ponce. *Computer Vision: A Modern Approach*. Prentice-Hall, 2002.
22. J.L. Gauvain, L. Lamel, and G. Adda. The limsi broadcast news transcription system. *Speech Communication*, 37(1–2):89–108, 2002.
23. A. Ghoshal, P. Ircing, and S. Khudanpur. Hidden markov models for automatic annotation and content based retrieval of images and video. In *The 28th International ACM SIGIR Conference*, Salvador, Brazil, August 15–19 2005.
24. E. Izquierdo and A. Dorado. Semantic labelling of images combining color, texture and keywords. In *Proceedings of the IEEE International Conference on Image Processing (ICIP2003)*, Barcelona, Spain, September 2003.
25. J. Jeon, V. Lavrenko, and R. Manmatha. Automatic image annotation and retrieval using cross-media relevance models. In *26th Annual International ACM SIGIR Conference*, pages 119–126, Toronto, Canada, July 28 – August 1 2003.
26. J. Jeon and R. Manmatha. Using maximum entropy for automatic image annotation. In *the Proceedings of the 3rd International Conference on Image and Video Retrieval (CIVR 2004)*, pages 24–32, Dublin City University, Ireland, July 21–23 2004.
27. R. Jin, J. Y. Chai, and S. Luo. Automatic image annotation via coherent language model and active learning. In *The 12th ACM Annual Conference on Multimedia (ACM MM 2004)*, New York, USA, October 10–16 2004.
28. V. Lavrenko, R. Manmatha, and J. Jeon. A model for learning the semantics of pictures. In *the Proceedings of the Seventeenth Annual Conference on Neural Information Processing Systems*, volume 16, pages 553–560, 2003.
29. J. Li and J.Z. Wang. Automatic linguistic indexing of pictures by a statistical modeling approach. *IEEE Transaction on Pattern Analysis and Machine Intelligence*, 25(9):1075–1088, September 2003.
30. C.-Y. Lin, B.L. Tseng, and J.R. Smith. Video collaborative annotation forum:establishing ground-truth labels on large multimedia datasets. In *NIST TREC-2003 Video Retrieval Evaluation Conference*, Gaithersburg, MD, November 2003.
31. D.G. Lowe. Distinctive image features from scale-invariant keypoints. *International Journal of Computer Vision*, 60(2), 2004.

32. O. Maron and A.L. Ratan. Multiple-instance learning for natural scene classification. In *The Fifteenth International Conference on Machine Learning*, 1998.
33. K. Mikolajczyk. Face detector. INRIA Rhone-Alpes, 2004. Ph.D Report.
34. F. Monay and D. Gatica-Perez. On image auto-annotation with latent space models. In *Proceedings of the ACM International Conference on Multimedia (ACM MM)*, Berkeley, CA, USA, November 2003.
35. F. Monay and D. Gatica-Perez. Plsa-based image auto-annotation: Constraining the latent space. In *Proceedings of the ACM International Conference on Multimedia (ACM MM)*, New York, October 2004.
36. Y. Mori, H. Takahashi, and R. Oka. Image-to-word transformation based on dividing and vector quantizing images with words. In *First International Workshop on Multimedia Intelligent Storage and Retrieval Management*, 1999.
37. F.J. Och and H. Ney. A systematic comparison of various statistical alignment models. *Computational Linguistics*, 1(29):19–51, 2003.
38. J.-Y. Pan, H.-J. Yang, P. Duygulu, and C. Faloutsos. Automatic image captioning. In *In Proceedings of the 2004 IEEE International Conference on Multimedia and Expo (ICME2004)*, Taipei, Taiwan, June 27–30 2004.
39. J.-Y. Pan, H.-J. Yang, C. Faloutsos, and P. Duygulu. Automatic multimedia cross-modal correlation discovery. In *Proceedings of the 10th ACM SIGKDD Conference*, Seatle, WA, August 22–25 2004.
40. S. Satoh and T. Kanade. Name-it: Association of face and name in video. In *Proceedings of IEEE Conference on Computer Vision and Pattern Recognition(CVPR)*, 1997.
41. A.W.M. Smeulders, M. Worring, S. Santini, A. Gupta, and R. Jain. Content based image retrieval at the end of the early years. *IEEE Transactions on Pattern Analysis and Machine Intelligence*, 22(12):1349–1380, 2000.
42. C.G.M. Snoek and M. Worring. Multimodal video indexing: A review of the state-of-the-art. *Multimedia Tools and Applications*, 25(1):5–35, January 2005.
43. R.K. Srihari and D.T Burhans. Visual semantics: Extracting visual information from text accompanying pictures. In *AAAI 94*, Seattle, WA, 1994.
44. P. Virga and P. Duygulu. Systematic evaluation of machine translation methods for image and video annotation. In *The Fourth International Conference on Image and Video Retrieval (CIVR 2005)*, Singapore, July 20–22 2005.
45. L. Wenyin, S. Dumais, Y. Sun, H. Zhang, M. Czerwinski, and B. Field. Semi-automatic image annotation. In *Proceedings of the INTERACT : Conference on Human-Computer Interaction*, pages 326–333, Tokyo Japan, July 9–13 2001.
46. J. Yang, M-Y. Chen, and A. Hauptmann. Finding person x: Correlating names with visual appearances. In *International Conference on Image and Video Retrieval (CIVR'04)*, Dublin City University Ireland, July 21–23 2004.
47. R. Zhao and W.I. Grosk. Narrowing the semantic gap: Improved text-based web document retrieval using visual features. *EEE Transactions on Multimedia*, 4(2):189–200, 2002.
48. W. Zhao, R. Chellappa, P.J. Phillips, and A. Rosenfeld. Face recognition: A literature survey. *ACM Computing Surveys*, 35(4):399–458, 2003.

Chapter 10
Machine Learning for Semi-structured Multimedia Documents: Application to Pornographic Filtering and Thematic Categorization

Ludovic Denoyer and Patrick Gallinari

Abstract We propose a generative statistical model for the classification of semi-structured multimedia documents. Its main originality is its ability to simultaneously take into account the structural and the content information present in a semi-structured document and also to cope with different types of content (text, image, etc.). We then present the results obtained on two sets of experiments:

- One set concerns the filtering of pornographic Web pages.
- The second one concerns the thematic classification of Wikipedia documents.

10.1 Introduction

The development of the Web and the growing number of documents available electronically has been paralleled by the emergence of semi-structured data models for representing textual or multimedia documents. The semi-structured documents can be described in a large variety of formats like HTML, XML, XHTML, PDF, etc. All these formats are based on the idea of modelling both the content information and the logical structure of the documents. The content information is mainly text or images while the logical structure represents the relations between the different document elements. For instance, XML and to a lesser extent HTML allow us to identify elements in a document (like its title or links to other documents) and to describe relations between those elements (we can identify the author of a specific part of the text). Figure 10.1 gives an example of such a document.

Given the growing amount of structured document collections, it is important to develop tools able to take into account the increased complexity of these

Ludovic Denoyer
LIP6, UPMC, Paris, France, e-mail: `ludovic.denoyer@lip6.fr`

Patrick Gallinari
LIP6, UPMC, Paris, France, e-mail: `patrick.gallinari@lip6.fr`

(**a**)

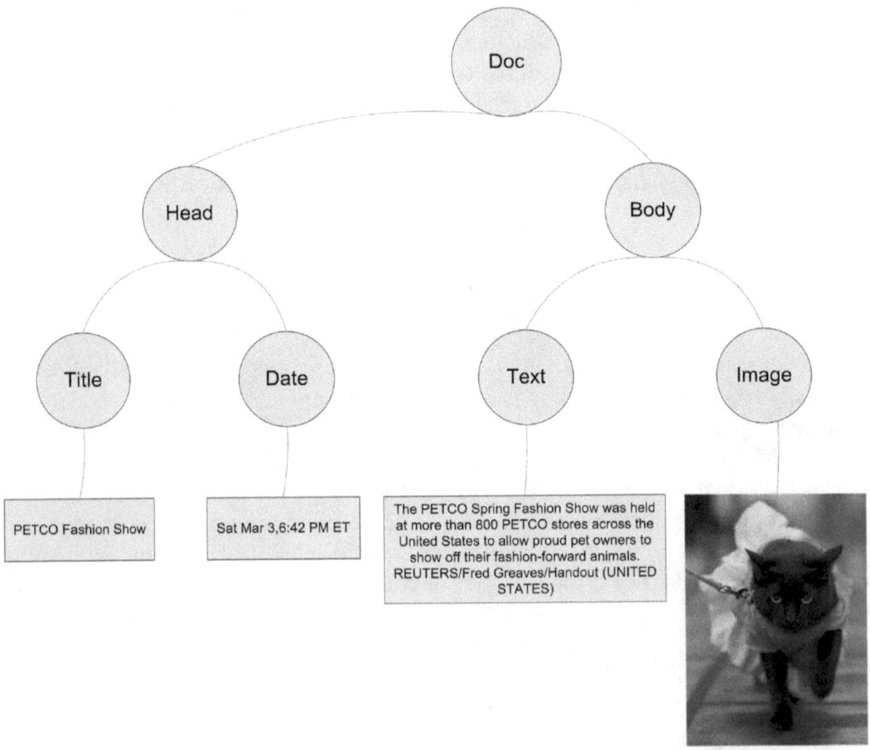

(**b**)

Fig. 10.1 An example of semi-structured multimedia document (**a**) and the corresponding labelled tree (**b**) representations. Up to now, information retrieval (IR) has mainly developed models for handling flat documents, and IR methods should now adapt to these new types of documents.

In this work, we focus on the task of *supervised classification* of semi-structured documents. This problem is a generic IR problem which have a lot of different applications: email filtering or classification, thematic classification of Web pages, document ranking, spam detection, etc. Although classification has been considered in IR for a long time, it is mainly since the 1990s that it has gained popularity and has developed as a sub-branch of the IR domain. Much progress in this area has been obtained through recent machine learning classification techniques. Most classification models for text or image have been developed before the emergence of structured documents and are devoted only to flat representations. Recently, some attempts have been made to adapt these techniques for the classification of complex documents, e.g. XML textual documents or multimedia documents.

We propose here a more principled approach to the problem of structured multimedia document classification. The model presented in this article is able to simultaneously take into account the structure and the content information. It can be seen as a *meta model* because it allows to combine existing models for flat documents. It is based on the use of dynamic belief networks which are classical statistical framework for learning dependences between random variables. The main advantage of this model is its low complexity which allows us to learn it and use it on a large number of documents.

The chapter is organized as follows: we first review in Sect. 10.2 existing work on information classification for structured documents. We then describe our model for classifying multimedia documents in Sect. 10.3. Finally, we describe the two corpora used for the experiments and explain the results obtained for the filtering and classification tasks.

10.2 Previous Work

A large amount of work has been dedicated over the last few years on document categorization, either text or image. However, only a few attempts deal with principled methods for structured and multimedia document classification. In the information retrieval community, it is often considered that structure should play an important role: for example, a word can have different meanings according to its location in the document or a document can be considered relevant for a class, if only one of its parts is relevant. On the other hand, most models have been developed for flat documents only. For example, in the text community, bag of word representations, which do not consider word ordering or document structure, are used in most models. We briefly review below recent work on structured and multimedia document classification.

10.2.1 Structured Document Classification

Generally speaking, classifiers fall into two categories: generative models which estimate class conditional densities $P(document|class)$ and discriminant models which directly estimate posterior probabilities $P(class|document)$. For instance, the Naive Bayes model [14] is a popular generative categorization model while support vector machines [11] has been widely used among discriminant models for the last few years. See [16] for a complete review of the categorization models which handle flat documents.

In machine learning, most document classifiers are designed for vector or sequence representations, and very few models consider content and structure information. The growing number of structured documents has recently motivated some work in this area. Some years ago, the development of the Web has created a need for classifying HTML pages. In an HTML page, the different parts of the text do not play the same role, and titles, links, text can be considered as different sources of information. Most of the techniques developed recently for this task use prior knowledge about the meaning of HTML tags to encode page structure using very simple schemes or to combine basic classifiers [9, 17]. These first attempts show that combining these different types of information may sometimes increase page categorization scores.

More recently, different techniques have been developed for general structured document classification. These models are not HTML specific and can be used for any type of structured documents, in particular XML. For example, Yi and Sundaresan [18] present an extension of the Naive Bayes model to semi-structured documents where essentially global word frequencies estimators are replaced with local estimators computed for each path element. A drawback of this technique is an important index growth which leads to poor estimation of probabilities. The hidden tree Markov model (HTMM) proposed by Diligenti et al. [8] extends classical HMM to semi-structured representations. Documents are represented by a tree and for each node, words are generated by a specific HMM. This model has been used for HTML document classification.

Dealing with XML document collections is a particularly challenging task for ML and IR. Compared to other domains where the structured data consists only of the "structure" part with no content this is much more complex and this had been addressed only for very specific problems in the ML community. Note that most existing ML methods can only deal with only one type of information (either structure or content). Some new methods for XML classification (see [6] for a summary of these methods) have been recently proposed in the context of the XML Document Mining Challenge.[1] Note that the Wikipedia corpus used in our article is very close to the one used in this track. It is not exactly the same because it has been adapted to Multimedia classification.

[1] http://xmlmining.lip6.fr.

10.2.2 Multimedia Documents

Many different methods have been proposed for the classification of multimedia documents. Most work in the multimedia area makes use of text information such as keywords for enhancing the performance of existing image, video or song classifiers. For example, Cascia et al. [3] propose a system that combines textual and visual statistics in a single index vector. Textual statistics are captured using latent semantic indexing (LSI) and visual ones are colour and orientation histograms. This approach allows improving performance in conducting content-based search. Barnard and Forsyth [1] present a generative hierarchical model, where the data is modelled as being generated by a fixed hierarchy of nodes. Each node in the tree has some probability of generating each word or image feature. This model could be useful for IR tasks such as database browsing and search for images based on text and image features. In [15], a document is represented as a collection of objects which themselves are represented as collections of features (words for text, colour and texture features for images). Several similarity measures over text and images are combined. More recently, a method has been proposed for using images for word sense disambiguation, which suggests that combining image features and text can outperform simple text classifiers [2]. Some work has been done in the field of Web filtering. Jones and Rehg [12] combine a naked people photo image detector with standard text analysis using an "OR" operator. Chan et al. [4] present an algorithm to identify images that contain large areas of skin. To identify pornographic pages, they combine this skin detector with a text analyser into a weighing scheme. They also claim that text-based approaches generally give poor results, whereas our structured model performs quite well on text-only documents. Most of these attempts rely on the combination of basic classifiers trained independently on the different information sources (e.g. text and images). They do not consider the global context of the document nor its logical organization, i.e. they ignore the relations between the different document parts.

The model we propose provides a general framework for the classification of structured multimedia documents. It can be used for any structural representation or language (e.g. HTML or XML). This model is an extension to multimedia data of the model proposed in [5] which operates only on textual structured document.

10.3 Multimedia Generative Model

10.3.1 Classification of Documents

We consider here the problem of *single label classification* where each document belongs to exactly one category. This problem corresponds to the classical filtering problem where we have to detect if a document belongs or not to a particular category (i.e. pornography for example). We consider here a set of categories denoted by $\mathscr{C} = \{c_i\}_{i \in [1,C]}$ where C is the number of categories. Let $\mathscr{D}$ be the set of documents.

We consider the *target function* $\Phi : \mathscr{D} \rightarrow \mathscr{C}$ which associates to each document d_j the real category of the document. The goal of our classification model is to learn a *decision function* $\Psi : \mathscr{D} \rightarrow \mathscr{C}$ that is a good approximation of the target function. The function Ψ will be learned over a set of labelled documents (i.e. training set).

10.3.2 Generative Model

We adopt a stochastic approach to the problem of classification. We consider a model that compute the probability of generating a specific document $d \in \mathscr{D}$ for a category $c \in \mathscr{C}$ denoted by $P(d|\theta_c)$, where θ_c is a set of parameters corresponding to the class c. These parameters will be learnt over the training set.

The score of a document for a class c will be computed using the Bayes rule:

$$P(c/d) = \frac{P(d|\theta_c)P(c)}{P(d)}. \tag{10.1}$$

We consider the *single label* classification task so we can rewrite $P(d)$ as

$$P(d) = \sum_{i=1}^{C} P(d|\theta_{c_i})P(c_i). \tag{10.2}$$

The probability that the document d belongs to the class c is then

$$P(c/d) = \frac{P(d|\theta_c)P(c)}{\sum\limits_{i=1}^{C} P(d|\theta_{c_i})P(c_i)}. \tag{10.3}$$

This probability defines the following *decision function*:

$$\begin{aligned}\psi(d) &= \arg\max_{cj} \frac{P(d|\theta_{c_j})P(c_j)}{\sum\limits_{i=1}^{C} P(d|\theta_{c_i})P(c_i)} \\ &= \arg\max_{cj} P(d|\theta_{c_j})P(c_j).\end{aligned} \tag{10.4}$$

In the following, we will denote the model parameters by θ corresponding either to $\theta_{c_1}, \theta_{c_2}, \ldots$.

10.3.3 Description

10.3.3.1 Informal Description

A generative stochastic model for a document corresponds to specific hypothesis about the physical generation of this document. Different hypotheses should be considered and the choice of a particular model most often corresponds to a compromise between an accurate representation of the document generation process and practical constraints depending on the task the model will be used for, the difficulty for accurately estimating model parameters from data collections, the availability of labelled corpus, etc. Different assumptions for structured textual document representations are discussed in [5] where it is shown that best performances for text classification are obtained with rather simple models of the dependencies between doxels. As it is often observed in document classification, more sophisticated models do not lead to increased performance. We adapt here one of these simple models to multimedia documents. The corresponding generative process is as follows: an author who wants to build a document about a specific topic (class) will first imagine the global logical structure of the document, once this structure is built he will then fill the content of each structural element. This content will depend on the type of the structural element: the process of writing a title will differ from the one for a paragraph, writing text is different from inserting an image or a piece of music, etc. This process is a simplified view of the reality and as will be seen below additional simplifying assumptions will be introduced in order to meet practical constraints. It embodies some crucial facts about structured documents: doxels are different depending on their type and the logical element they belong to, both logical structure and plain content of documents are essential for their descriptions, the topic of a document may influence both its logical structure and its content. The latter idea implies that the logical structure of a document may sometimes contain important information for characterizing the document class.

10.3.3.2 Dynamic Belief Network

We consider here that a semi-structured document is an ordered labelled tree where each node can contain or not a content information. A document of size N can be thus considered as an observation (or evidence) over a set of $N \times 2$ random variables:

$$P(D) = P(S^1, ..., S^N, T^1, ...T^N), \tag{10.5}$$

where S^i corresponds to the ith node of the document, $S^i \in \mathscr{S}$ where $\mathscr{S}$ is the set of all possible node labels (i.e. the XML tags). T^i corresponds to the content of the node i, $T^i \in \mathscr{T}$ where $\mathscr{T}$ is the set of all possible *pieces of content* (i.e. text or image) (Fig. 10.2).

A particular document d is a set of evidence over these random variables:

$$d = (s_d^1, ..., s_d^N, t_d^1, ...t_d^N), \tag{10.6}$$

where s_d^i is the label of the node i of d and t_d^i is the content of the node i of d.

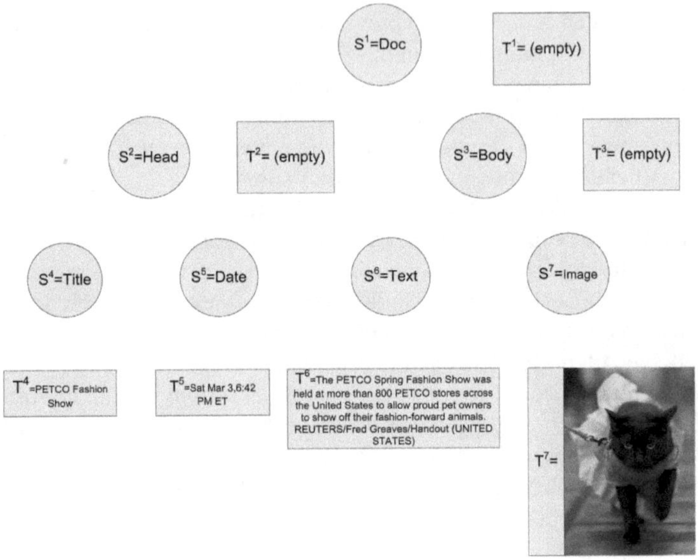

$$\mathscr{S} = \{\text{Doc}, \text{Head}, \text{Body}, \text{Title}, \text{Date}, \text{Text}, \text{Image}\}$$
$$N = 7$$

Fig. 10.2 The set of variables and evidence that correspond to the preceding document

Using (10.5), the probability of a document can be rewritten as

$$\begin{aligned} P(D=d|\theta) &= P(S^1 = s_d^1, ..., S^N = s_d^N, T^1 = t_d^1, ...T^N = t_d^N|\theta) \\ &= P(S^1 = s_d^1, ..., S^N = s_d^N|\theta) \\ &\quad \times P(T^1 = t_d^1, ...T^N = t_d^N|S^1 = s_d^1, ..., S^N = s_d^N, \theta). \end{aligned} \tag{10.7}$$

In the following, we will call $P(S^1 = s_d^1, ..., S^N = s_d^N|\theta)$ the *structural probability* and $P(T^1 = t_d^1, ...T^N = t_d^N|S^1 = s_d^1, ..., S^N = s_d^N, \theta)$ the *content probability*. The structural probability corresponds to the probability of a particular tree structure and the content probability corresponds to the probability of having a particular content information in conjunction with a known structure. In the following, in order to simplify the notations, we will write $P(s_d^1, ..., s_d^N|\theta)$ instead of $P(S^1 = s_d^1, ..., S^N = s_d^N|\theta)$ and $P(t_d^1, ...t_d^N|s_d^1, ..., s_d^N, \theta)$ instead of $P(T^1 = t_d^1, ...T^N = t_d^N|S^1 = s_d^1, ..., S^N = s_d^N, \theta)$.

We cannot compute directly this joint probability. We have to make different dependency assumptions over the random variables. In order to model these assumptions, we use the formalism of the belief networks that allows us to graphically represent these dependencies. Basically, a belief network is an oriented graph where the vertices correspond to the random variables and where an edge between two variables A and B explains that A is conditionally dependent on B.

10.3.3.3 Dependencies Between Structural Variables S_i.

The Fig. 10.3 shows different assumptions for a document with seven nodes. The number of random variables depends on the size (the number of nodes) of a document thus we are in the context of dynamic belief networks (see [10] for more

Belief network	Structural probability

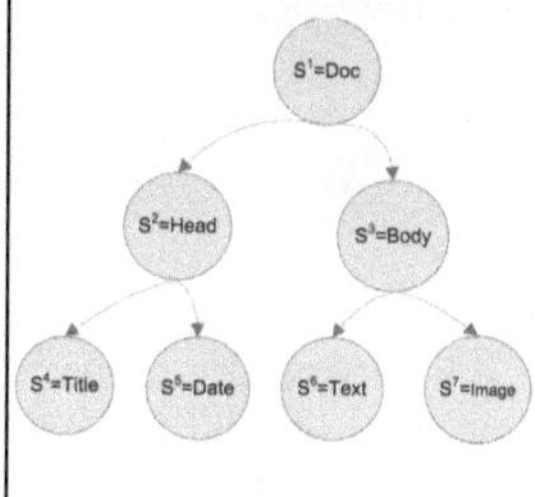

$$\begin{aligned} P(d) = & P(S^1|\theta)P(S^2|S^1,\theta)P(S^3|S^1,\theta)P(S^4|S^2,\theta) \\ & .P(S^5|S^2,\theta)P(S^6|S^3,\theta)P(S^7|S^3,\theta) \\ & = P(\text{Doc}|\theta)P(\text{Head}|\text{Doc},\theta)P(\text{Body}|\text{Doc},\theta) \\ & .P(\text{Title}|\text{Head},\theta)P(\text{Date}|\text{Head},\theta) \\ & .P(\text{Text}|\text{Body},\theta)P(\text{Image}|\text{Body},\theta) \end{aligned}$$

(a)

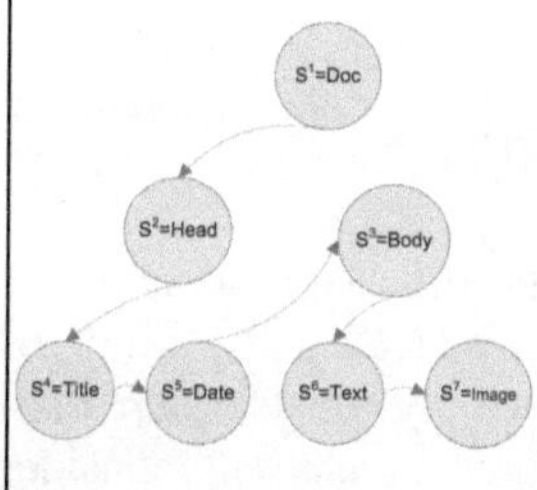

$$\begin{aligned} P(d) = & P(S^1|\theta)P(S^2|S^1,\theta)P(S^4|S^2,\theta)P(S^5|S^4,\theta) \\ & .P(S^3|S^5,\theta)P(S^6|S^3,\theta)P(S^7|S^6,\theta) \\ & = P(\text{Doc}|\theta)P(\text{Head}|\text{Doc},\theta)P(\text{Title}|\text{Head},\theta) \\ & .P(\text{Date}|\text{Title},\theta)P(\text{Body}|\text{Date},\theta) \\ & .P(\text{Text}|\text{Body},\theta)P(\text{Image}|\text{Text},\theta) \end{aligned}$$

(b)

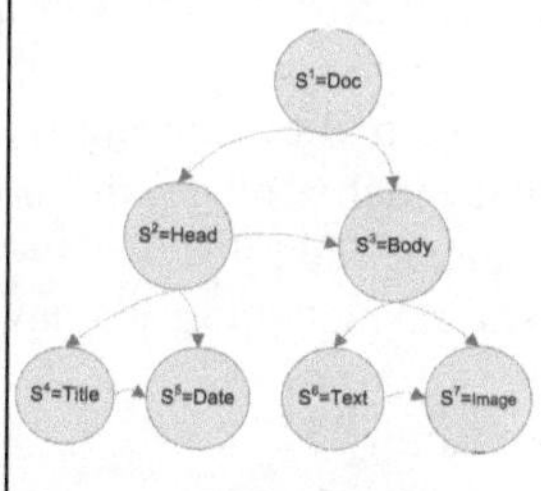

$$\begin{aligned} P(d) = & P(S^1|\theta)P(S^2|S^1,\theta)P(S^3|S^1,S^2,\theta)P(S^4|S^2,\theta) \\ & .P(S^5|S^2,S^4,\theta)P(S^6|S^3,\theta)P(S^7|S^3,S^6,\theta) \\ & = P(\text{Doc}|\theta)P(\text{Head}|\text{Doc},\theta)P(\text{Body}|\text{Doc},\text{Head},\theta) \\ & .P(\text{Title}|\text{Head},\theta)P(\text{Date}|\text{Head},\text{Title},\theta) \\ & .P(\text{Text}|\text{Body},\theta)P(\text{Image}|\text{Body},\text{Text},\theta) \end{aligned}$$

(c)

Fig. 10.3 Different belief networks and associated structural probability of a seven-node document. In **(a)**, the probabilistic dependencies follow the natural structure of the document. In **(b)**, the dependencies correspond to the sequence of the nodes. In **(c)** one considers both the parent information and the left sibling information. This last model is more complex than the other two. In this article, we have chosen to use the first model **(a)**

details). In the following, we consider only the case of a document with N nodes. This case can easily be extended to all the document sizes.

The structural variables denoted by S_i model the possible node labels. We consider here that the dependencies between structural variables follow the natural structure of the document (part *(a)* of Fig. 10.3). Note that other types of dependencies have been studied but gives quite similar results. We consider that a dependency from S_i to S_j exists if the node i is the parent of the node j in the document. Let us denote $pa_d(j)$ the *parent* function that corresponds to the index of the parent node of the node j in document d, the structural probability can be rewritten as

$$P(s_d^1,...,s_d^N|\theta) = \prod_{i=1}^{N} P(S^i = s_d^i|S^{pa_d(i)} = s_d^{pa_d(i)}). \tag{10.8}$$

For those who are not familiar with dynamic belief networks, Fig. 10.3 illustrates this process. The hypothesis here (a) is that a node label only depends on its parent label. For our generating process, this means that starting at the root, the first level of structural elements is built, after that the descendant of a structural node is built independent of the node brothers descendants. Here again, this can be viewed as a simplified process for defining the logical organization of a document.

10.3.3.4 Dependencies Between Content Variables T_i

As done in the previous section, we also have to consider different assumptions in order to simplify the computation of the content probability $P(t_d^1,...t_d^N|s_d^1,...,s_d^N,\theta)$. As the content depends on the structure of a document, we only consider here that a writer who writes a piece of content of a semi-structured document only considers the label of the part he is writing in order. Returning to our generation process, this means that once the document organization has been decided, each content node is filled independent of the others, by considering only the type of the structural element it belongs to. All content elements with the same type (paragraph, etc.) will share the same generative process. It could seem more natural to consider that content elements are filled in sequence, but early tests with such a model did not lead to improved results at the price of an increased complexity and this was then left out. Note that such simplifications are frequent in stochastic modelling and have led to very efficient models in different application areas.

With these assumptions, we can rewrite the content probability as

$$P(t_d^1,...t_d^N|s_d^1,...,s_d^N,\theta) = \prod_{i=1}^{N} P(t_d^i|s_d^i,\theta). \tag{10.9}$$

10.3.3.5 Final Probability

Finally, using (10.8) and (10.9), the joint probability corresponding to a document d is written as

$$P(d|\theta) = \prod_{i=1}^{N} P(S^i = s_d^i | S^{pa_d(i)} = s_d^{pa_d(i)}, \theta) \prod_{i=1}^{N} P(t_d^i | s_d^i, \theta). \quad (10.10)$$

The probabilities $P(S^i = s_d^i | S^{pa_d(i)_d} = s_d^{pa_d(i)})$ correspond to the probabilities of seeing a node with label s_d^i as child of a node with label $s^{pa_d(i)}$ (for example: probability of having a node with label *title* as child of a node with label *document*). This probabilities are defined on $\mathcal{S} \times \mathcal{S}$ and can easily be estimated using a training corpus (see Sect. 10.4).

The probabilities $P(t_d^i | s_d^i, \theta)$ correspond to the probability of seeing a particular content information t_d^i in a node with label s_d^i. The content information t_d^i can be of different nature (text, image, ...). In this work, we propose to model $P(t_d^i | s_d^i, \theta)$ using generative models called *local generative models*. Each node type in the structure will have its own generative content model. For example, the nodes with label *title* will use a local generative model of textual information (e.g. Naive Bayes for example), the node with labels *image* will use a local generative model of pictures (see Sect. 10.5), and so on. Our model can be seen as a *meta model* that allows us to use any kind of existing generative model, one generative model for each possible node label. Each local generative model will have its own parameters denoted by θ^l where l corresponds to the node label associated to this local generative model. With such an assumption, we can rewrite (10.10) as

$$P(d|\theta) = \prod_{i=1}^{N} P(s_d^i | s^{pa_d(i)}, \theta^s) \prod_{i=1}^{N} P(t_d^i | s_d^i, \theta^{s_d^i}), \quad (10.11)$$

where $\theta = \theta^s \bigcup \{\theta^l\}_{l \in \mathcal{S}} - \theta^s$ corresponds to the parameters for the structural probability (so called *structural parameters*).

As an example, for modelling the document in Fig. (10.1), we will use seven local generative models:

- A textual generative model of parameters θ^{Doc} for the text contained in tags Doc
- A textual generative model of parameters θ^{Head} for tags *Head*
- A textual generative model of parameters θ^{Title} for the text contained in tags Title
- A textual generative model of parameters θ^{Date} for tags *Date*
- A textual generative model of parameters θ^{Body} for the text contained in tags Body
- A textual generative model of parameters θ^{Text} for tags *Text*
- An image generative model of parameters θ^{Image} for the image contained in tag image

In fact, in this case, the models for Doc, Head and Body will not be created because these nodes do not contain any content information.

Rewriting (10.11) as

$$P(d|\theta) = \prod_{i=1}^{N} P(s_d^i | s_d^{pa_d(i)}, \theta^s) P(t_d^i | s_d^i, \theta^{s_d^i}), \quad (10.12)$$

we can see that our model corresponds to a *mixture of local generative models* where each local generative model is weighted by a structural probability $P(s_d^i|s_d^{pa_d(i)}, \theta^s)$.

In Sect. 10.5, we explain how we can model this probability using classical models from the IR community. Note that this model is able to deal with any kind of information (text, image, video, sound, etc.) if there exist a generative model for this kind of content which is often the case.

10.4 Learning the Meta Model

The model parameters will be learned by maximizing the data likelihood. Note that we expressed here how to train our model using a training set $\mathscr{D}_{\text{TRAIN}}$. For the categorization task, we will train a model for each class c using only the training documents that belong to this class c.

The log-likelihood of our training data is

$$
\begin{aligned}
L = & \sum_{d \in \mathscr{D}_{\text{TRAIN}}} \left\{ \sum_{i=1}^{|s_d|} \left(\log P(s_d^i|s_d^{pa_d(i)}, \theta^s) \right) \right. \\
& \left. + \sum_{i=1}^{|s_d|} \left(\log P(t_d^i|\theta_{s_d^i}) \right) \right\} \\
= & \left\{ \sum_{d \in \mathscr{D}_{\text{TRAIN}}} \sum_{i=1}^{|s_d|} \left(\log P(s_d^i|s_d^{pa_d(i)}, \theta^s) \right) \right\} + \\
& \left\{ \sum_{d \in \mathscr{D}_{\text{TRAIN}}} \sum_{i=1}^{|s_d|} \left(\log P(t_d^i|\theta_{s_d^i}) \right) \right\} \\
= & L_{\text{structure}} + L_{\text{content}}.
\end{aligned}
\tag{10.13}
$$

The maximization of L amounts to two separate maximizations on $L_{\text{structure}}$ and L_{content}. This is a classical optimization problem and the solution for the structural and content parameters is described in the following part.

10.4.1 Maximization of $L_{\text{structure}}$

We want to maximize

$$
\begin{aligned}
L_{\text{structure}} = & \sum_{d \in \mathscr{D}_{\text{TRAIN}}} \sum_{i=1}^{|s_d|} \log P(s_d^i|s_d^{pa_d(i)}, \theta^s) \\
= & \sum_{d \in \mathscr{D}_{\text{TRAIN}}} \sum_{i=1}^{|s_d|} \log \theta^s_{s_d^i, s_d^{pa_d(i)}}
\end{aligned}
\tag{10.14}
$$

under the constraint $\forall m \in \mathscr{S}, \sum_{l \in \mathscr{S}} \theta^s_{l,m} = 1$.

Using the Lagrange multipliers, for each $(n,m) \in \mathscr{S} \times \mathscr{S}$, we have

$$\frac{\partial(L_{\text{structure}} - \lambda_m(\sum_n \theta^s_{n,m} - 1)}{\partial \theta^s_{n,m}} = 0. \tag{10.15}$$

Let $N^d_{n,m}$ be the number of times a node of label n has its parent with label m in the document d, we solve

$$\frac{\sum_{d \in \mathscr{D}_{\text{TRAIN}}} N^d_{n,m}}{\theta^s_{n,m}} = \lambda_m. \tag{10.16}$$

The solution is

$$\theta^s_{n,m} = \frac{\sum_{d \in \mathscr{D}_{\text{TRAIN}}} N^d_{n,m}}{\sum_i \sum_{d \in \mathscr{D}_{\text{TRAIN}}} N^d_{i,m}}. \tag{10.17}$$

10.4.2 Maximization of L_{content}

We want to maximize

$$\begin{aligned} L_{\text{content}} &= \sum_{d \in \mathscr{D}_{\text{TRAIN}}} \sum_{i=1}^{|s_d|} \left(\log P(t^i_d | \theta^{s^i_d}) \right) \\ &= \sum_{l \in \mathscr{S}} \left(\sum_{d \in \mathscr{D}_{\text{TRAIN}}} \sum_{i=1/s^i_d=l}^{|s_d|} \log P(t^i_d | \theta^l) \right) \\ &= \sum_{l \in \mathscr{S}} L^l_{\text{content}}. \end{aligned} \tag{10.18}$$

This maximization is performed by maximizing the log-likelihood of each local generative model onto its own data. This means that learning our meta model is as simple as learning the local generative models separately.

10.5 Local Generative Models for Text and Image

We describe here the local generative models for text and image that have been used in our experiments. Note that more complex models could be used in order to improve the performances of the general model. Particularly, the image model used here is very simple (but quite well adapted to pornographic filtering) and the Wikipedia collection would need a more intelligent one.

10.5.1 Modelling a Piece of Text with Naive Bayes

If t_d^i corresponds to a textual information, the probability $P(t_d^i|s_d^i,\theta)$ can be easily computed using a textual generative model like Naive Bayes (see [13]), languages models, HMM, etc. In this article, we propose to use a simple Naive Bayes model (one Naive Bayes model for each possible node label). This model considers a probabilistic independence between the words of a text: each text is thus modelled as a bag of word. This model is known to be very robust when data belong to very high-dimensional spaces.

Let $t_d^i = (w_{d,1}^i, ..., w_{d,|t_d^i|}^i)$ be the textual content of node i in d, where $w_{d,k}^i$ represents the kth word of node i in d. $|t_d^i|$ is the number of words in node i.

NB computes the probability $P(t_d^i|\theta_{s_d^i})$ as

$$P(t_d^i|s_d^i,\theta^{s_d^i}) = \prod_{k=1}^{|t_d|} P(w_{d,k}^i|\theta^{s_d^i}). \tag{10.19}$$

Our model will use one Naive Bayes model for each label $l \in \mathscr{S}$. The model with parameters θ^l will be learned using the flat textual representation of all nodes with label l in the training set.

10.5.1.1 Learning Textual Models

In order to learn our textual Naive Bayes model, we have to maximize the log-likelihood for each possible node label. We do not detail the computation here but only give the results. Using a classical smoothing techniques, we obtain that for each possible word $w \in \mathscr{W}$ – $\mathscr{W}$ is called the word dictionary – and for each $l \in \mathscr{S}$, we have

$$P(w|l) = \frac{W_w^l + 1}{W_w + (\text{size of } \mathscr{W})}, \tag{10.20}$$

where W_w^l is the number of times the word w has been seen under in a node with label l in the training set and W_w is the number of times the word w has been seen in the train set.

10.5.2 Image Model

Before deciding on the image modelling, we made extensive preliminary experiments on the classification of pornographic images. As in [4], the conclusion was that the best workspace to detect pornographic images was the RGB colour space and that additional components like texture or shape did not improve performance. We then decided to represent images with a colour histogram in a normalized space. Note that this model is certainly not well adapted to Wikipedia.

Let t_d^i be an image, its histogram representation will be

$$t_d^i = (p_{d,1}^i, \ldots, p_{d,N_c}^i), \tag{10.21}$$

where $p_{d,k}^i$ is the number of pixels in the image with colour k. N_c represents the number of colours in the histogram. In order to keep image scores comparable, image size has been normalized to N_p pixels before computing the histogram. Under the independence hypothesis, we have

$$P(t_d^i|\theta_{s_d^i}) = \prod_{k=1}^{N_c} P(P_k = p_{d,k}^i|\theta_{s_d^i}), \tag{10.22}$$

where $P(P_k = p_{d,k}^i|\theta_{s_d^i})$ is the probability that there are $p_{d,k}^i$ pixels with colour k in image t_d^i.

This model is learned using a simple pixel count over all the images in the training set.

10.6 Experiments

10.6.1 Models and Evaluation

10.6.1.1 Models

We have made our experiments using different models:

- The *flat Naive Bayes* model is the reference baseline model on the textual information.
- The *flat TFIDF SVM* model is the other reference baseline model that corresponds to the use of a SVM with a linear kernel over TF-IDF vectors computed on flat documents.
- The *generative text model* corresponds to our generative model using only textual information
- The *generative text+image model* corresponds to our generative model using both textual and image information.

10.6.1.2 Evaluation Measures

As an evaluation measure, we use the recall obtained on the test corpus for each class. In order to give a synthetic recall value for each classifier, we compute for each of them their micro-average and macro-average recall. *Macro-average recall* is obtained by averaging recall values for the different classes. *Micro-average recall* is obtained by weighting the average by the relative size of each class.

10.6.2 Corpora

10.6.2.1 The Web Pornographic Corpus

We used for the experiments a database of about 100,000 HTML documents grabbed on the Web. These documents are labelled as pornographic or not. Note that the documents grabbed are expressed in different languages. In our experiments, we used half of each group for training and the remaining half for testing. Experiments have been made using the generative model and allow us to conclude that our meta model is able to mix the information from different media (text and image).

Preprocessing

For this corpus, textual parts of HTML documents have been cleaned by deleting figures, words smaller than three letters and all non-alphabetical symbols. The final vocabulary was composed of 20,834 terms. Images have been converted into features using colour histograms of size 100. All images have been projected in a 216 colour space.

10.6.2.2 The Wikipedia XML Corpus for Multimedia Classification

Wikipedia[2] is a well-known free content, multilingual encyclopaedia written collaboratively by contributors around the world. Anybody can edit an article using a wiki markup language that offers a simplified alternative to HTML. This encyclopaedia is composed of millions of articles in different languages. We have built a multimedia corpus for categorization using a subset of the whole Wikipedia XML corpus (see [7]). These corpus[3] is composed of 23,562 XML documents that contain images. These documents are associated with 21 different categories that correspond to Wikipedia Portals. Table 10.1 gives a summary of the document collection and Table 10.2 shows the distribution of the documents over the different categories.

Table 10.1 Description of the Wikipedia corpus for multimedia classification

Number of documents	11,830 (train) and 11,372 (test)
Number of XML nodes	1,496,290
Size of the vocabulary (without pruning)	180,445 words
Size of the vocabulary (with pruning at 3)	51,997 words
Number of node labels (XML tags)	3,767
Number of categories	21

[2] http://www.wikipedia.org.

[3] http://xmlmining.lip6.fr.

Table 10.2 Categories of the Wikipedia multimedia categorization corpus

Name of the category	Number of documents
Archaeology	1355
Art	7,624
Astronomy	1,105
Aviation	1,217
Chemistry	4,567
Christianity	4,671
Comics	600
Formula One	1,188
History	3,246
Law	24,213
Literature	16,929
Music	401
Physics	5,149
Pornography	458
Sexuality	2,402
Spirituality	2,704
Sports and games	14,595
Trains	953
University	605
War	2,217
Writing	412

In our experiment, we have used 50% of the documents for training and 50% for test. Results have been compared to classical text categorization techniques (Naive Bayes and SVM TF-IDF on flat documents).

Preprocessing

For this corpus, textual parts of XML documents have been preprocessed using a Porter Stemmer, words smaller than three letters and all non-alphabetical symbols have also been removed. We have kept words that appear at least three times. The final vocabulary was composed of 51,997 terms. Images have been converted into features using colour histograms of size 100. All images have been projected in a 64 colour space.

10.6.3 Results over the Pornographic Corpus

The baseline *Naive Bayes model* yields reasonably high micro-average and macro-average recall (88.4% and 89.9%) (Table 10.3). The *generative image only* model is lower and achieves only 82.7% on micro-average and 83% on macro-average. Our structured textual model has good results and is 4.5% better than Naive Bayes for the micro-average and 2.6% for the macro-average. The multimedia model is even

Table 10.3 The recall values for the four models using the pornographic Web corpus

Model	Porno	Not porno	Micro-average	Macro-average
Naive bayes	92.4	87.3	88.4	89.9
Generative text	91.8	93.1	92.9	92.5
Generative using only images	88.4	77.6	82.7	83.0
Generative text+image	**91.6**	**95.4**	**94.7**	**93.6**

better than the *generative text* model and achieves respectively 94.7% and 93.6% for the micro-average and the macro-average recall. These results are very encouraging and show that our structured approach is able to handle multimedia documents and to combine efficiently the information of simple generative models. Error rate is divided by 2 compared to NB and by 3 compared to *generative image only* model (Fig. 10.4).

10.6.4 Results over the Wikipedia Multimedia Categorization Corpus

The experiments made on the Wikipedia corpus help us to know how this model is able to deal with a difficult thematic categorization task. Table 10.4 shows the results of different models over this collection.

Text Only

We can see here that the baseline Naive Bayes model on text-onlyz information performs quite well on this corpus (micro-average recall is 74.2%) while due to

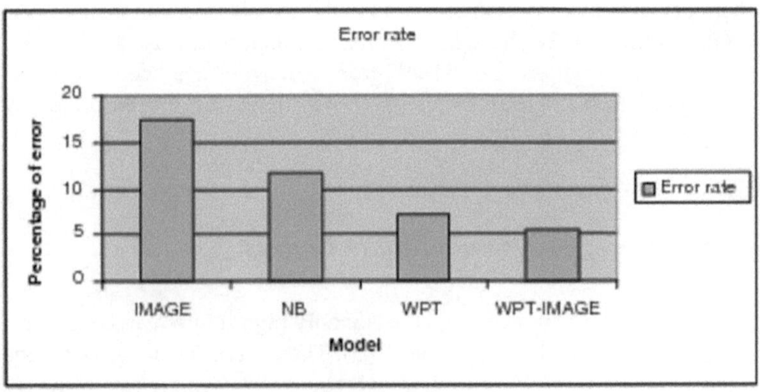

Fig. 10.4 The error rate for the four models for pornographic filtering

Table 10.4 Performance on the Wikipedia corpus

Model	Micro-average recall (%)	Macro-average recall (%)
Models on flat documents		
Naive Bayes text only	74.2	57.4
SVM TF-IDF	26.3	10.4
Models on structured documents		
Generative model text only (all tags)	49.7	18.0
Generative model text only (5 most frequent tags)	70.4	47.4
Generative model text+image (5 most frequent tags+image)	67.7	35.1
Image only	21.6	4.7

the size of the vector space and the low number of examples for each category, the SVM TF-IDF does not perform well. This is mainly due to over-fitting (this model performs 100% on the training set). Note that we have used the SVM models with a lot of different parameters and the performances proposed here are the best we obtained.

Text and Structure

The structured generative model using all the tags does not perform well (about 49.7% micro-average recall). It is mainly due to the number of parameters it has to learn (3767 tags × 51,997 words = about 200 million parameters) with respect to the small size of the corpus. We have performed experiments keeping only the five most frequent tags in the collection and obtained performances quite similar to the ones obtained with Naive Bayes (70.4% on micro-average). This model has already shown that it is able to have nice performances on XML classification on different corpora (see [5] and previous part for example of nice performances) but it seems that, on the Wikipedia corpus, it is not able to learn to deal correctly with the structure.

Text and Image

The classification of images in different thematics is a very hard task as it is shown by the results. The *image only* model only performs 5% macro-average recall which basically corresponds to a random model. These low performances decrease the results of the generative text+image model from 70.4% using only the textual information to 67.7% using both image and text. Note that the image model is – based on histogram – very simple and more experiments have to be made using a stringer model.

10.7 Conclusion

We have presented here a generative model that is able to model both the structure and the content of semi-structured – XML – multimedia documents. The main advantage of this model is that it is a *meta model* that is based on existing generative models. It allows us to use a large variety of methods and to mix them for the classification of complex documents. The model has shown nice performances on different XML classification tasks (see [den03b]) and particularly on the pornographic filtering task where the use of text, structure and image allows us to obtain a recall that is 6% better than a classical flat text classifier. On the other hand, this model which deals with many parameters has some difficulties to deal with small thematic corpus as presented and must be trained with a large amount of data.

References

1. K. Barnard and D. Forsyth. Learning the semantics of words and pictures. In *Proceedings of the 8th International Conference on Computer Vision*, volume 2, pages 408–415, 2001.
2. K. Barnard, M. Johnson, and D. Forsyth. Word sense disambiguation with pictures. In *Workshop on Learning word Meaning from Non-Linguistic Data*, 2003.
3. M. L. Cascia, S. Sethi, and S. Sclaroff. Combining textual and visual cues for content-based image retrieval on the world wide web. In *Proceedings of IEEE Workshop on Content-Based Access of Image and Video Libraries*, June 1998.
4. Y. Chan, R. Harvey, and D. Smith. Building systems to block pornography. In *Challenge of Image Retrieval*, 1999.
5. L. Denoyer and P. Gallinari. Using Belief Networks and Fisher Kernels for structured document classification. In *PKDD 2003*, 2003.
6. L. Denoyer and P. Gallinari. Report on the XML Mining Track at INEX 2005 and INEX 2006. In *Advances in XML Information Retrieval and Evaluation: Fifth Workshop of the INitiative for the Evaluation of XML Retrieval (INEX'06)*, 2007.
7. L. Denoyer and P. Gallinari. The Wikipedia XML Corpus. In *Advances in XML Information Retrieval and Evaluation: Fifth Workshop of the INitiative for the Evaluation of XML Retrieval (INEX'06)*, 2007
8. M. Diligenti, M. Gori, M. Maggini, and F. Scarselli. Classification of HTML documents by Hidden Tree-Markov Models. In *6th International Conference on Document Analysis and Recognition*, Seattle, WA, USA, August. 2001.
9. S. T. Dumais and H. Chen. Hierarchical classification of Web content. In N. J. Belkin, P. Ingwersen, and M.-K. Leong, editors, *Proceedings of SIGIR-00, 23rd ACM International Conference on Research and Development in Information Retrieval*, pages 256–263, Athens, GR, 2000. ACM Press, New York, US.
10. Z. Ghahramani. Learning Dynamic Bayesion Networks In Lecture Notes in Computer Science, pages 168–197, 1998
11. T. Joachims. Text categorization with support vector machines: learning with many relevant features. In C. Nédellec and C. Rouveirol, editors, *Proceedings of ECML-98, 10th European Conference on Machine Learning*, number 1398, pages 137–142, Chemnitz, DE, 1998. Springer Verlag, Heidelberg, DE.
12. M. J. Jones and J. M. Rehg. Detecting adult images. Technical report, 2002.
13. D. D. Lewis. *Representation and Learning in Information Retrieval.* PhD thesis, Department of Computer Science, University of Massachusetts, Amherst, US, 1992.

14. D. D. Lewis. Naive (Bayes) at forty: The independence assumption in information retrieval. In C. Nédellec and C. Rouveirol, editors, *Proceedings of ECML-98, 10th European Conference on Machine Learning*, number 1398, pages 4–15, Chemnitz, DE, 1998. Springer Verlag, Heidelberg, DE.
15. M. Ortega, K. Porkaew, and S. Mehrotra. Information retrieval over multimedia documents. In *the SIGIR Post-Conference Workshop on Multimedia Indexing and Retrieval (ACM SIGIR)*, 1999.
16. F. Sebastiani. Machine learning in automated text categorization. *ACM Computing Surveys*, 34(1), 2002.
17. Y. Yang, S. Slattery, and R. Ghani. A study of approaches to hypertext categorization. *Journal of Intelligent Information Systems*, 18(2–3):219–241, 2002.
18. J. Yi and N. Sundaresan. A classifier for semi-structured documents. In *Proceedings of the Conferance Knowledge Discovery in Data*, pages 190–197, 2000.

Chapter 11
Classification and Clustering of Music for Novel Music Access Applications

Thomas Lidy and Andreas Rauber

Abstract With an increasing number of people working with large music archives, advanced methods for automatic labeling and organization of music collections are required as these archives grow in size. Manual annotation and categorization is not feasible for massive collections of music. In the research domain of music information retrieval (MIR) a number of algorithms for the content-based description of music were developed, which perform the extraction of relevant features for the computation of similarity between pieces of music. This fundamental step enables a great range of applications for music retrieval and organization. With supervised machine learning, music can be classified into different kinds of categories, such as genres, artists or moods. Using unsupervised approaches such as the self-organizing map music can be clustered by similar style and visualized in a way that enables direct retrieval of similar music at a glance. In this chapter, we will review the most common audio feature extraction techniques, which serve as a basis for subsequent classification and clustering tasks. As an example, we will show how music is classified into a set of genres and how genre classification can be used for benchmarking. Moreover, the creation of the so-called "music maps" and their various visualizations is demonstrated, and an interactive application called "PlaySOM" is presented, with an interface which allows direct access to similar sounding pieces in a large music collection. Its mobile counterpart "PocketSOM-Player" allows direct playback from a music map on a mobile device without having to browse lists. Both allow the convenient interactive creation of situation-based playlists.

Thomas Lidy
Vienna University of Technology, Vienna, Austria, e-mail: `lidy@ifs.tuwien.ac.at`

Andreas Rauber
Vienna University of Technology, Vienna, Austria, e-mail: `rauber@ifs.tuwien.ac.at`

11.1 Introduction

Digital music databases are continuously gaining popularity both in the domains of professional repositories and personal audio collections. The increasing popularity and size of digital music archives drives the need for advanced methods to organize those archives. Traditional search based on file name, song title or artist does not meet the advanced requirements of people working with large music repositories, because it either presumes exact knowledge of these meta-data fields or involves browsing of long lists in the archive. Modern music information retrieval (MIR) systems offer search by content-based music similarity. They allow to formulate queries by audio examples – example songs, excerpts of recorded audio, hummed melodies, etc. – finding songs or artists similar to the query. Audio feature extraction, distance calculations in feature space and machine learning are the underlying techniques of modern MIR systems.

In the feature extraction domain there are two main directions: extraction from symbolic notations (e.g. MIDI files) and extraction from audio waveform signals. This chapter focuses entirely on systems that are based on extraction from waveform signals and describes applications that utilize machine learning techniques for accessing and interacting with music collections.

Research in recent years led to the development of a wealth of feature extraction algorithms, which are used to derive descriptors from audio content. Section 11.2 reviews a number of audio feature extraction techniques most commonly employed for content-based music retrieval.

Apart from retrieval by similarity, MIR systems offer a range of other advanced techniques for interaction with music repositories: The application of supervised machine learning techniques enables to classify pieces of music into a defined set of genres. Using these techniques, entire music archives can be classified and organized automatically. Some approaches and the accuracy of current music genre classification techniques are described in Sect. 11.3.

Moreover, this chapter explains how content-based audio description and unsupervised learning are employed to create interactive "music maps" for efficient music retrieval systems. Unsupervised learning approaches overcome genre boundaries and consider audio similarity independently from genre assignments. These approaches automatically build clusters of pieces of music which are similar according to the descriptors extracted from the acoustic content. Specifically, the self-organizing map (SOM) is employed in order to create virtual landscapes of music collections, the so-called "music maps". Advanced visualization methods are crucial for the presentation of music maps and are reviewed in Sect. 11.4, in which the creation of music maps is also illustrated.

The development of interfaces on top of the self-organizing map led to novel access methods and retrieval metaphors, which allow for intuitive browsing, discovering and searching of music. The map interface allows quick playlist creation by audio similarity or perceived music style. The PlaySOM application which offers many interaction facilities such as zooming into the music space and creating trajectories through the music map is described in detail in Sect. 11.5.

Mobile devices such as personal digital assistants (PDAs), mobile phones or portable audio players are in particular need for improved access to music collections, which motivated the implementation of music maps on handheld devices. The PocketSOMPlayer described in Sect. 11.6 allows for direct and intuitive access to music of the desired style and can play from the local storage of the portable device or via a streaming service. Its remote control function offers a convenient novel interface for controlling music in home environments.

11.2 Feature Extraction from Audio

Audio feature extraction algorithms perform an analysis of the content of audio, specifically music. While some of the techniques compute descriptors directly from the waveform of a piece of audio, the majority of feature extraction algorithms for audio content description are based on the spectral representation of the audio signal. In some form or another they use information of frequency bands, energy or statistical variations on the low level in order to be able to describe information necessary to detect rhythm, pitch, melody, timbre, etc. It was also shown that the incorporation of psycho-acoustics in order to be computationally close to human perception improves the information contained in the features. In this section a brief overview of some of the audio features most commonly employed in content-based audio retrieval, which also form the basis for subsequent analysis by machine learning techniques, is provided.

11.2.1 Low-Level Audio Features

The following features are low-level features employed in the context of many content-based audio retrieval projects, commonly in combination with other features or feature sets. They are for example also available in audio software frameworks such as MARSYAS [28], M2K [6] or CLAM [2].

11.2.1.1 Zero Crossing Rate

The Zero Crossing Rate (ZCR) is one of the features calculated directly from the audio waveform, i.e. in the time domain. It represents the number of times the signal crosses the 0-line, i.e. the signal changes from a positive to a negative value, within one second. It can be either a measure for the dominant frequency or the noisiness of a signal and serves as a basic separator of speech and music.

11.2.1.2 RMS Energy

Root Mean Square (RMS) energy is computed in time domain by calculating the mean of the square of all sample values in a time frame and taking the square root. Hence, it is a feature easy to implement. The RMS gives a good indication of loudness in a time frame and may also serve for higher-level tasks such as audio event detection, segmentation or tempo/beat estimation.

11.2.1.3 Low-Energy Rate

Low-energy rate is usually defined as the percentage of frames containing less energy than the average energy of all frames in a piece of audio. Energy is computed in time domain as RMS energy (see above). In [25] a frame is considered a low-energy frame when it has less than 50% of the average value within a one-second window.

11.2.1.4 Spectral Flux

Spectral flux is a frequency domain feature and is computed as the squared differences in frequency distribution of two successive time frames. It measures the rate of local change in the spectrum. If there is much spectral change between two frames the spectral flux is high.

11.2.1.5 Spectral Centroid

The spectral centroid is the center of gravity, i.e. the balancing point of the spectrum. It is the frequency where the energy of all frequencies below that frequency is equal to the energy of all frequencies above that frequency and is a measure of brightness and general spectral shape.

11.2.1.6 Spectral Rolloff

Another measure of spectral shape is the spectral rolloff which is the 90 percentile of the spectral distribution. It is a measure of the skewness of the spectral shape.

11.2.2 MPEG-7 Audio Descriptors

The Moving Picture Experts Group (MPEG) is a working group of ISO/IEC in charge of the development of standards for digitally coded representation of audio and video.[1] The MPEG-7 standard defines the multimedia content description interface and is the standard for description and search of audio and visual content. Part 4 of the MPEG-7 standard [1] contains a description of a number of low-level audio descriptors and some high-level description tools. The five defined sets for high-level audio description are partly based on the low-level descriptors and are intended for specific applications (description of audio signature, instrument timbre, melody, spoken content as well as for general sound recognition and indexing) and will not be further considered here.

The MPEG-7 low-level audio descriptors comprise 17 temporal and spectral descriptors, divided into seven classes. Some of them are based on basic waveform or spectral information while others use harmonic or timbral information. The following overview on the 17 descriptors is based on an MPEG-7 overview provided by the ISO on the web:[2]

11.2.2.1 Basic Temporal Descriptors

- *AudioWaveform* describes the audio waveform envelope (minimum and maximum).
- *AudioPower* is similar to RMS energy.

11.2.2.2 Basic Spectral Descriptors

The basic spectral descriptors are derived from the signal transformed into the frequency domain, similar to the spectral low-level features described in Sect. 11.2.1. However, instead of an equi-spaced frequency spectrum, a logarithmic frequency spectrum is used, where the resulting frequency bins are spaced by a power-of-two divisor or a multiple of an octave. This logarithmic spectrum is the common basis for the four MPEG-7 basic spectral audio descriptors:

- *AudioSpectrumEnvelope* computes the short-term power spectrum using the logarithmic frequency division and constitutes the evolution of the spectrum over time, hence a log-frequency spectrogram.
- *AudioSpectrumCentroid* represents the center of gravity of the log-frequency power spectrum.
- *AudioSpectrumSpread* describes the second moment of the log-frequency power spectrum, indicating whether the power spectrum is centered near the spectral centroid or spread out over the spectrum. This potentially enables discriminating between pure tone and noise-like sounds.
- *AudioSpectrumFlatness* describes the flatness of the spectrum of an audio signal for each of a number of frequency bands and is computed as the deviation of the power amplitude spectrum of each frame from a flat line.

---

[1] `http://www.chiariglione.org/mpeg/`

[2] `http://www.chiariglione.org/mpeg/standards/mpeg-7/mpeg-7.htm`

11.2.2.3 Signal Parameters

The two signal parameter descriptors estimate parameters of the signals which are fundamental for the extraction of other descriptors. The extraction of both of the following descriptors is possible, however, mainly from periodic or quasi-periodic signals:

- *AudioFundamentalFrequency* is the fundamental frequency of an audio signal.
- *AudioHarmonicity* represents the harmonicity of a signal, allowing distinction between sounds with a harmonic spectrum (e.g. musical tones or voiced speech, such as vowels), sounds with an inharmonic spectrum (e.g. metallic or bell-like sounds) and sounds with a non-harmonic spectrum (e.g. noise, unvoiced speech or dense mixtures of instruments).

11.2.2.4 Timbral Temporal Descriptors

- *LogAttackTime* characterizes the "attack" of a sound, i.e. the time it takes for the signal to rise from silence to the maximum amplitude.
- *TemporalCentroid* represents the point in time that is the center of gravity of the energy of a signal. (This descriptor may, for example, distinguish between a decaying piano note and a sustained organ note, when the lengths and attacks of the two notes are identical.)

11.2.2.5 Timbral Spectral Descriptors

The five timbral spectral descriptors are spectral features computed from a *linear-frequency* spectrum and are especially intended to capture musical timbre. The four harmonic spectral descriptors are derived from the components of harmonic peaks in the signal. Therefore, harmonic peak detection must be performed prior to feature extraction:

- *SpectralCentroid* is determined by the frequency bin where the energy in the linear spectrum is balanced, i.e. half of the energy is below that frequency and half of the energy is above it. This is equal to the SpectralCentroid explained in Sect. 11.2.1 and differs from the MPEG-7 AudioSpectrumCentroid descriptor by using a linear instead of a log-scale spectrum.
- *HarmonicSpectralCentroid* is the amplitude-weighted mean of the harmonic peaks of the spectrum. It has a similar semantic as the other centroid descriptors, but applies only to the harmonic parts of the musical tone.
- *HarmonicSpectralDeviation* indicates the spectral deviation of logarithmic amplitude components from a global spectral envelope.
- *HarmonicSpectralSpread* describes the amplitude-weighted standard deviation of the harmonic peaks of the spectrum, normalized by the HarmonicSpectral-Centroid.

- *HarmonicSpectralVariation* is the correlation of the amplitude of the harmonic peaks between two sequential time frames of the signal, normalized by the HarmonicSpectralCentroid.

11.2.2.6 Spectral Basis Descriptors

The two spectral basis descriptors represent projections of high-dimensional descriptors to low-dimensional space for more compactness, which is useful, e.g., for subsequent classification or indexing tasks:

- *AudioSpectrumBasis* is a series of (potentially time-varying and/or statistically independent) basis functions that are derived from the singular value decomposition of a normalized power spectrum.
- *AudioSpectrumProjection* used together with the AudioSpectrumBasis descriptor represents low-dimensional features of a spectrum after projection upon a reduced rank basis.

11.2.2.7 Silence Descriptor

- *Silence* detects silent parts in audio and attaches this semantic to an audio segment.

For further details on the MPEG-7 descriptors refer to the MPEG-7 standard [1] or the overview provided on the web: `http://www.chiariglione.org/mpeg/standards/mpeg-7/mpeg-7.htm`.

11.2.3 MFCCs

Mel frequency cepstral coefficients (MFCCs) originated in research for speech processing and soon gained popularity in the field of music information retrieval [11]. A cepstrum is defined as the inverse Fourier transform of the logarithm of the spectrum. If the Mel scale is applied to the logarithmic spectrum before applying the inverse Fourier transform the result is called Mel frequency cepstral coefficients.

The Mel scale is a perceptual scale found empirically through human listening tests and models perceived pitch distances. The reference point is 1000 Mels, equating a 1000 Hz tone, 40 dB above the hearing threshold. With increasing frequency, the intervals in Hertz which produce equal increments in perceived pitch get larger and larger. Thus, the Mel scale is approximately a logarithmic scale, which corresponds more closely to the human auditory system than the linearly spaced frequency bands of a spectrum. In MFCC calculation often the discrete cosine transform (DCT) is used instead of the inverse Fourier transform for practical reasons.

From the MFCCs commonly only the first few (between 5 and 20) coefficients are used as features.

11.2.4 MARSYAS Features

The MARSYAS system is a software framework for audio analysis, feature extraction, synthesis and retrieval and contains a number of extractors for the following feature sets.

11.2.4.1 STFT Spectrum-Based Features

MARSYAS implements the standard temporal and spectral low-level features described in Sect. 11.2.1: spectral centroid, spectral rolloff, spectral flux, RMS energy and zero crossings. Also, MFCC feature extraction is provided, cf. Sect. 11.2.3.

11.2.4.2 MPEG Compression-Based Features

Tzanetakis and Cook presented an approach which extracts audio features directly from MPEG compressed audio data (e.g. from mp3 files) [29]. The idea was that in MPEG compression much of the analysis is done already in the encoding stage, including a time–frequency analysis. The spectrum is divided into 32 equally spaced sub-bands via an analysis filterbank. Instead of decoding the information and again computing the spectrum this approach computes features directly from the 32 sub-bands in the MPEG data. Consequently, the derived features are called MPEG centroid, MPEG rolloff, MPEG spectral flux and MPEG RMS and are computed similar as their non-MPEG counterparts (cf. Sect. 11.2.1).

11.2.4.3 Wavelet Transform Features

The wavelet transform [12] is an alternative to the Fourier transform which overcomes the issue of the trade-off between time and frequency resolution. For high-frequency ranges, it provides low-frequency resolution but high time resolution, whereas in low-frequency ranges, it provides high frequency and lower time resolution. This is a closer representation of what the human ear perceives from sound.

The wavelet transform features represent "sound texture" by applying the wavelet transform and computing statistics over the wavelet coefficients:

- mean of the absolute value of the coefficients in each frequency band
- standard deviation of the coefficients in each frequency band
- ratios of the mean absolute values between adjacent bands

These features provide information about the frequency distribution of the signal and its evolution over time.

11.2.4.4 Beat Histograms

The calculation of this set of features includes a beat detection algorithm which uses a wavelet transform to decompose the signal into octave frequency bands followed by envelope extraction and periodicity detection. The time domain amplitude envelope of each band is extracted separately which is achieved by full-wave rectification,[3] low-pass filtering and downsampling. After removing the mean of each band signal these envelopes are summed together and the autocorrelation of the resulting envelope is computed. The amplitude values of the dominant peaks of the autocorrelation function are then accumulated over the whole song into a beat histogram. This representation captures not only the dominant beat in a sound, like other automatic beat detectors, but more detailed information about the rhythmic content of a piece of music. The following set of features are derived from a beat histogram:

- relative amplitude (divided by the sum of amplitudes) of the first and second histogram peak
- ratio of the amplitude of the second peak to the amplitude of the first peak
- period of the first and second beat (in beats per minute)
- overall sum of the histogram, as indication of beat strength

11.2.4.5 Pitch Histograms

For the pitch content features, the multiple pitch detection algorithm described by [27] is utilized. The signal is decomposed into two frequency bands (below and above 1000 Hz) and amplitude envelopes are extracted for each of them using half-wave rectification[4] and low-pass filtering. The envelopes are then summed up and an enhanced autocorrelation function is used – similar to beat histograms, but within smaller time frames (about 23 ms) – to detect the main pitches of the short sound segment. The three dominant peaks are then accumulated into a pitch histogram over the whole sound file. Each bin in the histogram corresponds to a musical note. Subsequently, a folded version of the pitch histogram can also be created by mapping the notes of all octaves onto a single octave. The unfolded version contains information about the pitch range of a piece of music and the folded version contains information about the pitch classes or the harmonic content. The following features are derived from pitch histograms:

---

[3] Full-wave rectification in the digital world means that each sample value is transformed into its absolute value.

[4] Half-wave rectification in the digital world means that each sample value < 0 is converted to 0.

- amplitude of the maximum peak of the folded histogram (i.e. magnitude of the most dominant pitch class)
- period of the maximum peak of the unfolded histogram (i.e. octave range of the dominant pitch)
- period of the maximum peak of the folded histogram (i.e. main pitch class)
- pitch interval between the two most prominent peaks of the folded histogram (i.e. main tonal interval relation)
- overall sum of the histogram (i.e. measure of strength of pitch detection)

MARSYAS can be applied to extract individual feature sets; however, a set of combinations of them is defined and has been applied successfully in music genre classification. It also consists of additional tools for classification and subsequent processing. For detailed descriptions of all the features available in MARSYAS refer to [28].

11.2.5 Rhythm Patterns

A Rhythm Pattern [17, 22, 23], also called fluctuation pattern, is a matrix representation of fluctuations on critical bands. Parts of it describe rhythm in the narrow sense. The algorithm for extracting a Rhythm Pattern is a two-stage process: First, from the spectral data the specific loudness sensation in Sone is computed for critical frequency bands. Second, the critical band scale sonogram is transformed into a time-invariant domain resulting in a representation of modulation amplitudes per modulation frequency. The block diagram for the entire approach of Rhythm Patterns extraction is provided in Fig. 11.2, steps of the first part are denoted with an "S" and steps of the second part with an "R".

In a pre-processing step the audio signal is converted to a mono signal (if necessary) and segmented into chunks of approximately 6 seconds.[5] Usually not every segment is used for audio feature extraction; the selection of segments, however, depends on the particular task. For music with a typical duration of about 4 minutes, frequently the first and last one or two (up to four) segments are skipped and from the remaining segments every third one is processed.

For each segment the spectrogram of the audio is computed using the short time Fast Fourier Transform (STFT). The window size is set to 23 ms[6] and a Hanning window is applied using 50% overlap between the windows.

The Bark scale, a perceptual scale which groups frequencies to critical bands according to perceptive pitch regions [35], is applied to the spectrogram, aggregating it to 24 frequency bands. A spectral masking spreading function is applied to the signal [26], which models the occlusion of one sound by another sound.

---

[5] The segment size is 2^{18} samples with a sampling frequency of 44 kHz, 2^{17} for 22 kHz and 2^{16} for 11 kHz, i.e. about 5.9 seconds.

[6] 1024 samples at 44 kHz, 512 samples at 22 kHz, 256 samples at 11 kHz.

The Bark scale spectrogram is then transformed into the decibel scale. Further psycho-acoustic transformations are applied: Computation of the Phon scale incorporates equal loudness curves, which account for the different perception of loudness at different frequencies [35]. Subsequently, the values are transformed into the unit Sone, reflecting the specific loudness sensation of the human auditory system. The Sone scale relates to the Phon scale in the way that a doubling on the Sone scale sounds to the human ear like a doubling of the loudness.

In the second part, the varying energy on a critical band of the Bark scale sonogram is regarded as a modulation of the amplitude over time. Using a Fourier transform, the spectrum of this modulation signal is retrieved. In contrast to the time-dependent spectrogram data the result is now a time-invariant signal that contains magnitudes of modulation per modulation frequency per critical band. The occurrence of high amplitudes at the modulation frequency of 2 Hz on several critical bands for example indicates a rhythm at 120 beats per minute. The notion of rhythm ends above 15 Hz, where the sensation of roughness starts and goes up to 150 Hz, the limit where only three separately audible tones are perceivable. For the Rhythm Patterns feature set usually only information up to a modulation frequency of 10 Hz is considered. Subsequent to the Fourier transform, modulation amplitudes are weighted according to a function of human sensation depending on modulation frequency, accentuating values around 4 Hz. The application of a gradient filter and Gaussian smoothing may improve similarity of Rhythm Patterns which is useful in classification and retrieval tasks.

A Rhythm Pattern is usually extracted per segment (e.g. 6 seconds) and the feature set is computed as the median of multiple Rhythm Patterns of a piece of music. The dimension of the feature set is 1440, if the full range of frequency bands (24) and modulation frequencies up to 10 Hz (60 bins at a resolution of 0.17 Hz) are used.

Figure 11.1 shows examples of Rhythm Patterns of a classical piece, the "Blue Danube Waltz" by Johann Strauß, and a rock piece, "Go With The Flow" by The Queens Of The Stone Age. While the rock piece shows a prominent rhythm at a modulation frequency of 5.34 Hz, both in the lower critical bands (bass) as well as in higher regions (percussion, e-guitars), the classical piece does not show a distinctive rhythm but contains a "blobby" area in the region of lower critical bands and low modulation frequencies. This is a typical indication of classical music.

11.2.6 Statistical Spectrum Descriptors

Statistical Spectrum Descriptors (SSD) [10] are based on the first part of the Rhythm Patterns algorithm, namely the computation of a psycho-acoustically motivated Bark scale sonogram. However, instead of creating a pattern of modulation frequencies, an SSD intends to describe fluctuations on the critical frequency bands in a more compact representation, by deriving several statistical moments from each critical band. A block diagram of SSD computation is given in Fig. 11.2.

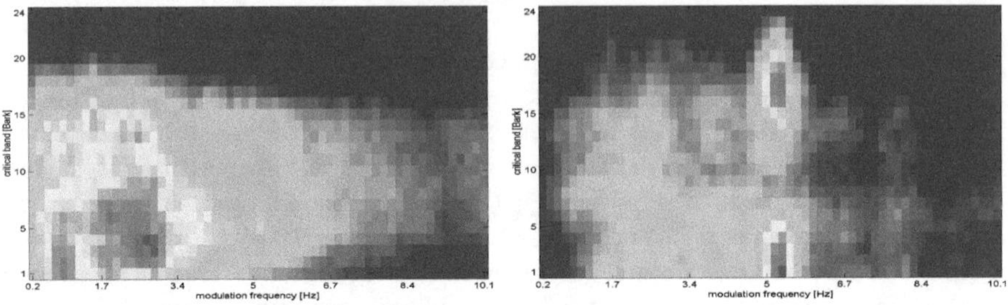

(a) Classical (b) Rock

Fig. 11.1 Rhythm patterns

The specific loudness sensation on different frequency bands is computed analogously to Rhythm Patterns (cf. Sect. 11.2.5): A short-time FFT is used to compute the spectrum. The resulting frequency bands are grouped to 24 critical bands, according to the Bark scale. Optionally, a spreading function is applied in order to account for spectral masking effects. Successively, the Bark scale spectrogram is transformed into the decibel, Phon and Sone scales. This results in a power spectrum that reflects human loudness sensation – a Bark scale sonogram.

From this representation of perceived loudness a number of statistical moments are computed per critical band, in order to describe fluctuations within the critical bands extensively. Mean, median, variance, skewness, kurtosis, min- and max-value are computed for each band, and a Statistical Spectrum Descriptor is extracted for each selected segment. The SSD feature vector for a piece of audio is then calculated as either the mean or the median of the descriptors of its segments.

Statistical Spectrum Descriptors are able to capture additional timbral information compared to Rhythm Patterns, yet at a much lower dimension of the feature space (168 dimensions).

11.2.7 Rhythm Histograms

Rhythm Histogram features are a descriptor for general rhythmic characteristics in a piece of audio. A modulation amplitude spectrum for critical bands according to the Bark scale is calculated, equally as for Rhythm Patterns (see Sect. 11.2.5 and Fig. 11.2). Subsequently, the magnitudes of each modulation frequency bin of all 24 critical bands are summed up, to form a histogram of "rhythmic energy" per modulation frequency. The histogram contains 60 bins which reflect modulation frequency between 0.17 and 10 Hz[7] (cf. Fig. 11.3). For a given piece of audio, the Rhythm

[7] Using the parameters given in Footnotes 5 and 6, the resolution of modulation frequencies is 0.17 Hz.

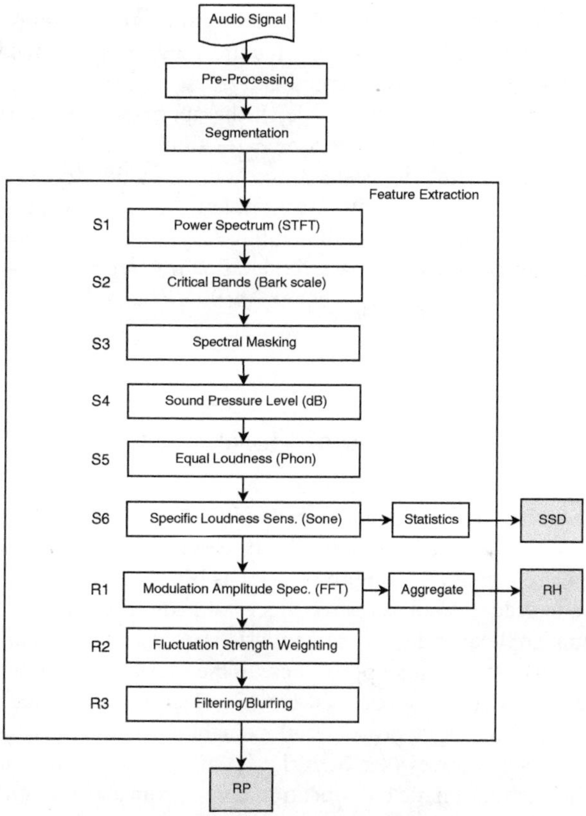

Fig. 11.2 Feature extraction process for Statistical Spectrum Descriptors (SSD), Rhythm Histograms (RH) and Rhythm Patterns (RP)

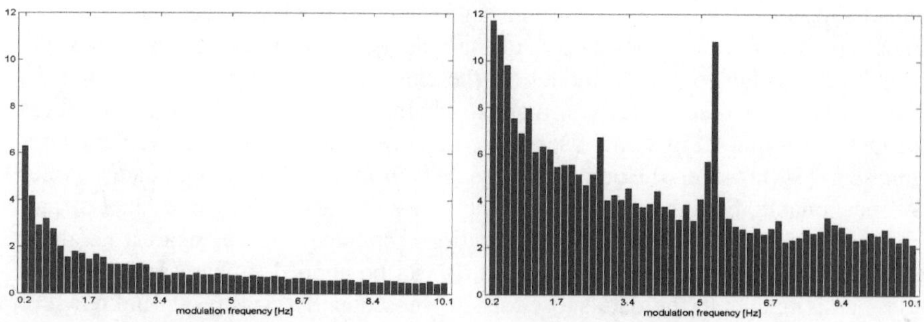

(**a**) Classical (**b**) Rock

Fig. 11.3 Rhythm Histograms

Histogram feature set is calculated by taking the median of the histograms of every 6 second segment processed. The resulting feature vector has 60 dimensions.

The Rhythm Histograms are similar in their representation to the beat histograms introduced by Tzanetakis (cf. Sect. 11.2.4.4); the approach, however, is different: The beat histogram approach uses envelope extraction and autocorrelation and accumulates the histogram from the peaks of the autocorrelation function.

Figure 11.3 compares the Rhythm Histograms of a classical piece and a rock piece (the same example songs as for illustrating Rhythm Patterns have been used). The rock piece indicates a clear peak at a modulation frequency of 5.34 Hz while the classical piece generally contains less energy, having most of it at low modulation frequencies.

11.3 Automatic Classification of Music into Genres

Audio descriptors computed in feature extraction processes build the basis for a range of different retrieval tasks. For similarity-based search and retrieval different distance metrics are applied to compute distances between the numeric descriptors in feature space in order to rank retrieved pieces of audio according to their similarity. Besides being applicable directly in similarity-based searches, audio descriptors are utilized for machine learning approaches. In the music information retrieval domain, both supervised and unsupervised machine learning techniques are applied. While unsupervised learning approaches are valuable in automatic organization and visualization of music archives (see Sect. 11.4), supervised machine learning techniques are applied for automatic classification tasks. From a given number of examples the computer learns how to classify music pieces into a number of previously defined classes. The taxonomy can be defined according to specific task requirements. Common scenarios are the classification of music into genres or genre hierarchies or according to artists.

Supervised classifiers are able to learn from the genre assignments to a rather small subset of the data and can predict the genre labels for unseen data. Given a music repository, one needs to provide sample songs per music genre in order to enable a classifier to build a model for the classification of further songs into the desired list of genres. Usually, a manual labeling/assignment of classes to pieces of music is required, in some cases genre meta-data can also be retrieved from online services (however, the credibility of labels from user-driven meta-data services is questionable, because there is often no consensus about the genre label of particular music). Any supervised classification technique, such as nearest neighbor or support vector machines (see Sect. 2.6), can be applied to learn a model from class assignments to the data subset and to subsequently classify yet unknown or unlabeled music. Differences in the classification results for various classifiers are dependent on the type and dimensionality of the feature set used. For feature sets with high dimension, such as for instance the Rhythm Patterns feature set, support vector machines usually achieve better results than other classifiers.

11.3.1 Evaluation Through Music Classification

The advantage of music classification through supervised machine learning is that – provided that ground truth is available – the results of the classification process can be directly measured, facilitating (1) the comparison of feature sets and (2) the assessment of the suitability of particular classifiers for specific feature sets. Direct evaluation is not possible in unsupervised learning tasks, where the result of the clustering is of subjective nature.

Feature sets and classification techniques are evaluated using a range of measures. The most commonly used one is accuracy. In a two-class problem accuracy is defined as

$$A = \frac{\mathrm{TP}+\mathrm{TN}}{N}. \tag{11.1}$$

TP is the number of true positives, TN is the number of true negatives, i.e. the two cases where the classifier predicted the correct class label. A false negative (FN) is when the classifier prediction was "false" while it should have been "true". A false positive (FP) appears when the classifier assigns an item that is actually labeled as "false" to the class "true". N is the number of items in the collection.

In music classification usually more than two classes exist, thus accuracy is computed as the sum of all correctly classified songs, divided by the total number of songs in a collection:

$$A = \frac{\sum_{i=1}^{|C|} \mathrm{TP}_i}{N}. \tag{11.2}$$

The determination of the number of correctly classified songs implies the availability of genre (class) labels for all songs in the collection, i.e. for evaluation a music collection with 100% annotated data is needed, which is called ground-truth data.

Besides accuracy, precision and recall are further performance measures often reported from classification tasks:

$$\pi_i = \frac{\mathrm{TP}_i}{\mathrm{TP}_i+\mathrm{FP}_i}, \quad P^{\mathrm{M}} = \frac{\sum_{i=1}^{|C|} \pi_i}{|C|}. \tag{11.3}$$

π_i is the precision per class, where TP_i is the number of true positives in class i and FP_i is the number of false positives in class i, i.e. songs identified as class i but actually belonging to another class. $|C|$ is the number of classes in a collection and P^{M} is macro-averaged precision. Precision measures the proportion of relevant pieces to all songs retrieved. Macro-averaging computes the precision per class first and then averages the precision values over all classes, while micro-averaged precision is computed by summing over all classes. In micro-averaging large classes are over-emphasized:

$$\rho_i = \frac{\mathrm{TP}_i}{\mathrm{TP}_i + \mathrm{FN}_i}, \qquad R^{\mathrm{M}} = \frac{\sum_{i=1}^{|C|} \rho_i}{|C|}. \tag{11.4}$$

ρ_i is the recall per class, where FN_i is the number of false negatives of class i, i.e. songs belonging to class i, but which the classifier assigned to another class. R^{M} is macro-averaged recall; micro-averaged recall is computed, analogously to micro-averaged precision, directly by summing over all classes. Recall measures the proportion of relevant songs retrieved out of all relevant songs available.

An additional performance measure is the F-measure, which is a combined measure of precision and recall, computed as their weighted harmonic mean. The most common F-measure is F_1-measure, which weights precision and recall equally:

$$F_1 = \frac{2 \cdot P^{\mathrm{M}} \cdot R^{\mathrm{M}}}{P^{\mathrm{M}} + R^{\mathrm{M}}}. \tag{11.5}$$

The general definition for the F-measure is

$$F_\alpha = \frac{(1+\alpha) \cdot P^{\mathrm{M}} \cdot R^{\mathrm{M}}}{\alpha \cdot P^{\mathrm{M}} + R^{\mathrm{M}}}, \tag{11.6}$$

where α influences the weighting of precision and recall.

As the performance measures may fluctuate depending on the particular partitioning of the data collection into subsets used for training and testing, usually a cross-validation approach is chosen in order to get a more stable assessment of the classifier results. A standard form of cross-validation is 10-fold cross-validation: The music collection is divided into 10 subsets; in each of 10 iterations a different subset is chosen as test data and the other 90% of the data is used for the training process. The cross-validation result is the average of the 10 classification runs.

11.3.2 Benchmark Data Sets for Music Classification

In order to compare different approaches, benchmark data sets with annotated ground truth for evaluation are needed. However, as music information retrieval is a rather young research field, standard benchmark collections are rare. Nevertheless, there are three music collections that are repeatedly used in evaluations and experiments reported in scientific papers, which allow for comparison of different feature extractors, similarity measures and/or classifiers.

Collection A This music collection has been used in the evaluation of the MARSYAS system (cf. Sect. 11.2.4). The collection and the experiments for genre discrimination are described in [28]. It consists of 1000 pieces of music equi-distributed among 10 popular music genres (blues, classical, country, disco, hip hop, jazz, metal, pop, reggae, rock).

Collection B Music collection B has been used in the ISMIR 2004 Rhythm classification contest [8], where the task was the distinction of Latin American and ballroom dance rhythms. It stems from ballroomdancers.com and consists of 698 excerpts of 8 genres from ballroom dance music (Cha-Cha-Cha, Jive, Quickstep, Rumba, Samba, Slow Waltz, Tango, Viennese Waltz).

Collection C Music collection C is a part of the free repository at magnatune.com and has been used in the ISMIR 2004 Genre classification contest [8]. It contains 1458 complete songs, the pieces being unequally distributed over 6 genres (classical, electronic, jazz&blues, metal&punk, rock&pop, world music).

Based on these data sets, a comparison of three feature sets and two classifiers was done, using 10-fold cross-validation. As classifiers, a k-nearest neighbor classifier with $k = 5$ and a linear support vector machine with pairwise classification have been chosen. The implementations of the Weka machine learning software [34] were used with default parameter settings. The following feature sets are being compared: Rhythm Patterns (RP), Rhythm Histograms (RH) and Statistical Spectrum Descriptors (SSD) (see Sect. 11.2). Table 11.1 presents the results of this comparison experiment in terms of accuracy in percent.

The results show that in these settings linear support vector machines generally outperform nearest neighbor classifiers. The performance of Rhythm Histograms is in most cases lower than for the other two feature sets, except for k-NN classifier on collection B. It is noticeable that the Rhythm Histogram features achieve a comparable performance to Rhythm Patterns on collection B, although the dimensionality of Rhythm Histograms is only 1/24 of that of rhythm patterns. Rhythm Patterns perform very well on collection B; obviously those two feature sets discriminate the different rhythms of the Latin American and ballroom dances very well. Contrarily, Statistical Spectrum Descriptors apparently describe differences in popular music genres well, probably due to different instrumentation and timbre. In [10] it was shown that a combination of several feature sets improves the classification results. Alternatively, different feature sets can be classified separately and the classification result, i.e. the predicted class label, is determined by a majority vote on the predictions of the single classifiers. A combination of several classifiers is also the idea of ensemble techniques, which significantly improve the results compared to single classifiers (see Sect. 2.5).

Table 11.1 Music genre classification: comparison of three feature sets and two classifiers using 10-fold cross-validation and three music test collections (accuracy in %)

Classifier	k-NN, $k = 5$			Linear SVM		
Feature set	RP	RH	SSD	RP	RH	SSD
Collection A (10 genres)	50.7	41.2	67.1	64.6	47.1	73.2
Collection B (8 genres)	79.7	79.8	50.3	90.0	85.2	60.9
Collection C (6 genres)	69.8	63.0	77.7	73.5	62.9	78.8

For a simulation of real world scenarios, huge music collections would be needed, but objective ground truth for such collections is difficult to find [3]. With the ISMIR 2004 Audio description contest, a standard evaluation forum for the music information retrieval research community has been established. Since 2005, the annual evaluation exchange is called MIREX [5] and is also performed in parallel to the International Conference on Music Information Retrieval (ISMIR). The evaluation is open to any groups who desire a comparison of their approach(es) to current state-of-the-art algorithms. In 2005, participants could submit their algorithms to the following list of tasks [14]:

- audio artist identification
- audio drum detection
- audio genre classification
- audio melody extraction
- audio onset detection
- audio tempo extraction
- audio and symbolic key finding
- symbolic genre classification
- symbolic melodic similarity

As the list shows, there are numerous problems regarding the extraction of semantics from music addressed in the MIR community and both the symbolic and audio-based music information retrieval domains are represented.

From the list of tasks, the audio and symbolic genre classification tasks are the ones corresponding to the classification problem described before: to extract suitable features from a benchmark music collection and to classify the pieces of music into a given list of genres. The results of the MIREX 2005 audio genre classification task indicate the performance of state-of-the-art systems for classification of audio-based music collections (such as personal mp3 collections) at that time. In the MIREX 2005 evaluation, two benchmark data sets have been involved: the Magnatune data set, comprising 1515 audio files from 10 music genres and the USPOP 2002 data set, comprising 1414 songs from 6 genres. Results have been evaluated separately on these data sets; however, an overall score has been calculated from both. Table 11.2 shows the overall results of the 2005 audio genre classification task.

Bergstra, Casagrande & Eck extracted a relatively large number of timbre features (MFCCs, RCEPS, ZCR, LPC, rolloff, among others) at an intermediate time scale (every 13.9 seconds) and calculated mean and variance of the features for each segment. They used AdaBoost.MH [7] for classification. In variant (1) they boosted decision stumps and in variant (2), which was ranked first in the MIREX 2005 evaluation, they boosted 2-level trees. They classified the features extracted from each time segment independently and determined the class label for a song by averaging the outputs of the meta-feature classifiers. Further details about the MIREX 2005 results and data sets as well as the implementations of the approaches involved are available on the MIREX 2005 website [14].

Table 11.2 Audio genre classification in the MIREX 2005 evaluation of state-of-the-art algorithms (mean accuracy in % from two benchmark data sets)

Rank	Participant	Result (%)
1	Bergstra, Casagrande & Eck (2)	82.34
2	Bergstra, Casagrande & Eck (1)	81.77
3	Mandel & Ellis	78.81
4	West, K.	75.29
5	Lidy & Rauber (SSD+RH)	75.27
6	Pampalk, E.	75.14
7	Lidy & Rauber (RP+SSD)	74.78
8	Lidy & Rauber (RP+SSD+RH)	74.58
9	Scaringella, N.	73.11
10	Ahrendt, P.	71.55
11	Burred, J.	62.63
12	Soares, V.	60.98
13	Tzanetakis, G.	60.72

11.4 Creating and Visualizing Music Maps Based on Self-organizing Maps

There are numerous clustering algorithms that can be employed to organize audio by sound similarity based on a variety of feature sets. One model that is particularly suitable is the self-organizing map (SOM), an unsupervised neural network that provides a mapping from a high-dimensional input space to a usually two-dimensional output space [9]. Section 3.4 provides a detailed review of the SOM training algorithm. In this section we show how the self-organizing map is used to create music similarity maps.

In 2001 the first map visualization of a music archive based on the self-organizing map (SOM) has been presented [21]. Meanwhile a number of other research projects focus on the creation of SOM-based interfaces to music archives [15, 18, 24]. Commonly, a rectangular map is chosen, but also toroidal maps are common, which avoid saturations at the borders of the map, but usually need unfolding for display on a 2D screen. With the MnemonicSOM [13], the algorithm has been modified so that maps with virtually arbitrary shapes can be created. The SOM is initialized with an appropriate number i of units, proportional to the number of tracks in the music collection. In case of a rectangular SOM the units are arranged in a two-dimensional grid. The result of the SOM training procedure is a topologically ordered mapping of the presented input signals in the two-dimensional space. The SOM associates patterns in the input data with units on the grid, hence similarities present in the audio signals are reflected as faithfully as possible on the map, using the feature vectors extracted from audio.

The result is a similarity map, in which music is placed according to perceived similarity: Similar music is located close to each other, building clusters, while pieces with more distinct content are located farther away. If the pieces in the music collection are not from clearly distinguishable genres the map will reflect this by placing pieces along smooth transitions.

Due to the fact that the clusters and structures found by the trained music map are not inherently visible, several visualization techniques have been developed. Choosing appropriate visualization algorithms and appealing color palettes facilitate insight into the structures of the SOM from different perspectives. The influence of color palettes is important, as the different views on the data can be interpreted accordingly as mountains and valleys, islands in the sea, etc. For the following visualization examples, collection B described in Sect. 11.3.2 was used to train a rectangular map with 20×14 units. As feature set for music similarity Rhythm Pattern features (cf. Sect. 11.2.5) have been used. The collection and hence the map contains music from eight different Latin American and Ballroom dances: Cha-Cha-Cha, Tango, Jive, Samba, Rumba, Quickstep, Slow Waltz and Viennese Waltz. In order to elicit the cluster information a number of visualization techniques have been devised to analyze the map's structures. An additional visualization which is *not* based on the map structures but on external meta-data is the class visualization. In order to give an overview of the music map used in the examples the class visualization will be described first. An interactive web demo of a music map trained from the same music collection is available on the Web.[8]

11.4.1 Class Visualization

While the self-organizing map does not rely on any manual classification, nevertheless there are often genre labels available for the music titles (or at least for the artists). Also, music collections sometimes have been manually sorted into different classes. The availability of genre or class information allows the creation of a visualization which assists the analysis of the clustered structures on the map. External genre information allows to color-code the genres and to overlay the genre information onto other existing visualizations. Commonly, pie charts (cf. Fig. 11.4(a)) are used to visualize the distribution of the classes within a map unit, thus, one pie chart is placed on every unit of the map. Alternatively, if a full-area visualization is preferred, a Voronoi-like approach (cf. Fig.11.4(b)) is taken to fill regions covered by a single genre with its respective color. If units contain a mixed set of classes, an approach similar to dithering is used to represent multiple classes within the area covered by that unit [4]. The additional clues provided by genre or class visualization vastly facilitate the description of regions or clusters identified by other visualizations, such as smoothed data histograms.

Describing the map (cf. Fig. 11.4), the cluster on the top left contains Cha-Cha-Cha, Samba was clustered bottom left, with quickstep just to its right. Jive music is found in a cluster slightly left of the center, Rumba at the bottom, Slow Waltz at the outer right, with Viennese Waltz to its left. Tango has been partitioned into three

[8] `http://www.ifs.tuwien.ac.at/mir/playsom/demo/`

(a) Pie chart overlay (b) Voronoi-like tessellation

Fig. 11.4 Class visualization. Classes: Cha-Cha-Cha (red), Tango (cyan), Jive (blue), Samba (orange), Rumba (yellow), Quickstep (green), Slow Waltz (pink), Viennese Waltz (magenta)

clusters, one at the center left, one at the top and another part below Viennese Waltz. Parts of Quickstep are also found left of the latter two Tango clusters.

11.4.2 Hit Histograms

A hit histogram visualization depicts the distribution of data over the SOM. For each unit, the number of items (songs) mapped is counted to compute the hit histogram. Different visualizations of hit histograms are possible: Likewise to other visualizations, the number of mapped songs can be visualized by different colors (Fig. 11.5(a)). Another variant is the use of markers (circles, bars, etc.) on top of the map grid, where the number of hits for each unit is reflected by the size of the marker (Fig. 11.5(b)). Hit histograms are good for getting an overview of the map and an indication where much of the data is concentrated. In Fig. 11.5(a) and (b) for example the highest peaks correspond to the Cha-Cha-Cha, Samba and Slow Waltz clusters, while the visualization also exhibits lower peaks for other clusters.

The approach is also very useful if not the entire data set is to be visualized, but only a part of it. With the same techniques it is possible to display (statistical) information about a subset of the data collection. This is used, for instance, to visualize results of a query to the music map: Querying the map with the name of a certain artist, a hit histogram is constructed by counting the number of times this artist is present with a piece of music on every unit. Only a part of the map will have hits on that query, the histogram values of the respective units are increased by one for each hit, while the values of the remaining units are set to zero. The visualization provides an immediate overview of where the resulting items of the query are located. Depending on the graphical result, one receives an indication of whether the music of a given artist is rather distributed or aggregated in a certain area of the map. Hit histograms can be employed in a number of situations: If genre or class labels of the

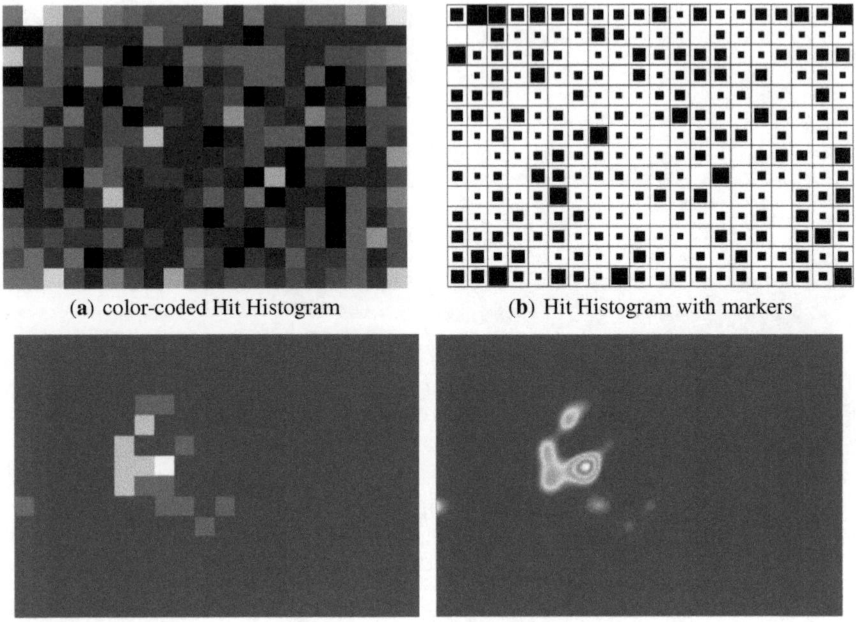

(**a**) color-coded Hit Histogram (**b**) Hit Histogram with markers

(**c**) Hit Histogram of a query for *Jive* music (flat) (**d**) Hit Histogram of a query for *Jive* music (spline interpolated)

Fig. 11.5 Hit histograms

songs are available, the distribution of a particular genre can be visualized with hit histograms. Figure 11.5(c) shows a hit histogram of the distribution of "Jive" music within the example music collection in a color-coded representation similar to the one in Fig. 11.5(a). In Fig. 11.5(d) spline interpolation has been used in order to improve the intuitive perception of clusters in the query results. Another possibility would be the visualization of cover songs having the same title, but different performers. Hit histograms may be a useful visualization for virtually any sort of query in which frequency counts are involved.

11.4.3 U-Matrix

One of the first and most prominent SOM visualizations developed is the U-matrix [33]. The U-matrix visualizes distances between the weight vectors of adjacent units. The local distances are mapped onto a color palette: small distances between neighboring units are depicted with another color than large distances, the color gradually changes with distance. As a consequence the U-matrix reveals homogeneous clusters as areas with one particular color, while the cluster boundaries having larger distances are visualized with another color. In Fig. 11.6(a) the cluster boundaries are shown by bright colors. There is, e.g., a particularly strong boundary between Tango and Samba music as well as between Quickstep and Rumba. With an appropriate color palette one can give this visualization the metaphor of mountains and valleys: Large distances are visualized with brown or even white color (for the mountain tops), lower distances are visualized with different shadings of green color. The mountains then indicate the barriers between homogeneous (flat) regions. As the U-matrix is computed from distances between adjacent units, the visualization result is depicted at a finer level as other (per-unit) visualizations. The U-height-matrix depicted in Fig. 11.6(b) is a unit-wise aggregation of adjacent U-matrix values.

11.4.4 P-Matrix

The P-matrix visualization shows local relative data densities based on an estimated radius around the unit prototype vectors of the SOM. First, the so-called Pareto radius is determined as a quantile of the pair-wise distances between the data vectors [31]. Then, for each map unit, the number of data points within the sphere of the

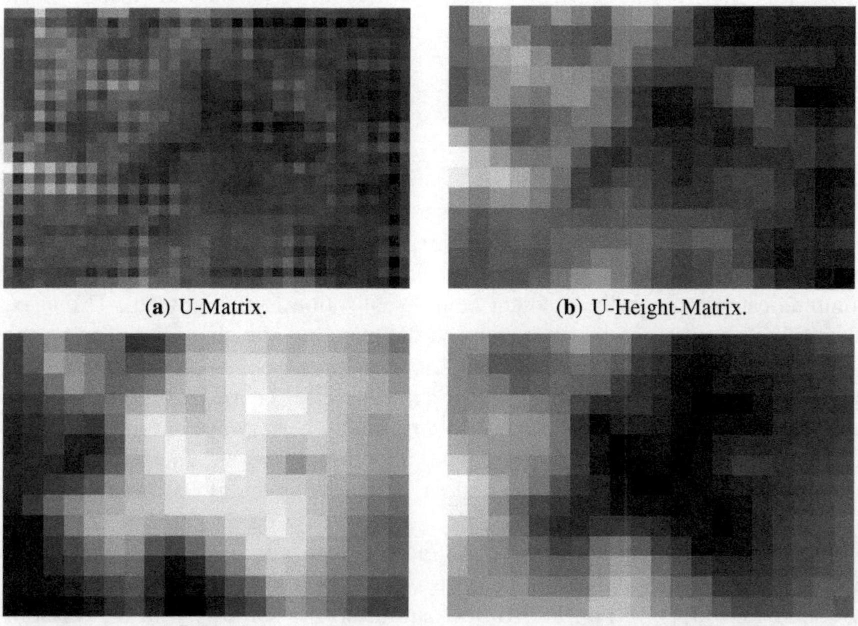

(**a**) U-Matrix. (**b**) U-Height-Matrix.

(**c**) P-Matrix. (**d**) U*-Matrix

Fig. 11.6 U-matrix, U-height-matrix, P-matrix and U*-matrix previously calculated radius is counted and visualized on the map grid. The purpose of this visualization is to show the relative density of the map units. The map nodes in the center of the map usually have a higher P value than the ones at the border, which is also the case in Fig. 11.6(c).

11.4.5 U*-matrix

The U*-matrix aims at showing cluster boundaries taking both the local distances between the unit vectors and the data density into account [32]. It is derived from the P-matrix and the U-matrix. This is performed by weighting the U-matrix values according to the P-matrix values: Local distances within a cluster with high density (high P-matrix values) are weighted less than distances in areas with low density, resulting in a smoother version of the U-matrix. The intention is to reduce the effects of inhomogeneous visualization in actually dense regions. Comparing Fig. 11.6(b)–(d)), it can be seen that the dense region exhibited in the P-matrix is visualized more homogeneously in the U*-matrix visualization than in the U-matrix. Consequently, the smaller distances in the dense areas have vanished while the cluster boundaries between Tango and Samba as well as Quickstep and Rumba are depicted more clearly. The U*-matrix has been particularly designed for emergent SOMs [30], i.e. very large SOMs, where the number of units on the map is much larger than the number of input data.

11.4.6 Gradient Fields

The gradient field visualization [20] aims at making the SOM readable for persons with engineering background who have experience with flow and gradient visualizations. It is displayed as a vector field overlay on top of the map. The information communicated through the gradient field visualization is similar to the U-matrix, identifying clusters and coherent areas on the map, but allowing for extending the neighborhood width, and thus showing more global distances. Another goal is to make explicit the direction of the most similar cluster center, represented by arrows pointing to this center. The method turns out to be very useful for SOMs with a large number of map units. The neighborhood radius is an adjustable parameter: a higher radius has a smoothing effect, emphasizing the global structures over local ones. Thus, this parameter is selected depending on the level of detail one is interested in. Fig. 11.7(a) depicts a gradient field with a neighborhood radius of 2, with the arrows indicating the direction to the center of each genre cluster (compare Fig. 11.4). In Fig. 11.7(b) the parameter was set to 7, with the result of the arrows pointing mostly to the most salient cluster peaks exhibited in the hit histogram (cf. Fig. 11.5(a) and (b)).

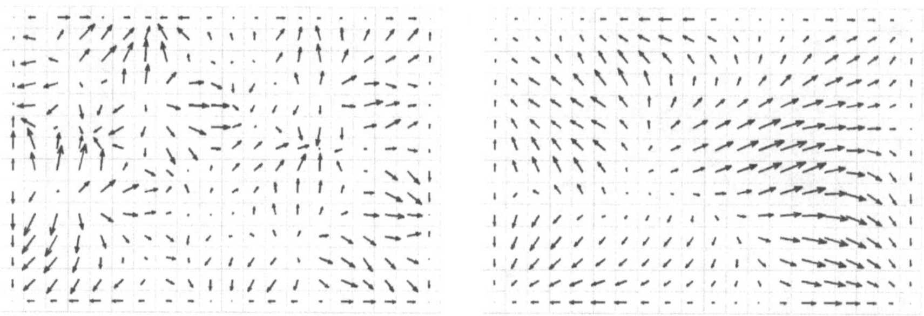

(a) $\sigma = 2$ (b) $\sigma = 7$

Fig. 11.7 Gradient field visualization with neighborhood parameter σ set to different values

11.4.7 Component Planes

Component planes visualize the distribution of particular features or attributes (components) of the feature set. A single component of the unit weight vectors is used to create the visualization allowing to investigate the influence of a particular feature (such as the zero crossing rate or a modulation frequency on a specific band within the Rhythm Patterns feature set) to the mapping of certain pieces of music on particular regions of the map. For the component planes visualization, each unit on the map is color-coded, where the color reflects the magnitude of a particular component of the weight vector of each unit. With the appropriate color palette, this visualization is comparable to "weather charts" [19].

When maps are created using feature sets with large dimensions, the visualization of every component of the feature set is probably not desired. Especially feature sets that allow the aggregation of attributes to semantic subsets are suitable to create a "weather chart"-like visualization, permitting the description of map regions by comprehensive terms. For this purpose subsets of feature vector components are being accumulated. Particularly for the rhythm pattern feature set (cf. Sect. 11.2.5) four "weather chart" visualizations have been created reflecting the psycho-acoustic characteristics inherent in the feature set:

- Maximum fluctuation strength is calculated as the highest value contained in the Rhythm Pattern. Its weather chart indicates regions with music dominated by strong beats.
- Bass denotes the aggregation of the values in the lowest two critical bands with a modulation frequency higher than 1 Hz indicating music with bass beats faster than 60 beats per minute.
- Non-aggressiveness takes into account values with a modulation frequency lower than 0.5 Hz of all critical bands except the lowest two. The respective weather chart indicates rather calm songs with slow rhythm.
- Low frequencies dominant is the ratio of the five lowest and highest critical bands and measures in how far the low frequencies dominate.

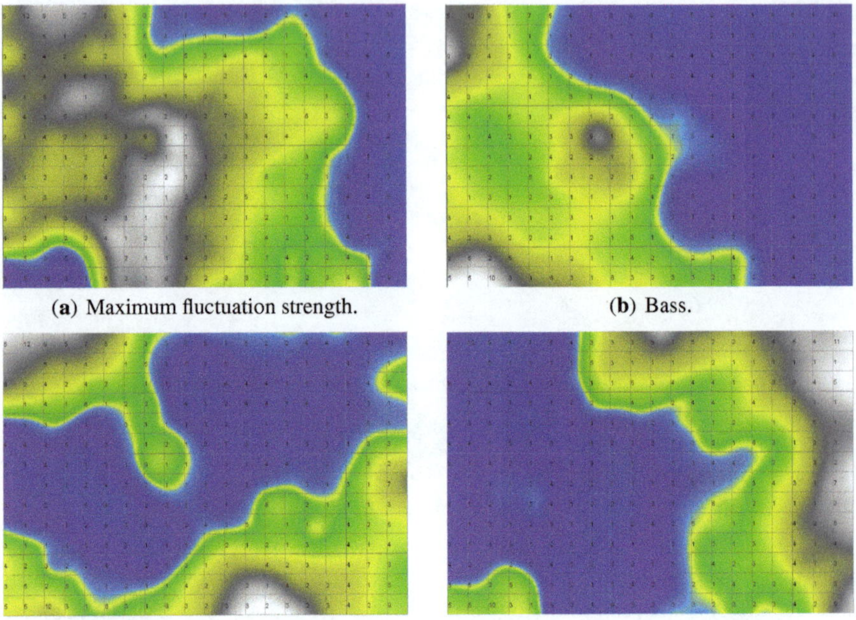

(**a**) Maximum fluctuation strength. (**b**) Bass.

(**c**) Non-aggressiveness. (**d**) Low frequencies dominant.

Fig. 11.8 Component planes: weather chart visualizations of characteristics inherent in the Rhythm Patterns feature set

As these examples show, component planes provide an intuitive explanation of the map, its regions and the underlying features. Figure 11.8 shows examples of the four Rhythm Pattern Weather Chart visualizations explained. Regarding the figures we see that the maximum magnitude of fluctuation strength corresponds to quickstep and Jive music. Bass covers the genres Cha-Cha-Cha, Jive, Samba, Quickstep and Rumba and partly Tango. Cha-Cha-Cha and Rumba have been identified to have the least aggressive rhythm, while in slow Waltz and Viennese Waltz the low frequencies dominate.

11.4.8 Smoothed Data Histograms

Detecting and visualizing the actual cluster structure of a map is a challenging problem. The U-matrix described above visualizes the distances between the model vectors of units which are immediate neighbors, aiming at cluster boundary detection. Smoothed data histograms [23] are an approach to visualize the cluster structure of the data set in a more global manner. The concept of this visualization technique is basically a density estimation and resembles the probability density of the whole data set on the map. When a SOM is trained, each data item is assigned to the map unit which best represents it, i.e. the unit which has the smallest distance between its model vector and the respective feature vector. However, by continuation of these distance calculations it is also possible to determine the second best, third best and so on, matching units for a given feature vector. A voting function is introduced using a robust ranking, which assigns points to each map unit: For every data item, the best matching unit gets n points, the second best $n-1$ points, the third $n-2$ and so forth, for the n closest map units, where n is the user-adjustable smoothing parameter. All votes are accumulated resulting in a histogram over the entire map. The histogram is then visualized using spline interpolation and appropriate color palettes. Depending on the palette used, map units in the centers of clusters are represented by mountain peaks while map units located between clusters are represented as valleys. Using another palette the SDH visualization creates the *islands of music* [17] metaphor, ranging from dark blue (deep sea), via light blue (shallow water), yellow (beach), dark green (forest), light green (hills), to gray (rocks) and finally white (snow).

The SDH visualization, contrary to the U-matrix, offers a sort of hierarchical representation of the cluster structures on the map. On a higher level the overall structure of the music archive is represented by large continents or islands. These larger genres or styles of music might be connected through land passages or might be completely isolated by the sea. On lower levels the structure is represented by mountains and hills, which can be connected through a ridge or separated by valleys. For example, there might be an island (or even a "continent") comprising non-aggressive, calm music without strong beats. On this island there might be two mountains, one representing classical and the other one orchestral film music, which is somewhat more dynamic. Another example might be an island comprising electronic music and the hills and mountains on it representing sub-genres with different rhythm or beat. This does not imply that the most interesting pieces are always located on or around mountains; interesting pieces might also be located between two strongly represented distinctive groups of music and would thus be found either in the valleys between mountains or even in the sea between islands, in the case of pieces which are not typical members of the main genres or music styles represented by the large islands (clusters) on the map.

The parameter n mentioned before, which determines the number of best matching units to be considered in the voting scheme for the SDH, can be adjusted by the user to interactively change the appearance of the SDH visualization. A low value of n creates more and smaller clusters (islands) on the map; with an increasing value of n the islands grow and eventually merge building greater islands or continents. Analogous to the hierarchical representation described before, this enables the user of the map to create a cluster structure visualization at different levels, which depends on whether a more general aggregation of the data or a more specialized one is desired.

Figure 11.9 shows two SDH visualizations of the music collection described at the beginning of this section: in the left visualization, the smoothing parameter was set to 12, showing the dominant clusters of Cha-Cha-Cha, Samba and slow Waltz, as

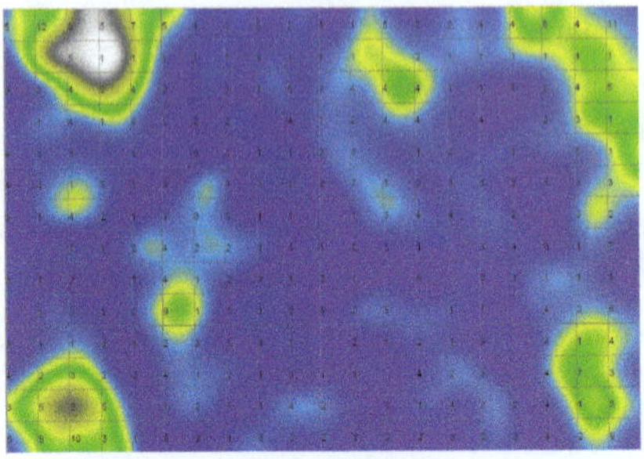

(**a**) smoothing factor $n = 12$

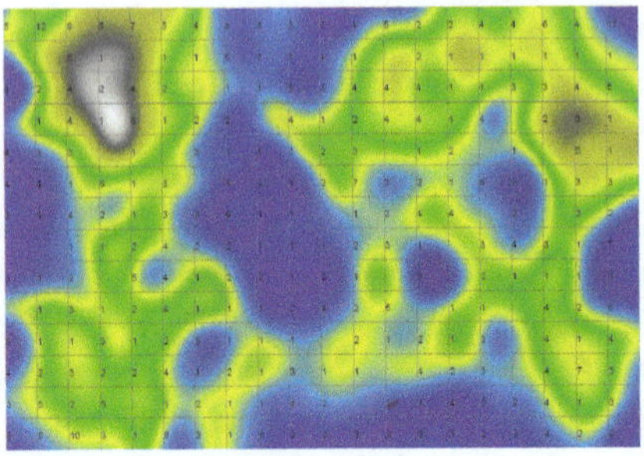

(**b**) smoothing factor $n = 33$

Fig. 11.9 Smoothed data histograms

well as the peaks of Tango and Quickstep. On the right image the smoothing parameter was set to 33, showing large clusters which are beginning to merge, joining also the two parts of the Cha-Cha-Cha cluster and the slow Waltz clusters which were separated in Fig. 11.9(a). Moreover, Jive music is found by a "sea-ground level" cluster in the center of the map, surrounded by "islands". Jive was the genre with the lowest number of pieces in the collection which is probably the reason why the SDH visualization does not show an "island" cluster of Jive as well.

11.5 PlaySOM – Interaction with Music Maps

Music maps provide a convenient overview of the content of music archives. Yet, their advantages are augmented by the PlaySOM application, which enriches music maps with facilities for interaction, intuitive browsing and exploration, semantic zooming, panning and playlist creation. This moves the SOM from a purely analytical machine learning tool for analysing the high-dimensional feature space to an actual and direct application platform.

11.5.1 Interface

The main PlaySOM interface is shown in Fig. 11.10. Its largest part is covered by the interactive map on the right, where squares represent single units of the SOM. At the outmost zooming level, the units are labeled with numbers indicating the quantity of songs per unit. The left-hand side of the user interface contains

- a bird's eye view showing which part of the potentially very large map is currently displayed in the main view
- the color palette used in the currently active visualization

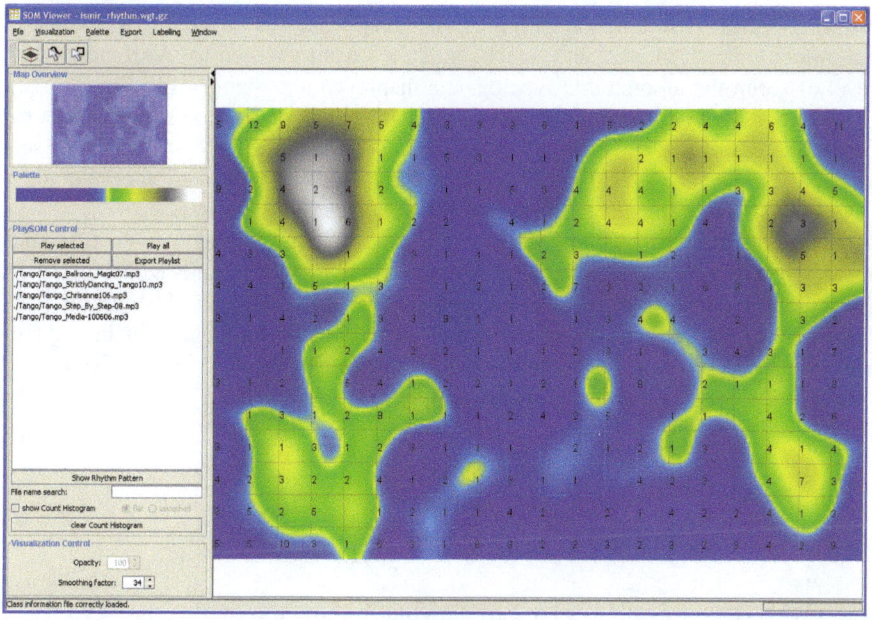

Fig. 11.10 PlaySOM desktop application: main interface

- the playlist containing titles of the last selection made on the map, alongside buttons to refine, play or export the playlist
- search fields for queries to the map
- a visualization control to influence the parameters of the currently active visualization

The menu bar at the top of the window contains menus for additional settings, for switching between the visualization, changing the palette and for exporting the map into different formats, including the PocketSOMPlayer format. Also, if genre tags are available for the music titles, the distribution of genres on the music map can be displayed as colored overlay as an additional clue. A toolbar allows the user to switch between the two different selection models and to automatically zoom out to fit the map to the current screen size.

11.5.2 Interaction

PlaySOM allows the user to select from and to switch between the different visualizations described in the previous section. The weather charts visualization for instance, indicating particular musical attributes (see Sect. 11.4.7), aids the user in finding the music of a particular genre or style. With the SDH visualization creating an *Islands of Music* interface a metaphor for a geographic map is offered, which allows for intuitive interaction with the music map. Users can move across the map, zoom into areas of interest and select songs they want to listen to. With increasing level of zoom the amount and type of data displayed are changed (cf. Fig. 11.11), providing more details about the items on the units. The interaction model allows to conveniently traverse and explore the music map. At any level of detail users can select single songs and play them or create playlists directly by selection on the map. Playlists can either be played immediately or exported for later use.

The application also offers traditional search by artist name or song title and locates the retrieved titles on the map by marking the respective units with a different color. Alternatively, the results of a query are visualized using hit histograms (see Sect. 11.4.2), showing the distribution of the search results on the map including the number of hits per unit. From the retrieved locations it is easy to browse for and to discover similar (yet unknown) music simply by selecting the SOM units close to the marked location.

11.5.3 Playlist Creation

The two selection models offered by PlaySOM allow the user to directly create playlists by interacting with the map. Thus the application not only allows for convenient browsing of music collections containing hundreds or thousands of songs, but also enables the creation of playlists based on real music similarity instead of albums or meta-data. This relieves users from traditional browsing through lists and hierarchies of genres and albums, which often leads to rather monotonous playlists consisting of complete albums from a single artist. Instead of the burdensome compilation of playlists title by title, PlaySOM allows to directly select a region of the map with the music style of interest. Moreover, by drawing tra-

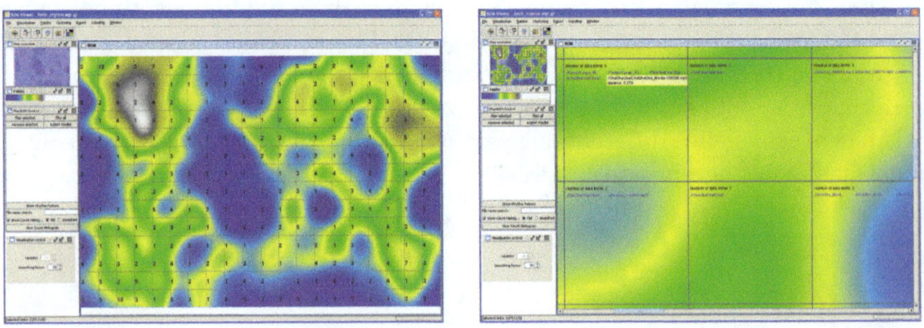

(**a**) Low zooming level: the number of songs mapped on the units are indicated.

(**b**) High zooming level: song titles and additional information are displayed on the respective units.

Fig. 11.11 Semantic zooming and its influence on the amount of information displayed jectories on the map, it is possible to generate playlists which are traversing multiple genres. Figure 11.12 depicts the playlist creation models that are supported by PlaySOM. The rectangular selection model (cf. Fig. 11.12(a)) allows the user to drag a rectangle and select the songs belonging to units inside that rectangle without preserving any order of the selected tracks. This model is used to select music from one particular cluster or region on the map and is useful if music from a particular genre or sub-genre well-represented by a cluster is desired. The path selection model allows users to draw trajectories and select all songs belonging to units beneath that trajectory. Figure 11.12(b) shows a trajectory that moves from one music cluster to another one, including the music that is located on the transition between those clusters. Paths can be drawn on the map, for instance, starting with slow Waltz music going via Tango to Jive music and back to slow Waltz via Viennese Waltz. It can be fixed from the beginning that such a "tour" should take, for example, two hours. The PlaySOM application then automatically selects music along these path lines, or, optionally, plays music randomly from within the trajectory drawn. Such an approach offers a wonderful possibility to quickly prepare a playlist for particular situations (party, dinner, background music, etc.). Once a user has selected songs on the map the playlist element in the interface displays the list of selected titles. It is possible to play the music in the list directly or to refine the list by manually dropping single songs from the selection. The playlist can also be exported for later use on the desktop computer or on other devices like mobile phones, PDAs or audio players. The music can be played either locally or, if the music collection is stored on a server, via a streaming environment. Furthermore, the PlaySOM application can be conveniently and efficiently used on a Tablet PC (see Fig. 11.13), because its interface is easily controllable via pen input. It is even usable as a touch screen application.

Summarizing, the PlaySOM interface allows the interactive exploration of entire music archives and the creation of personal playlists directly by selecting regions of

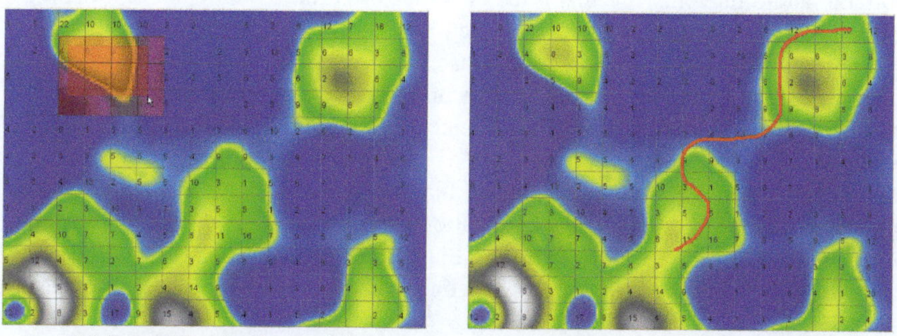

(**a**) Rectangle selection model. (**b**) Trajectory selection model.

Fig. 11.12 Different models of selecting music on the PlaySOM music map

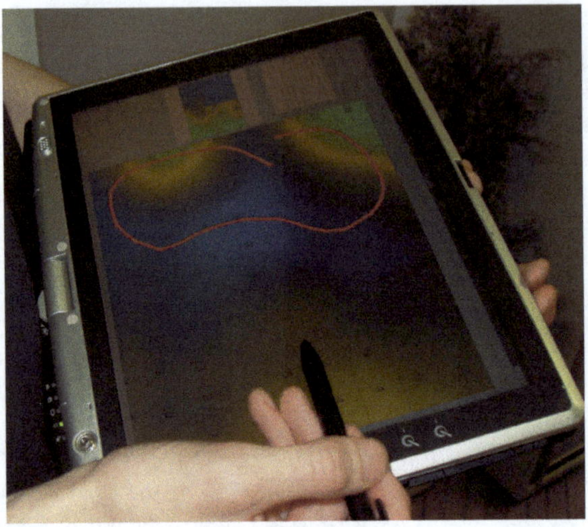

Fig. 11.13 PlaySOM running on a Tablet PC with pen input

one's personal taste, without having to browse a list of available titles and manually sorting them into playlists. Thus, the map metaphor constitutes a completely novel experience of music retrieval by navigation through "music spaces".

11.6 PocketSOMPlayer – Music Retrieval on Mobile Devices

Traditional selection methods such as browsing long lists of music titles or selecting artists from alphabetical lists or entering queries into a search field are particularly cumbersome when used on mobile devices, such as PDAs, mobile phones or portable audio players. Yet, this issue becomes even more annoying with people's music collections constantly getting larger.

The need for improved access to music collections on portable devices motivated the implementation of music maps on those devices, allowing for direct and intuitive access to the desired music. Like for PlaySOM, a self-organizing map builds the basis for intuitive visualizations and forms the application interface. A lightweight application has been created that runs on Java-enabled PDAs and mobile phones (see Fig. 11.14). The application takes an image export (e.g. an islands of music (SDH) visualization) from the PlaySOM application for its interface, i.e. currently the PlaySOM application is needed to create a map for the PocketSOMPlayer.

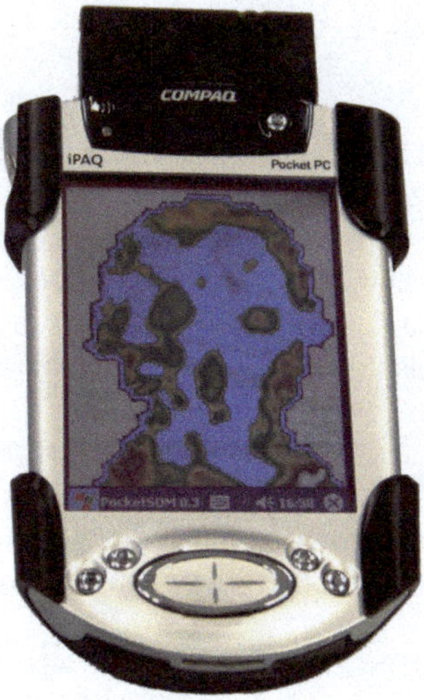

(**a**) on a PDA (iPAQ)

(**b**) on a PDA phone (BenQ P50)

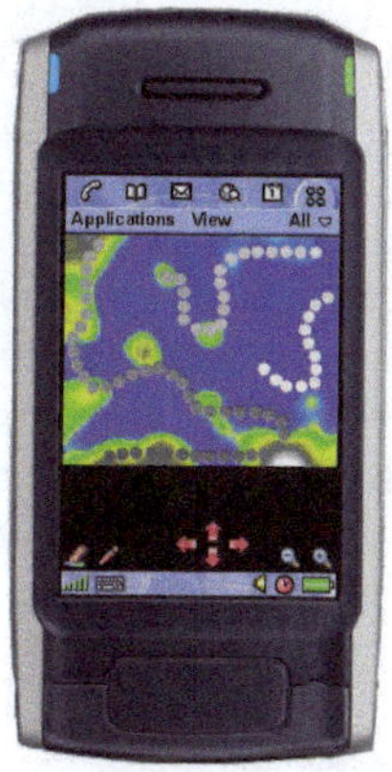

(**c**) on a mobile phone (Sony Ericsson emulator)

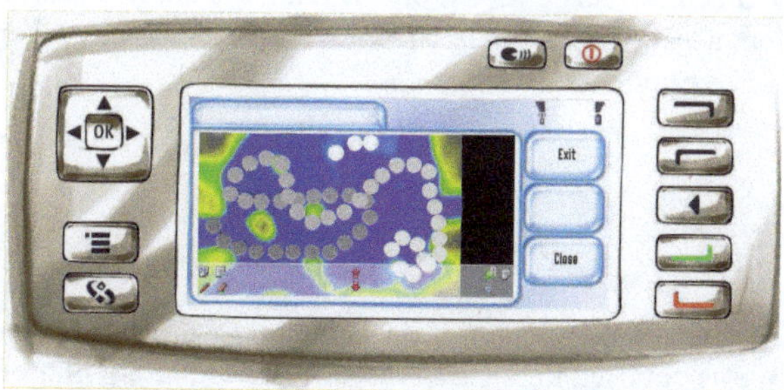

(**d**) on a multimedia phone (Nokia 7710 emulator)

Fig. 11.14 Different implementations of the PocketSOMPlayer

11.6.1 Interaction

The interface offered by the PocketSOMPlayer [16] is similar to its desktop counterpart PlaySOM. It also offers exploring a music map by zooming and selection and playlists are created by drawing paths with a pen on the screen (provided the device supports pen input). Due to the limitations in screen size, playlists are displayed on the full screen after a selection is made, offering the choice of fine-tuning the playlist.

11.6.2 Playing Scenarios

Several play modes exist for the PocketSOMPlayer:

First, if the device has sufficient capacity to store entire music collections on it, music can be played directly from the device.

Alternatively, the PocketSOMPlayer can also be used for streaming one's personal music collection from the desktop computer at home. A connection is opened from the mobile device to one's personal computer and each time a playlist is created by drawing a path on the mobile device, the PocketSOMPlayer starts to stream the music from the desktop computer to the handheld device.

Instead of streaming the music to the mobile device, the PocketSOMPlayer can also be used as convenient remote control to select the music one wants to listen to in one's living room. After selecting a path or an area on the music map on the PDA or mobile phone a playlist is sent to the desktop computer which then plays the music.

With an active connection to the Internet the PocketSOMPlayer is able to stream the music on the selected map trajectory from a server. Thus, while travelling around, with this technology one can access a music repository from wherever one has access to the Internet, be it via GPRS, UMTS or wireless LAN. This also enables the idea of a central music repository with a huge archive of music in it and a multitude of users accessing this music from wherever they are, offering room for portal-based service providers.

11.6.3 Conclusion

Selecting music via drawing trajectories on a touch screen is straightforward, easy to learn and intuitive as opposed to clicking through hierarchies of genres or interprets. The PocketSOMPlayer offers a convenient alternative to traditional music selection and may also constitute a new model of how to access a music collection on portable audio players.

11.7 Conclusions

Machine learning has a nearly indefinite amount of application areas. One of them is music information retrieval, where several different techniques are employed for a wealth of retrieval and organization tasks. Supervised machine learning is applied to recognize artists, to classify music into genres or to categorize music collections into entire genre hierarchies. In order to achieve this, various methods for audio description by feature extraction have been developed which serve as the basis for all subsequent machine learning tasks. Unsupervised learning techniques are employed to organize entire music archives fully automatically. One particular approach that has been chosen is the use of self-organizing maps to create maps of music archives. These music maps not only allow for deeper investigation of clusters of similar music within the music repository, but also enable a great number of different visualization methods which enhance the view of the cluster structures inherent in the data set. Moreover, the interactive applications that have been developed on top of SOM-based music maps permit exploration of and interaction with the maps by browsing and zooming, navigation through the map and ad hoc creation of playlists, based on music similarity measures extracted from audio. Implementations are also available for PDAs and mobile phones and facilitate access to music collections on portable devices, also enabling stream-based services. The application of machine learning techniques thus constitutes the foundation for a range of novel music access models.

References

1. *Information technology - Multimedia content description interface - Part 4: Audio. ISO/IEC 15938-4:2002*. International Organisation for Standardisation, 2002.
2. Xavier Amatriain, Pau Arum', and David Garcia. CLAM: A framework for efficient and rapid development of cross-platform audio applications. In *Proceedings of ACM Multimedia 2006*, Santa Barbara, CA, USA, 2006.
3. Jean-Julien Aucouturier and Francois Pachet. Representing musical genre: A state of the art. *Journal of New Music Research*, 32(1):83–93, 2003.
4. Taha Abdel Aziz. Coloring of the SOM based on class labels. Master's thesis, Vienna University of Technology, October 2006.
5. J. Stephen Downie. The music information retrieval evaluation exchange (MIREX). *D-Lib Magazine*, 12(12), December 2006.
6. J. Stephen Downie, Andreas F. Ehmann, and Xiao Hu. Music-to-knowledge (M2K): A prototyping and evaluation environment for music digital library research. In *Proceedings of the Joint Conference on Digital Libraries (JCDL)*, page 376, Denver, CO, USA, June 7–11 2005.
7. Yoav Freund and Robert E. Schapire. A decision-theoretic generalization of on-line learning and an application to boosting. In *Proceedings of the European Conference on Computational Learning Theory (EUROCOLT)*, pages 23–37, 1995.
8. ISMIR 2004 Audio Description Contest. Website, 2004. `http://ismir2004.ismir.net/ISMIR_Contest.html`.
9. Teuvo Kohonen. *Self-Organizing Maps*, volume 30 of *Springer Series in Information Sciences*. Springer, Berlin, 3rd edition, 2001.
10. Thomas Lidy and Andreas Rauber. Evaluation of feature extractors and psycho-acoustic transformations for music genre classification. In *Proceedings of the International Conference on Music Information Retrieval (ISMIR)*, pages 34–41, London, UK, September 11–15 2005.
11. Beth Logan. Mel frequency cepstral coefficients for music modeling. In *Proceedings of the International Symposium on Music Information Retrieval (ISMIR)*, Plymouth, MA, USA, October 23–25 2000.
12. Stéphane Mallat. *A Wavelet Tour of Signal Processing*. Academic Press, New York, 2nd edition, 1999.
13. Rudolf Mayer, Dieter Merkl, and Andreas Rauber. Mnemonic SOMs: Recognizable shapes for self-organizing maps. In *Proceedings of the Workshop On Self-Organizing Maps (WSOM)*, pages 131–138, Paris, France, September 5–8 2005.
14. 2nd annual Music Information Retrieval Evaluation eXchange. Website, 2005. `http://www.music-ir.org/mirex2005/index.php/Main_Page`.

15. Fabian Mörchen, Alfred Ultsch, Mario Nöcker, and Christian Stamm. Databionic visualization of music collections according to perceptual distance. In *Proceedings of the International Conference on Music Information Retrieval (ISMIR)*, London, UK, September 11–15 2005.
16. Robert Neumayer, Michael Dittenbach, and Andreas Rauber. PlaySOM and PocketSOM-Player – alternative interfaces to large music collections. In *Proceedings of the International Conference on Music Information Retrieval (ISMIR)*, pages 618–623, London, UK, September 11–15 2005.
17. Elias Pampalk. Islands of music: Analysis, organization, and visualization of music archives. Master's thesis, Vienna University of Technology, December 2001.
18. Elias Pampalk, Simon Dixon, and Gerhard Widmer. Exploring music collections by browsing different views. In *Proceedings of the International Conference on Music Information Retrieval (ISMIR)*, pages 201–208, Baltimore, MD, USA, October 26–30 2003.
19. Elias Pampalk, Andreas Rauber, and Dieter Merkl. Content-based organization and visualization of music archives. In *Proceedings of ACM Multimedia 2002*, pages 570–579, Juan-les-Pins, France, December 1–6 2002.
20. Georg Pölzlbauer, Michael Dittenbach, and Andreas Rauber. Advanced visualization of self-organizing maps with vector fields. *Neural Networks*, 19(6–7):911–922, July–August 2006.
21. Andreas Rauber and Markus Frühwirth. Automatically analyzing and organizing music archives. In *Proceedings of the European Conference on Research and Advanced Technology for Digital Libraries (ECDL)*, Darmstadt, Germany, September 4–8 2001.
22. Andreas Rauber, Elias Pampalk, and Dieter Merkl. Using psycho-acoustic models and self-organizing maps to create a hierarchical structuring of music by musical styles. In *Proceedings of the International Conference on Music Information Retrieval (ISMIR)*, pages 71–80, Paris, France, October 13–17 2002.
23. Andreas Rauber, Elias Pampalk, and Dieter Merkl. The SOM-enhanced JukeBox: Organization and visualization of music collections based on perceptual models. *Journal of New Music Research*, 32(2):193–210, June 2003.
24. Markus Schedl, Peter Knees, and Gerhard Widmer. Using CoMIRVA for visualizing similarities between music artists. In *Proceedings of the 16th IEEE Visualization Conference (Vis 2005)*, Minneapolis, MN, USA, 23–28 October 2005.
25. Eric Scheirer and Malcolm Slaney. Construction and evaluation of a robust multifeature speech/music discriminator. In *Proceedings of the International Conference on Acoustics, Speech and Signal Processing (ICASSP'97)*, pages 1331–1334, Munich, Germany, 1997.
26. Manfred R. Schröder, Bishnu S. Atal, and Joseph L. Hall. Optimizing digital speech coders by exploiting masking properties of the human ear. *Journal of the Acoustical Society of America*, 66:1647–1652, 1979.
27. Tero Tolonen and Matti Karjalainen. A computationally efficient multipitch analysis model. *IEEE Transactions on Speech and Audio Processing*, 8(6):708–716, November 2000.
28. George Tzanetakis. *Manipulation, Analysis and Retrieval Systems for Audio Signals*. PhD thesis, Computer Science Department, Princeton University, 2002.
29. George Tzanetakis and Perry Cook. Sound analysis using MPEG compressed audio. In *Proceedings of the International Conference on Audio, Speech and Signal Processing (ICASSP)*, Istanbul, Turkey, 2000.
30. Alfred Ultsch. Data mining and knowledge discovery with emergent self-organizing feature maps for multivariate time series. In *Kohonen Maps*, pages 33–45, Elsevier, Amsterdam, 1999.
31. Alfred Ultsch. Pareto density estimation: A density estimation for knowledge discovery. In *Innovations in Classification, Data Science, and Information Systems – Proceedings 27th Annual Conference of the German Classification Society (GfKL)*, pages 91–100, Berlin, Heidelberg, 2003.
32. Alfred Ultsch. U*-matrix: A tool to visualize clusters in high dimensional data. Technical report, Department of Mathematics and Computer Science, Philipps-University Marburg, 2003.
33. Alfred Ultsch and H. Peter Siemon. Kohonen's self-organizing feature maps for exploratory data analysis. In *Proceedings of the International Neural Network Conference (INNC'90)*, pages 305–308, Dordrecht, Netherlands, 1990.

34. Ian H. Witten and Eibe Frank. *Data Mining: Practical machine learning tools and techniques.* Morgan Kaufmann, San Francisco, CA, 2nd edition, 2005.
35. Eberhard Zwicker and Hugo Fastl. *Psychoacoustics – Facts and Models*, volume 22 of *Springer Series of Information Sciences*. Springer, Berlin, 1999.

Index

Cognitive Technologies

H. Prendinger, M. Ishizuka (Eds.)
Life-Like Characters
Tools, Affective Functions, and Applications
IX, 477 pages. 2004

H. Helbig
Knowledge Representation and the Semantics of Natural Language
XVIII, 646 pages. 2006

P.M. Nugues
An Introduction to Language Processing with Perl and Prolog
An Outline of Theories, Implementation, and Application with Special Consideration of English, French, and German
XX, 513 pages. 2006

W. Wahlster (Ed.)
SmartKom: Foundations of Multimodal Dialogue Systems
XVIII, 644 pages. 2006

B. Goertzel, C. Pennachin (Eds.)
Artificial General Intelligence
XVI, 509 pages. 2007

O. Stock, M. Zancanaro (Eds.)
PEACH — Intelligent Interfaces for Museum Visits
XVIII, 316 pages. 2007

V. Torra, Y. Narukawa
Modeling Decisions: Information Fusion and Aggregation Operators
XIV, 284 pages. 2007

P. Manoonpong
Neural Preprocessing and Control of Reactive Walking Machines
Towards Versatile Artificial Perception–Action Systems
XVI, 185 pages. 2007

S. Patnaik
Robot Cognition and Navigation
An Experiment with Mobile Robots
XVI, 290 pages. 2007

M. Cord, P. Cunningham (Eds.)
Machine Learning Techniques for Multimedia
Case Studies on Organization and Retrieval
XVI, 290 pages. 2008

MIX
Papier | Fördert gute Waldnutzung
FSC® C083411

Kolibri 360 GmbH
Ferdinand-Jühlke-Straße 7
99095 Erfurt, Deutschland
produktsicherheit@kolibri360.de